Contemporary Issues in Digital Communication, Media, and Culture

Kostas Karpouzis
Panteion University of Social and Political Sciences, Greece

Ioannis Skarpelos
Panteion University of Social and Political Sciences, Greece

IGI Global
Scientific Publishing
Publishing Tomorrow's Research Today

Vice President of Editorial	Melissa Wagner
Director of Acquisitions	Mikaela Felty
Director of Book Development	Jocelynn Hessler
Production Manager	Mike Brehm
Cover Design	Jose Rosado

Published in the United States of America by
IGI Global Scientific Publishing
701 East Chocolate Avenue
Hershey, PA, 17033, USA
Tel: 717-533-8845 | Fax: 717-533-7115
Website: https://www.igi-global.com E-mail: cust@igi-global.com

Library of Congress Cataloging-in-Publication Data

LCCN: 2024058311 (CIP Data Pending)
ISBN13: 9798369387290
Isbn13Softcover: 9798369387306
EISBN13: 9798369387313

British Cataloguing in Publication Data
A Cataloguing in Publication record for this book is available from the British Library.

Table of Contents

Detailed Table of Contents

The integration of Artificial Intelligence (AI) into journalism constitutes a transformative development in the evolution of news reporting and the broader media landscape. This chapter explores the emergence of AI in the Greek media system through qualitative analyses of journalists and CEOs, complemented by an examination of five case studies. The findings suggest that the majority of media organizations in Greece adopt a cautious "wait-and-see" approach before fully incorporating AI tools into their workflows. Nevertheless, they are taking steps in order to prepare for the integration of AI into their value chain. Journalists are experimenting with AI tools, using them primarily as Intelligent Assistants (IA) while in certain cases, AI is employed for specific tasks. These varied approaches are shaped by concerns about AI's limitations, such as linguistic challenges, ethical considerations, intellectual property issues, and potential job displacement. Moreover, the digital divide and a lack of technical skills emerge as significant barriers to the adoption of AI.

Contemporary platform society necessitates fact-checking initiatives to combat fake news proliferation and enhance the credibility of digital discourse. While automated fact-checking experiments show promise in improving verification efficiency, they also raise concerns regarding the diminishing authority of human fact-checkers. Based on the theories of "disruptive technologies" and "affordances", this study explores the perception and application of automated technologies in the routines of fact-checkers across multiple countries, using data from an online survey of thirty-two fact-checkers and interviews with seven editors-in-chief of fact-checking organisations affiliated with the International Fact-Checking Network. Findings indicate that, while fact-checkers recognise the utility of AI tools for specific tasks, they resist full automation of the fact-checking process, highlighting the necessity of human oversight. AI is viewed primarily as an aid in the fight against disinformation rather than a complete solution due to inherent limitations that human capabilities surpass.

This study presents a computational and behaviorally grounded approach to health communication, employing audience segmentation as a strategy to optimize the dissemination of public health messages during the COVID-19 pandemic. Drawing on a national-level database of the Mexican population (INEGI, 2020), we conducted an exploratory factor analysis and subsequent statistical modeling to identify three distinct attitudinal segments: indifferent and carefree, positive but evasive, and concerned and involved. These segments reflect heterogeneous emotional, cognitive, and behavioral patterns in response to the pandemic. The findings underscore the relevance of gender as a determinant of segment distribution, suggesting the need for differentiated messaging strategies. This research contributes to the development of more precise, culturally resonant, and evidence-based social marketing campaigns that leverage data analytics to promote preventive health behaviors in crisis contexts.

Malware detection is a critical aspect requiring robust and adaptive techniques to identify evolving threats. The rapid proliferation of IoT devices has led to exponential growth. Machine learning methods, though effective, often struggle to capture the complex spatial. This study proposed a hybrid deep learning model that integrates CNN with Gated Recurrent Units (GRU) to enhance the classification performance on IoT datasets. The architecture leverages the spatial feature extraction capabilities of CNNs and the temporal sequence modeling strength of GRUs. The preprocessing pipeline includes data cleaning, feature extraction, and feature selection, followed by a data-splitting strategy to ensure robust training and evaluation. The proposed model achieves superior classification accuracy Furthermore, performance evaluation using training/testing accuracy and loss metrics demonstrates the model's stability, efficiency, and generalization ability. The findings suggest that the proposed method offers a highly reliable and scalable solution for IoT data analysis.

This chapter explores how digital transformation (DT) affects both performance and customer experience (CX) in small and medium-sized enterprises (SMEs) in Vietnam. It examines the role of information technology and innovation, such as artificial intelligence, big data, and cloud computing, in promoting economic growth and improving customer satisfaction. While DT offers SMEs opportunities for growth and greater flexibility, many of them still face obstacles such as limited funding and a lack of technical expertise. The research highlights how Vietnamese SMEs can use DT to increase their efficiency and strengthen their competitive position in today's digital economy.

The global fashion system changes decisively due to the impact of the growing technological innovation and social changes. Fashion is a method of self-expression that gives people the opportunity to be their true selves.The purpose of this study is to present a new perspective about the combination of fashion and gaming. The connection between the two seemingly different "worlds" is more relevant than ever considering that over three billion people worldwide are actively participate in-game experience. Further research seems necessary to explore how virtual fashion is used for new forms of self-identity. To develop a research model for approaching this aim, the study employs four theoretical approaches. So, this systematic literature review investigates the evolution of avatar fashion, the innovative ways that virtual environment used for forms of social identity, the collaboration between fashion-gaming industries and the Metaverse as a digital universe of avatars.

Visual Politics has its roots in the history of political communication, and it takes on particular relevance in the contemporary media scenario, together with the assumption of centrality of social media. The chapter begins with a brief history of the role of image in electoral campaigns. The contribution of Visual Politics properly so-called is presented through the evolution of the theoretical framework and exemplary research conducted over time, first in the television field then in the social media field. The social media environment is then identified as an exemplary place where Visual Politics can express its potential in the hybridization of research methods belonging to the sociology of communication and computer sciences. Some first evidence on the contribution of Computer Vision to the study of the 2024 European Elections, provided by two joint research programs in which the authors are involved, are presented.

Preface

INTRODUCTION

Rapid technological innovation is currently impacting the disciplines of communication, media, and culture: the emergence of Web 2.0, the rise and spread of social media, participatory journalism, and the growing influence of Artificial Intelligence in fields beyond Computer Science have collectively redefined how we produce, distribute, and engage with content. These developments have not only altered public discourse but have also compelled researchers, practitioners, and educators to reconsider the frameworks through which we understand these evolving dynamics.

At the heart of these transformations is a methodological shift sparked by the advent of the Digital Humanities, Cultural Analytics, Social Informatics, and Computational Social Sciences. These emerging fields challenge traditional disciplinary boundaries and encourage new forms of inquiry, both qualitative and quantitative, that are better suited to the complexity of our digital age, and to the opportunities that modern tools offer.

Contemporary Issues in Digital Communication, Media, and Culture explores this evolving landscape through an interdisciplinary lens, focusing on the convergence of Computer Science and the Social Sciences and Humanities. The chapters within this volume critically engage with how computational methods can be harnessed to analyze, interpret, and influence human communication and cultural phenomena. By combining theoretical discussions with applied case studies, the volume serves as both a conceptual guide and a practical resource for understanding the multifaceted nature of digital communication today.

This book aims to:

- Advance interdisciplinary research by promoting collaboration between computer scientists and social scientists, encouraging projects that bridge technical and humanistic approaches;
- Innovate research methodologies by introducing novel computational tools and techniques that enhance analytical rigor and broaden the scope of inquiry;
- Inform policy and practice by offering insights relevant to media regulation, cultural preservation, ethical technology use, and digital governance;
- Enhance educational curricula by providing a foundational text for courses integrating computational methods into the study of communication, media, and culture.

Our intended audience includes researchers and professionals across a wide range of disciplines: communication, media and journalism, cultural studies, computing, data science, artificial intelligence, and beyond. Practitioners involved in the development of computational applications in journalism, social media, content creation, advertising, cultural management, and the creative industries will also find the book a valuable resource.

By promoting dialogue between the computational and social sciences, we hope this volume contributes meaningfully to ongoing conversations around the future of communication, media, and culture in an increasingly digital world.

ORGANIZATION OF THE BOOK

This volume brings together a collection of original and timely research contributions that examine the interplay between digital technologies, communication practices, and cultural formations. Ranging from algorithmic governance and media ecosystems to digital fashion and computational health campaigns, each chapter offers a focused investigation into the dynamic spaces where technology and society converge. Below, we provide an overview of each chapter to guide readers through the book's thematic and methodological diversity.

Chapter 1. Building Culturally Aware Information Technology

This opening chapter lays the groundwork for a deeper understanding of how culture and information technology shape each other. Employing a Cultural Epistemology framework, the authors articulate four key dimensions, i.e., information sharing, digital access, algorithmic fairness, and AI understanding, through which culture and IT intersect. Each dimension is explored through critical questions that offer insight into the nuanced relationship between cultural diversity and technological design. This chapter is especially valuable for policymakers, system developers, and researchers working toward more inclusive, equitable digital infrastructures.

Chapter 2. Algorithmic Decision-Making and the Ambiguity of Transparency in EU Digital Regulation

Focusing on the European regulatory landscape, this chapter interrogates the ambiguity surrounding algorithmic transparency and explainability. By examining EU legislation, key judicial decisions from Italy, and definitional challenges, the authors highlight the tension between legal mandates and technological complexity. The chapter makes a case for the creation of standardized, measurable criteria for algorithmic transparency, particularly in public administration, and underscores the need for interdisciplinary collaboration in shaping fair, accountable AI systems.

Chapter 3. The Integration and Impact of AI in the Greek Media Ecosystem: Challenges, Opportunities, and the Digital Divide

This chapter examines the cautious integration of AI within the Greek media landscape through interviews and case studies involving media professionals. It reveals a fragmented approach to AI adoption, shaped by ethical concerns, technical limitations, and fears of job displacement. While AI is beginning to assist journalistic tasks, full integration remains hindered by the digital divide and insufficient technical skills. The chapter contributes valuable regional insight into the global conversation on AI in journalism, offering implications for both policy and practice.

Chapter 4. Automated Fact-Checking in Combating Mis-/Disinformation: An Investigation of Fact-Checkers' Perceptions in 18 Countries

This cross-national study investigates how fact-checkers perceive and implement automated technologies in their verification work. Grounded in theories of disruptive innovation and technological affordances, the chapter draws from survey data and in-depth interviews to show that while AI tools are increasingly used, human judgment remains indispensable. The findings highlight the benefits of automation for efficiency, yet stress the irreplaceable value of editorial oversight in safeguarding credibility, underscoring the limits of full automation in combating disinformation.

Chapter 5. Segmenting Health Audiences through Computational Analysis: A Social Marketing Approach During the COVID-19 Pandemic in Mexico

By integrating computational analysis with behavioral segmentation, this chapter presents a novel approach to public health communication in crisis contexts. Using large-scale demographic data from Mexico, the authors identify three audience segments with varying attitudes toward COVID-19. The study emphasizes the importance of gender and cultural context in shaping communication strategies and demonstrates how data-driven segmentation can lead to more targeted, resonant public health messaging, offering a valuable model for social marketers and public health officials.

Chapter 6. Hybrid CNN-GRU Architecture for Malware Detection Using Memory Forensic Datasets

This technically robust chapter introduces an innovative hybrid deep learning model combining Convolutional Neural Networks (CNN) and Gated Recurrent Units (GRU) for malware detection in IoT environments. With a detailed account of the preprocessing and model evaluation process, the authors demonstrate high classification accuracy and generalizability. This contribution stands out for its practical significance, offering a scalable solution to cybersecurity threats and advancing methodological approaches in memory forensics and IoT analytics.

Chapter 7. Customer Experience and the Impact of Digital Transformation on Firm Performance in SMEs: A Comparative Analysis of Vietnam

Exploring the intersection of customer experience and digital transformation, this chapter focuses on small and medium-sized enterprises (SMEs) in Vietnam. It discusses how emerging technologies like AI, big data, and cloud computing are being leveraged to improve operational performance and consumer engagement. Despite these benefits, many SMEs face barriers such as funding limitations and digital skill gaps. The chapter provides actionable insights into how digital transformation can serve as a competitive advantage when supported by appropriate policies and investment.

Chapter 8. Digital Fashion and Social Identity

This chapter describes the emerging relationship of fashion, gaming, and identity formation in virtual environments: through a systematic literature review and a multi-theoretical approach, the authors examine how virtual fashion is becoming a medium for self-expression and social signaling. The discussion includes the evolution of avatar aesthetics, fashion-industry collaborations, and the growing relevance of the Metaverse. This interdisciplinary investigation opens new avenues for understanding how digital environments are reshaping the construction and performance of identity.

Chapter 9. Visual Politics at the Time of Platform Society

Concluding the volume, this chapter situates visual political communication within the broader context of platform society. Beginning with a historical overview of imagery in electoral politics, it traces the evolution of visual politics from traditional media to social media platforms. The chapter introduces hybrid methodological frameworks combining sociology and computer science, including early findings from recent European election studies utilizing computer vision. It offers a compelling case for expanding the analytical lens on political visuals in an increasingly digitized and image-saturated public sphere.

Together, these chapters provide a deep and multifaceted exploration of how digital technologies are transforming communication, media, and culture across contexts. From the algorithmic to the aesthetic, the practical to the theoretical, the local to the transnational, this volume aims to both document and shape ongoing conversations at the intersection of computational methods and social inquiry.

CONCLUSION

As editors of this volume, we are pleased to present a body of work that reflects the growing complexity, diversity, and interdependence of digital communication, media, and culture in a rapidly evolving technological range of disciplines. The chapters collected here demonstrate that the convergence of computational tools and social inquiry is not merely a methodological development, but a necessary and transformative shift in how we understand the world.

One of the central themes that emerges across the volume is the tension between innovation and responsibility; as digital tools become more powerful and ubiquitous, questions of fairness, accessibility, ethics, and human agency grow more urgent. Whether exploring AI's impact on media practices, developing strategies for public health communication, or investigating the implications of visual politics in the platform society, our contributors remind us that technology is never neutral: it is embedded in cultural contexts, shaped by human choices, and capable of reinforcing or disrupting existing power structures.

We hope that *Contemporary Issues in Digital Communication, Media, and Culture* serves as a basis for further inquiry, dialogue, and collaboration. We envision this volume as both a reflective space and a practical resource for researchers seeking to deepen their methodological toolkit, for educators designing interdisciplinary curricula, and for practitioners navigating the complexities of digital transformation in their fields. As the boundaries between disciplines continue to blur and the digital continues to redefine our cultural and communicative landscapes, this volume illustrates the importance of critical, context-aware, and forward-looking scholarship.

Chapter 1
Building Culturally Aware Information Technology

Vijaya Kittu Manda
https://orcid.org/0000-0002-1680-8210
PBMEIT, India

Srinivas Bikkina
https://orcid.org/0000-0002-5411-7146
Vingas Industries Private Limited, India

U. V. Adinarayana Rao
GITAM University, India

Dina Darwish
https://orcid.org/0000-0002-6708-649X
Ahram Canadian University, Egypt

Theodore Tarnanidis
https://orcid.org/0000-0002-4836-3906
International Hellenic University, Greece

ABSTRACT

Culture and Information Technology (IT) influence each other. This chapter uses cultural epistemology (CE) perspectives to explain the four dimensions of Culture and IT that influence each other. First, a discussion on the impact of Culture on Information Sharing and Communication is included. Then, Culture and its impact on the Digital Divide and Global Access to Information are discussed. Information Bias and Algorithm Fairness are vital considerations for IT developers when building culture-rich applications. Finally, a discussion on Artificial Intelligence and Cultural Understanding is covered to catch up with the emerging theme. Each dimension raises

DOI: 10.4018/979-8-3693-8729-0.ch001

three questions, which are discussed using the CE approach. This study benefits many stakeholders by providing new insights into the relationship between Culture and IT. This knowledge can be used to develop more effective policies, practices, and systems that promote the equitable use of IT and address the digital divide.

INTRODUCTION

The global rise of Information Technology (IT) has highlighted the importance of intercultural communication (Zhang & Zhou, 2023). IT allows individuals from diverse backgrounds to connect, interpret, and share information. It allows them to spread ideas and promote collaboration among global teams (Gibbs et al., 2021). It helps build trust across cultures on the booming and evolving online platforms (Chameroy et al., 2024). However, identifying and minimizing cultural differences is required to ensure proper understanding among stakeholders. Global information technologies can promote intercultural communication skills (Kumari & Nirban, 2018). With this, they can bridge cultural divides. They can help create a more inclusive online space that is diverse because all voices can be heard and understood. Technologies require cultural consideration to ensure they are not biased and can be used by everyone daily. The list includes Artificial Intelligence and Media and communication platforms like social media. Overall, it is clear that digital humanities and intelligence computing are playing a key role in cultural heritage (Wang et al., 2024).

This chapter adopts Cultural Epistemology (CE) as its guiding philosophical framework to explore the complex relationship between culture and information technology systematically. CE provides a lens through which we can examine how culture shapes knowledge and its technological representations. It also examines how technology, in turn, influences cultural understanding and values. The following section introduces the core concepts of CE. It presents the relevance of the design and use of IT systems. Eventually, it sets the foundation for the subsequent analysis.

Cultural Epistemology Approach in IT

Cultural Epistemology (CE) is a branch of Philosophy that focuses on how cultural factors influence knowledge and knowing (Sulistyono et al., 2023). It discusses how culture constructs, validates, and transmits knowledge (Liu, 2010). It is a way to understand how people in different cultural contexts see knowledge. It discusses how the views of the people shape the acquisition, use, and evaluation of knowledge. CE is a tool for understanding the underlying forces of culture, knowledge, and power. It studies how cultural contexts shape knowledge systems, belief structures, and ways of better understanding the world.

Cultural Epistemology is used in this study to understand how culture influences Information Technology (IT) and vice versa. This philosophical angle is used in this chapter to assess how IT platforms and tools are (to be) designed to reflect the values and assumptions of culture. By building culturally sensitive platforms, developers can promote inclusiveness and be more effective (Passey, 2015). They will be meaningful because they help in intercultural communication, collaboration, and knowledge sharing (Kampermann et al., 2021). Table 1 shows a few examples of how the IT design process can incorporate cultural epistemology.

Table 1. Examples of cultural epistemology in the IT design process

Application Area	Description
Development of culturally sensitive search engines	They should consider the different ways people from different cultures search for information.
Design of online learning platforms	The online learning platforms support different learning styles and cultural perspectives.
Development of social media platforms	The platforms facilitate dialogue and collaboration between people from different cultures.

Source: Author's creation

This section establishes Cultural Epistemology as a foundational framework. Existing research on the intersection of culture and IT lacks a comprehensive, philosophically grounded approach. Digital humanities and AI studies have made significant improvements. However, critical gaps remain in how cultural dimensions are integrated into technological design and evaluation. The following section outlines these gaps. The chapter addresses this by applying CE to key IT development and use domains. It provides insights and perspectives on Culture and IT that can benefit stakeholders such as Academicians, Researchers, Educators, Policymakers, IT professionals, Cultural organizations, and the public.

Literature Gap

Existing research has already acknowledged that the humanities and information technology are rich topics for research and that digital humanities (DH) is an institutional endeavor (Svensson, 2016). A literature study showed a lack of a comprehensive and culturally sensitive approach to understanding the relationship between Culture and Information Technology (IT). This chapter identifies and addresses the following gaps in the literature:

1. There is a need to provide a philosophical framework for understanding the relationship between Culture and IT through the lens of Cultural Epistemology (CE).
2. It is essential to identify the major dimensions in which Culture and IT influence each other.
3. Cultural sensitivity is vital in designing and using IT systems and is worth putting effort into. It includes the need for ethical considerations and the potential for IT to perpetuate cultural stereotypes.
4. There is a need to highlight the role of IT in promoting intercultural understanding and bridging cultural divides, including language translation tools and cultural heritage preservation initiatives. The study can connect with emerging trends such as Artificial Intelligence (AI) and Individual Personal Data Privacy practices.

While current models often focus on accommodating cultural preferences, this study uses CE as a philosophical framework. The idea is to understand how culture fundamentally shapes knowledge and knowing in the context of technology. It allows assessing the design of IT platforms and how they reflect the values and assumptions of different cultures. This focus on the philosophical framework of CE helps fill a gap in the literature that lacks a comprehensive and culturally sensitive approach to the relationship between Culture and IT. It also allows for a deeper exploration of ethical considerations. For example, it speaks about the potential for IT to continue cultural biases or, conversely, to promote intercultural understanding.

This chapter proposes a focused research agenda in response to the identified literature gaps. Grounded in Cultural Epistemology, this agenda aims to deepen our understanding of how culture and IT mutually shape one another. The following section outlines the four core thematic areas that structure this inquiry.

Research Agenda

The Research Agenda of this chapter is to connect and study the interaction of Culture and Information Technology from a Cultural Epistemology perspective. With this goal, this chapter proposes four key sections under the following heads:

1. Impact of Culture on Information Sharing and Communication
2. Digital Divide and Global Access to Information
3. Information Bias and Algorithmic Fairness
4. Artificial Intelligence and Cultural Understanding

The first dimension of the research agenda examines how cultural norms and values influence how individuals share information and communicate through digital platforms. This section is a starting point for understanding the micro-level interactions underpinning broader cultural-technological dynamics. With this backdrop, the study begins to see how deeply culture is rooted in everyday technological use. The research goals can be addressed by analyzing online communication patterns across cultures.

IMPACT OF CULTURE ON INFORMATION SHARING & COMMUNICATION

Humans, over the millennia, have created, shared, and transformed information networks. Culture influences how people share information and communicate online (Ardichvili et al., 2006). Cultural norms vary substantially regarding an individual's self-disclosure, privacy, and etiquette. These differences sometimes lead to misunderstandings and challenges in cross-cultural interactions (Korkut et al., 2018). For example, some cultures may value more open and expressive communication online. Some other cultures prefer more reserved and formal styles. Communication between people from these cultures would lead to misunderstandings.

Technology adoption sometimes gets delayed (Weber & Kauffman, 2011), perhaps because some cultures do not readily accept new machines and technologies such as Robots and Artificial Intelligence (Yam et al., 2023). These delays lead to delays in intercultural exchange. Designing IT platforms to be culturally sensitive can address these issues (Häkkilä et al., 2020). The platforms should incorporate features and interfaces that accommodate diverse cultural preferences and offer a space for mutual understanding. IT developers should propose culturally sensitive communication strategies allowing stakeholders to express their concerns freely and, if required, voice their concerns. Their voice should be allowed to be heard and understood by all other stakeholders. Ongoing research and user testing of applications should include checkpoints to identify and mitigate cultural biases in technology design. Intercultural collaboration and communication have become a new challenge (Papadima-Sophocleous, 2003).

With this backdrop, this section attempts to discuss three questions:

1. How do cultural norms and values influence how people share information online and through social media?
2. What challenges arise in intercultural communication due to differences in information technology use and expectations?

3. How can IT platforms and tools be designed to be more culturally sensitive and promote understanding across cultures?

Cultural Norms and Values' Influence on Online Information Sharing

This research theme attempts to study and answer the question:

How do cultural norms and values influence how people share information online and through social media?

The rise of social media and online platforms is a significant trend in online environments (Guţă, 2021). These tools changed the way people communicate. The increasing number of online interactions shows that they have become standard tools for information dissemination in the 21st century (Sabedini, 2017). However, beneath the "borderless world" lies the persistent influence of cultural norms and values. Understanding these aspects is crucial for building a culturally aware information technology (IT).

Cultural norms and values shape how people use and perceive information sharing through online platforms and social media. Studies have already shown how communication styles, privacy concerns, views on self-expression, and community greatly vary across cultures. Culture influences an individual's perception and the choice of communication media (Kampermann et al., 2021). Some cultural dimensions influencing how individuals share information online include Individualism versus Collectivism, Power Distance, Uncertainty Avoidance, and Masculinity/Femininity.

Beliefs and behaviors of different cultures influence how information is shared online and how privacy-conscious the users are (Li et al., 2022). Research on East Asian harmonious group interactions indicates that Asian students socialize predominantly with others from the same racial, cultural, and national background (Mahaffey et al., 2011). These harmonious group interactions are far more than in Western cultures. The East Asian groups prioritize independence and individualism. Social media engagement patterns clearly show this. East Asians are more likely to share updates focused on families and communities (Gravili, 2016).

On the other hand, Americans and Europeans tend to share personal achievements, opinions, or lifestyle content. Like in North America, individualistic cultures prioritize self-expression and personal branding online. Sharing achievements, opinions, and curated experiences becomes a form of self-presentation. Collectivistic cultures are prevalent in Asian countries such as China and South Korea. These cultures emphasize group harmony, interconnectedness, and social respect (Hyun & Yoan, 2018). Here, information sharing is filtered based on its potential impact on the collective. Dissonance or criticism may be expressed indirectly or in veiled language to avoid disrupting group dynamics.

Privacy attitudes also differ. East Asians, especially Chinese and Koreans, are generally more circumspect about sharing private details due to cultural emphasis on restraint and modesty (Broeder & Lee, 2016). For example, sharing personal information in some cultures is considered rude, while it is acceptable to others. In order to communicate freely online, it is essential to be aware of people's different cultural norms and values.

Japan is an example of a strong culture and high-context communication style. The country emphasizes implicit (hidden) meanings and nonverbal signals (Broeder, 2021). Much information is conveyed indirectly through shared understanding (Izotova, 2012). Public online information might be sparse, with emphasis on in-group communication channels. Conversely, low-context cultures, like the United States, tend towards explicit communication. The social media profiles of people from the US will be more detailed. The information they share online will usually be more precise and direct (Gravili, 2016).

Cultural conceptions of truth and knowledge also play a significant role. Cultures prioritizing logical reasoning and empirical evidence might value factual accuracy and online data-driven information. On the other hand, cultures with a more general view of knowledge accept subjective experiences and emotional narratives alongside objective facts. Algorithmic bias within IT can worsen these disparities. They make access to specific knowledge systems a privilege and marginalize others.

IT and Intercultural Communication Challenges

This research agenda deals with the following:

What challenges arise in intercultural communication due to differences in information technology use and expectations?

Intercultural communication will be challenging from a cultural epistemology point of view. Differences exist in IT usage, value, and expectations of what they deliver (Ebrahimi et al., 2010). These differences lead to misunderstandings, conflict, and even discrimination. Another reason could be because of access to and familiarity with information technology by different cultures. People of some cultures have access to a wide range of information. They are also very comfortable using it to communicate. They see technology as a means of empowerment and liberation. However, people in some other cultures may have limited access to IT and related infrastructure. While some cannot afford the technology (Reddick et al., 2020), others may not be as comfortable using it even if they have access to it. They may use technology as a tool for surveillance and control. Examples are technology tools such as social media, email, and devices such as smartphones. This difference at a basic level of access and familiarity can lead to misunderstandings and exclusion.

A culture that values verbal (oral) tradition is less likely to embrace technology than one that prioritizes written communication (Xudoyqulov & Abdurazzoqov, 2022). Similarly, a culture that values direct and explicit communication may struggle to adapt to technology that favors indirect and implicit communication (Gildert et al., 2019).

Also, different cultures have different expectations about the role of IT in communication. People in some cultures are expected to be available always to communicate. They have higher expectations for speed and immediacy in communication. For example, in many business environments in the United States, there is an expectation for employees to be readily available via email or messaging platforms during work hours. In some other cultures, people indulge in communicating only during certain times of the day or week. They are not fast-paced and usually think deeply before expressing their views. For example, in some European countries, like Germany, there is a stronger emphasis on work-life balance. It is common for employees to have specific hours during which they are available for communication. Expecting and waiting for their replies can lead to frustration, misunderstandings, and conflict. These hurt more when the expectations are unmet. Some cultures view technology as a tool to facilitate communication. Some others view it as a barrier to authentic communication.

Culturally Sensitive IT Design for Cross-Cultural Understanding

This research agenda discusses the following question:

How can IT platforms and tools be designed to be more culturally sensitive and promote understanding across cultures?

Designing culturally sensitive IT platforms and tools is a step towards advancing diverse cultures. As a part of this, developers must identify and appreciate how knowledge, values, and norms are constructed and negotiated within different cultural contexts.

IT platforms generally have inherent biases and assumptions from the design and development processes. Recognizing and acknowledging the presence of these is a first step towards this. These biases often come from the designers' cultural backgrounds and their personalities. They inadvertently prioritize specific cultural perspectives over others, which must be mitigated. A concerted effort should involve diverse cultural perspectives and epistemologies throughout the design process. The process should promote an inclusive and collaborative approach.

IT developers must also acknowledge and allow societal changes. There are three ways in which these changes appear:

1. Changes in communication styles
2. Changes in language preferences
3. Changes in symbolic representations.

These elements are essential and will be present in various cultures. IT developers can take the help of experts to identify cultural changes. Then, a list of changes that are to be made to the applications is to be prepared. Once the changes are made, they will be reviewed again to ensure that the cultural changes are reflected in their applications and that no new undesired outcomes are introduced. Finally, these subtle societal changes appear in the application in the form of:

1. User interfaces
2. Content
3. Interaction modalities.

The IT application should have sufficient provisions allowing its users to customize. The documentation should clearly explain the steps required to make the changes. This way, users from more cultural contexts can use the application comfortably. Users will be able to engage with the technology in a manner that aligns with their cultural norms and expectations. For example, India is a culturally rich country. GPay from Google is a popular mobile payment application in India. It uses pictures and graphics that connect well with Indian culture. Some examples of this are the use of pictures of indigenous toys, coconut trees, and pull carts seen in rural India. Payment systems in India reflect cultural preferences, affecting inclusion and digital participation (M. Manda et al., 2024).

IT platforms and tools should provide ample intercultural dialogue and knowledge exchange opportunities. They should promote an environment where diverse cultural perspectives can be shared, understood, and appreciated. Respectful discourse can be encouraged as a feature of the application. The features facilitate cultural learning and allow users to express and explore their identities and worldviews.

IT systems should include some culturally relevant content. The examples used in them should resonate with diverse audiences. They increase the relevance and accessibility of the technology. It also promotes a deeper understanding. It promotes appreciation of cultural differences, challenging ethnocentric biases, and cultural sensitivity. Japanese messaging app Line, for example, resonates with its diverse user base. The app's emojis, regional dialects, cultural references, and festival observations reflect this.

Ethical considerations should always be at the forefront of the design process. They help the IT platforms and tools respect and uphold all users' privacy, dignity, and cultural values. IT developers should implement robust data protection measures.

They should provide transparency regarding data collection and usage practices. They act as cultural contexts where privacy and data sovereignty are highly valued.

Figure 1 lists ten areas where IT platforms and tools can be culturally sensitive. Some famous examples are Google Translate, Wikipedia, and The Rosetta Project. These IT platforms and tools are designed to be more culturally sensitive and promote understanding across cultures.

Figure 1. Making IT platforms & tools culturally sensitive

(Author's creation)

While cultural norms shape how individuals communicate online, access to the technologies enabling such communication is unevenly distributed. The insights drawn from this section on cultural influences highlight the need to examine structural inequalities in access and participation. It naturally leads to the next dimension of our research agenda - the role of culture in shaping—and being shaped by—the global digital divide.

DIGITAL DIVIDE AND GLOBAL ACCESS TO INFORMATION

Cultural factors worsen the digital divide. They can create unequal access to information and communication technologies. Some factors limiting some populations' ability to participate in the digital world include differences in technological literacy, access to technology infrastructure, economic resources, and social norms. IT can play a crucial role in bridging the digital divide. IT should be accomplished by improving connectivity, developing culturally relevant educational content, and creating inclusive online platforms.

Technology is essential to accessing information. It is a communication tool that creates opportunities for cultural exchange. Marginalized communities need it to stay in sync with the broader society. Culturally appropriate digital resources should be co-created with local stakeholders (Häkkilä et al., 2020). Digital transformation must align with local cultural practices to ensure inclusive adoption (Mampilly & Manda, 2023). Such resources empower underserved groups and help achieve greater global understanding. Addressing the digital divide requires giving priority to cultural sensitivity and equity. For example, in some cultures, women and girls have less access to ICT than men and boys. The difference in access could be due to gender stereotypes and discrimination. Such things reinforce the digital divide, which is not apt for the progress of society.

Considering this backdrop, this section raises and attempts to discuss the following three questions:

1. How do cultural factors contribute to the digital divide and the gap in access to information and communication technologies between different populations?
2. What role can IT play in reducing this gap and promoting cultural exchange opportunities for all?
3. How can culturally appropriate educational resources be developed and disseminated through IT platforms?

Cultural Factors and Digital Divide

The central question of this research agenda is as follows:

How do cultural factors contribute to the digital divide and the gap in access to information and communication technologies between different populations?

Cultural factors sometimes promote the digital divide. So, it is a dual-sided sword. Culture's positive or negative impact on the digital divide depends on how society uses it. Culture influences Information and Communication Technologies (ICTs) and vice-versa (Ladislav, 2009). The ICT gaps between different populations can be attributed to the five important reasons:

1. **Perceiving and Valuing IT:** Culture shapes how people perceive and value ICTs. In some cultures, ICTs are seen to be beneficial tools. They help in education, economic development, and social progress. In other cultures, ICTs are felt to be luxury tools. These cultures would see ICT as a potential threat to their traditional values. These different perceptions of ICTs can lead to significant gaps, eventually impacting their access and use.
2. **Cultural Hesitance**: In some societies, there is deep skepticism or resistance to new technologies. It is primarily due to their traditional beliefs, values, or practices. This cultural hesitation can prohibit or delay the introduction and use of ICT. It widens the gap between those who readily embrace digital tools and

those who remain on the periphery. In some cultures, indigenous knowledge systems (IKS) are not well aligned with digital (cyber) trends (Adrian, 2012). It leads to tension and resistance to adopting ICTs (Winschiers-Theophilus et al., 2021). So, IKS and ICT could become incompatible with cultural traditions or worldviews. Similarly, some societies highly value face-to-face interactions. Such societies are less inclined to adopt online communication tools.

3. **Resource & Opportunity Distribution**: Culture influences the distribution of resources and opportunities (Joshua, 2011). There is a digital divide between urban and rural areas, rich and poor, and men and women in many societies. This divide is often due to insufficient investment in ICT infrastructure and training in underserved communities. Access to education, particularly in STEM fields, is very limited in some cultures. Access is only for specific groups based on gender, ethnicity, or socio-economic status. This lack of exposure and training in technological domains creates barriers to fully participating in the digital environment. Further, historical experiences influence digital access. Formerly colonized countries are still struggling with infrastructural inequalities inherited from colonial eras.
4. **IT Interactions**: Culture can shape how people interact with ICTs and vice versa. Users are increasingly involved in electronic cultural activities as consumers and producers. ICT plays a significant role in enhancing interaction between citizens and cultural institutions. It provides greater access to cultural resources (Ardielli, 2017). For example, some cultures have strong norms around privacy and data protection. People in these cultures are reluctant to use ICTs for specific purposes. These cultures still follow traditional banking and healthcare methods.
5. **Design and Development**: Culture influences how ICTs are designed and developed (Gajendran & Brewer, 2012). For example, Western cultural bias can be in many ICT designs. People from other cultures find it difficult to use such ICT tools. A linguistic divide can hamper effective engagement with digital resources and services.

Figure 2 illustrates the reasons behind the ICT gaps between different populations.

Figure 2. Reason for ICT gaps between different populations

(Author's creation)

IT as a Bridge for Cultural Exchange & Inclusion

This research agenda's central question is as follows:

What role can IT play in reducing this gap and promoting cultural exchange opportunities for all?

Information Technology (IT) has the potential to bridge cultural gaps (McKay, 2025). It helps in promoting cultural exchange opportunities across diverse backgrounds. IT tools and platforms can help in three significant ways: Intercultural dialogue, Cross-cultural learning, and Understanding and appreciation of different cultures.

Virtual Tools and Online Connectivity

IT helps create virtual platforms and online communities. These platforms are primarily established for educational purposes as technology becomes accessible (Garber, 2004). These tools span geographical boundaries and allow individuals from different cultural backgrounds to connect, interact, and share their perspectives. Online platforms and forums, social media, video conferencing tools, virtual meetings, and webinars provide spaces for cultural exchange. These instruments help individuals to engage in meaningful conversations. They can share their experiences and learn from one another. They help share perspectives and gain insights into cultural practices, beliefs, and traditions. These platforms promote a sense of global interconnectedness. They break down cultural barriers, promote empathy and mutual understanding, and make digital humanities an intersectional meeting place (Svensson, 2016).

Repository

IT provides access to a vast repository of cultural content (Tsipi et al., 2023). The content can be literature, music, art, and film. Online libraries, databases, and, more recently, streaming services are tools that allow access to the content. These tools allow individuals to connect with cultural expressions from around the world. This exposure to different cultural objects and perspectives widens horizons. They challenge stereotypes and cultivate a deeper appreciation for the richness and diversity of human culture. Online museums, art galleries, and cultural repositories make heritage accessible to everyone, regardless of location (Khosrow-Pour, 2018).

Dissemination

IT can facilitate the distribution of multimedia content. Some common forms of such multimedia content include documentaries, interactive exhibitions, and educational resources. Traditional tools allow individuals to share their cultural experiences, traditions, and stories with a global audience (Shao & Crook, 2015). Famous tools are blogs, podcasts, videos, and e-books. These multimedia tools can effectively convey the richness and complexity of diverse cultures because of colorful content, audio and music, and video and graphical elements. Immersive technologies, such as Virtual Reality (VR) and Augmented Reality (AR), are the latest tools in the market. They offer interactive ways to experience and explore different cultural landscapes. It can promote empathy and cultural awareness. VR experiences provide valuable knowledge acquisition, especially in cultural education (Christopoulos et al., 2024) and virtual tours of historical sites, landmarks, and cultural events. People can use them to explore distant places without leaving their homes.

Information Technology empowers individuals to document and share their cultural traditions, practices, and narratives. Some tools that allow marginalized voices to be heard and celebrated are Digital Storytelling, Online Exhibitions, and Community-driven Archives.

They promote cultural preservation and revitalization. An example is the depiction of the Indian historical and religious figure Hanuman. Comics, television, and online content widely portray Hanuman. The medium thus bridges tradition and modernity, contributing to cultural preservation and appreciation. IT tools allow individuals to share their unique perspectives. Hence, IT tools create platforms for cultural expression and help build an inclusive and diverse culture-rich society.

Culturally Appropriate Educational Resources through IT platforms

This research theme discusses the following question:

How can culturally appropriate educational resources be developed and disseminated through IT platforms?

Culturally Relevant Pedagogy (CRP) is a term coined by Gloria Ladson-Billings. It is a famous academic teaching approach that helps students learn with their cultural strengths, backgrounds, and experiences (Bolander, 2024). It follows an approach that helps students relate to experiences from their own lives. IT developers can also use this academic approach to disseminate culturally appropriate educational resources through their IT platforms. The traditional CRP is built on five fundamental principles (Brown-Jeffy & Cooper, 2011) as illustrated in Figure 3.

Figure 3. The five principles of traditional CRP

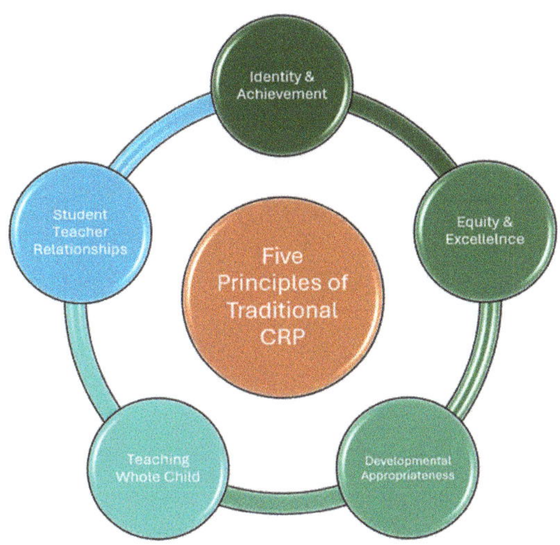

Author's creation with inputs from Brown-Jeffy and Cooper (2011)

Understanding Local Context

Developers are required to understand the local context where teaching and learning occur. As a part of this, they need to learn and understand the concepts of cultural norms, values, and linguistic nuances. They must engage with community members and cultural experts and take their guidance. With the help of collaboration, developers can identify specific educational needs and preferences. An example is the collaborative innovation between India and Canada, wherein knowledge transfer and translation through ICT were done (Gupta & Pal, 2020). Based on these inputs, they can build culture-rich applications. Application development is best done in a collaborative mode, which encourages teamwork. Then, developers from outside the team should be engaged to conduct an independent code review. Such practices ensure that cultural elements are rightfully added. It must be understood that educational content cannot be made relevant and practical without this knowledge and effort.

Flexible Methodological Approaches

CRP encourages flexibility in teaching methods. Educators should be able to change their teaching approaches according to the tastes and needs of diverse learning styles and cultural backgrounds. Similarly, IT tools should be built to be flexible and customizable. This flexibility ensures that educational technology caters to a broader audience. More people from different cultures will be able to use it.

Language Adaptation

Educational applications should use language that aligns with the linguistic proficiency of their user (the student). Content developers should avoid jargon and complex terminology to the extent possible. Developers should build applications that include accessibility features. Using familiar themes, colors, aesthetic styles, and cultural references also helps.

Design Principles

Developers must learn, practice, and use culturally sensitive design principles. Educational materials should reflect the target culture's visual aesthetics, language, and storytelling traditions. Such applications improve engagement. It makes the learning process more meaningful for learners.

Blended Learning

Combining face-to-face instruction with technology creates a better dialogue between teachers and students. Blended learning allows for personalized interactions while utilizing what IT platforms offer. The Smart Classroom of Singapore and MathCloud in Bhutan are examples of this.

Equitable Treatment

CRP emphasizes treating students equitably irrespective of their competency level. Developers can create inclusive resources that empower all learners by addressing cultural and linguistic barriers. Further, the platforms must be user-friendly and accessible to all learners, regardless of their technical proficiency. Although the cultural deficit situation of the 1960s was the prime reason for the wide use of CRP, recent literature has begun to stress that culture can be used as an equity strategy. This has caused research to focus on culturally relevant teaching as something that a teacher can 'do' instead of a critical stance that a teacher takes (Schmeichel, 2012).

Ongoing Evaluation and Feedback

These mechanisms help assess the effectiveness of the resources. It allows developers to make necessary adjustments. It is an iterative process and requires much patience. The result ensures that the resources remain relevant and responsive to the community's evolving needs.

Beyond access, the quality and fairness of information experienced by users depend heavily on the algorithms that govern digital platforms. Even when individuals gain access to technology, they may encounter systems that reflect and reinforce cultural biases. This brings us to the third dimension of our research agenda: understanding how cultural biases become embedded in algorithms and what ethical implications arise from such biases.

INFORMATION BIAS AND ALGORITHMIC FAIRNESS

Digital technologies are regarded as 'Products of Culture' (Wesner, 2022). However, the algorithms that power many digital technologies can perpetuate and amplify cultural biases, often subtly. The data and assumptions used to train these algorithms may reflect the perspectives and experiences of their creators. Consequently, the search results will be skewed. The content recommendations on online and OTT platforms appear to be incorrect. Incorrect suggestions given by the system will complicate the decision-making process. This can lead to several ethical implications (Martin, 2019). It limits access to information, opportunities, and fair treatment to marginalized groups. IT developers must prioritize algorithmic fairness and diversity to address this challenge (Grgić-Hlača et al., 2017). The data sources used for the application should be critically evaluated. The applications need to be exclusively tested for biases. The design process should incorporate input from diverse cultural perspectives. Information bias can be countered to an extent by using proactive measures. All these steps can help create more inclusive and equitable digital experiences.

Considering this backdrop, this section deals with the following three questions:

1. How do cultural biases get embedded in algorithms used in search engines, social media platforms, and other IT applications?
2. What are the ethical implications of these biases for access to information and opportunities?
3. How can IT be developed to promote cultural diversity and counter information bias?

Cultural Bias in Algorithms, Platforms, and Applications

This theme deals with the following question:

How do cultural biases get embedded in algorithms used in search engines, social media platforms, and other IT applications?

Cultural biases can quickly come into digital tools through various routes and mechanisms (Litts et al., 2021). The most common forms of bias entering into IT systems are through data, design, and societal context (Ntoutsi et al., 2020). Here are some ways by which this happens:

1. **Training Data**: Algorithms learn from historical data. The data fed to the algorithm might already have cultural biases, societal biases, stereotypes, and discriminatory patterns. The algorithm will inherit those biases if training data predominantly represents certain groups or perspectives. For example, suppose the training data is primarily taken from websites written by people from a particular cultural background. In that case, the algorithms will learn to prioritize results that align with that culture's perspectives. Similarly, the data is sometimes skewed or lacks representation from specific demographics. This results in the final models inevitably perpetuating these imbalances.
2. **Algorithm Design**: Design choices can inadvertently lead to cultural biases. These can be in the form of feature selection, model architecture, and optimization criteria. For instance, if an algorithm prioritizes engagement metrics, it may amplify sensational or polarizing content. Sometimes, the algorithms themselves can be designed with inherent biases. For example, consider an algorithm that ranks search results like a search engine. Assume that the results are based on popularity. This will result in the search engine showing content that favors the majority culture. This can happen even when the content does not accurately represent the diversity of perspectives of society.

3. **Implicit Assumptions**: IT Developers and Engineers make assumptions about user behavior, relevance, and quality. These assumptions are influenced by their cultural norms, values, experiences, and unconscious biases. The assumptions influence the design and implementation of these systems and result in biased outcomes. Audits of live development environments help. Such audits revealed that even minor, seemingly harmless design choices make a huge difference. The design choices can be as simple as the selection of variables or the prioritization of specific outcomes. Even these minor choices reflect cultural assumptions and may ultimately cause undesirable outcomes.
4. **Feedback Loops**: Algorithms learn and transform themselves from user interactions. The algorithm may even support biased content if it is (statistically proven) engaging. The optimization objectives and performance metrics used to train and evaluate algorithms may prioritize outcomes that align with dominant cultural norms and values. This further increases bias. For example, maximizing user engagement is a goal for website user interface designers. Similarly, a marketer monitors click-through rates as part of the marketing campaign. By focusing too much on performance metrics, these professionals could inadvertently build content that could harm stereotypes or marginalize certain groups.
5. **Lack of Diversity**: Large IT consulting companies developing software and services have global teams. Homogeneous development teams are groups of people similar in race, gender, ethnicity, and other demographic characteristics. These groups may overlook cultural nuances. Cultural nuances are subtle differences in behavior, communication, and values (Cao et al., 2024). They can vary from one culture to another. The opposite of a Homogeneous development team is a Diverse development team. Members of diverse teams come from different backgrounds and experiences (Lazar et al., 2020). It should be remembered that diversity is not just about race, gender, and ethnicity. It can also include diversity of thought, experience, and perspective. Diverse perspectives are crucial for identifying and addressing biases.
6. **Contextual Blind Spots**: Algorithms lack contextual understanding. They are designed to follow the rules or instructions based on which they are coded and tuned. They are less able to understand the context or meaning of the data they are processing. While technological advances are reducing the gap between human and machine thinking, it is still a work in progress. Till then, it is assumed that algorithms may misinterpret cultural references, slang, or regional variations.
7. **Societal Norms**: Algorithms reflect societal norms. This includes stereotypes and discriminatory beliefs. Biased training data reinforces (emphasizes) these norms. The way the algorithms are used can lead to perpetuating cultural biases. For instance, consider a social media platform that uses an algorithm to personalize user content. It may reinforce existing biases by showing users content

that aligns with their beliefs and preferences. Doing this limits their exposure to alternative perspectives.

8. **Data Labeling**: Data labeling is the process of tagging images or audio clips with labels (Chaki & Dey, 2021). It is typical for human annotators to be engaged in adding these labels. These human annotators could introduce their biases when labeling data during the system training. These biases propagate to the algorithm. Engaging a diversified set of human annotators can partially reduce this bias (Fan et al., 2022).
9. **Ethical Considerations**: Ethical guidelines during algorithm development are essential. Without them, biases continue to keep coming into the system. IT development teams should have members trained to monitor ethical considerations in the project (Becker et al., 2023).
10. **Feedback Mechanisms**: Machine learning algorithms require feedback mechanisms to control bias. This can be done with the help of regular audits, maintaining transparency levels, and getting user feedback. Internal or external auditors can conduct audits. Several different techniques are used in the audit process. Some famous methods are statistical analysis, case studies, and user surveys. Transparency levels are to be maintained by algorithm developers. They should document the processes to the best of their ability. User feedback mechanisms are often built into the IT application, though user surveys are also done outside the application environment.

Addressing these biases is critical and requires Interdisciplinary efforts, Ethical awareness, and Ongoing vigilance. If systems and applications with biases are developed, it will result in skewed search results, disproportionate content moderation, or discriminatory decision-making processes.

Biased systems will lead to systemic inequalities. They shape societal narratives to make the dominant cultural perspectives stand out. Hence, developers should be careful to ensure that algorithms are used in a way that promotes equity and inclusion.

Ethical Implications of Algorithmic Biases

This theme deals with the following question:

What are the ethical implications of these biases for access to information and opportunities?

Algorithms pose significant ethical implications of cultural biases. The biases lead to systemic discrimination (Bigman et al., 2023). Underrepresented groups would get marginalized. This further accelerates existing societal inequalities (Stypinska, 2023). Sometimes, specific demographics are excessively excluded or misrepresented within the digital ecosystem. Their voices, narratives, perspectives, and lived experiences get ignored. These biases can be divided into two dimensions.

1. **Access to Information:** This leads to limited exposure to diverse perspectives and the silencing of marginalized voices (such as gendered language).
2. **Access to Opportunities:** This leads to:
   a. Discrimination in hiring and lending (job opportunities)
   b. Reduced access to education (such as in college admissions)
   c. Healthcare (healthcare disparities)

Figure 4. Addressing ethical implications of cultural biases in algorithms

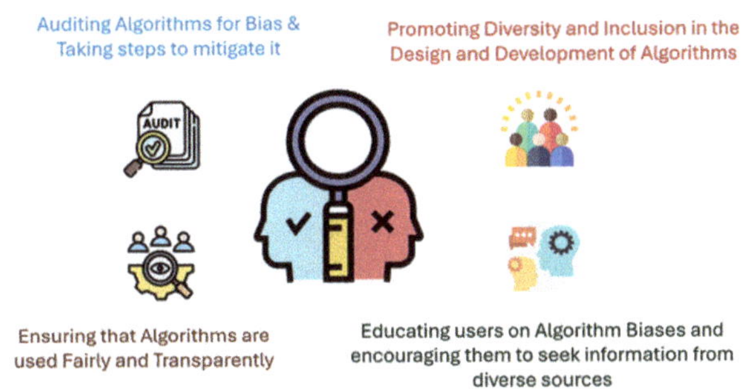

(Author's creation)

Various ways of addressing the ethical implications of cultural biases in algorithms are shown in Figure 4. Addressing the ethical implications can help ensure that algorithms promote equity and inclusion. Failure to address them will lead to the strengthening of existing biases and inequalities.

Countering Information Bias

This theme deals with the following question:

How can IT be developed to promote cultural diversity and counter information bias?

Information Technology (IT) can be developed to promote cultural diversity and to counter information bias. IT systems and platforms should be designed to promote diverse cultural narratives and perspectives. They should develop ways for users to get to know the global culture.

Promoting Cultural Diversity through Design

The first step towards promoting cultural diversity is recognizing and accommodating diverse cultural perspectives. As a part of this, IT systems should be built to recognize the cultural embeddedness of knowledge. They should support diverse knowledge representations and communications. They should incorporate cross-cultural design to the extent possible. To do this, IT developers should have some basic understanding of culture. Hence, they should be trained in cultural epistemologies, worldviews, and communication styles.

Culturally Responsive Design (CRD) principles can help account for various cultural groups' unique worldviews, values, and communication styles. This could involve collaborating with cultural experts and community representatives. This ensures that user interfaces, content curation algorithms, and data collection methods are inclusive and culturally sensitive.

Global trends are towards decentralization (Babarykina et al., 2022), such as decentralized governance, decentralized financial products such as cryptocurrencies, and decentralized insurance (V. K. Manda, Abukari, et al., 2024). Keeping in line with the global trends, IT developers should create decentralized applications and community-driven platforms. The platforms should empower marginalized cultures. They should allow the communities to curate and disseminate their narratives. When marginalized cultures can use the platforms, they challenge the dominance of mainstream media and information sources. These platforms use disruption technologies such as the Blockchain, Metaverse, and Web3 to improve the experience while reflecting the evolving cultural values around ownership and digital identity (Khaliq & Manda, 2023). Some early research shows that Blockchain adoption, though in its early stages, can potentially enhance public cultural services (Rubino et al., 2025). Technology brings trust, transparency, accountability, and equitable representation.

Integrating Artificial Intelligence (AI) and Machine Learning (ML) algorithms in IT systems can help. They should be accompanied by strong measures to detect and mitigate inherent biases. AI and ML systems should use diverse and representative training data. They can use bias-mitigation techniques such as adversarial debiasing and causal reasoning. They help test, measure, and eliminate potential bias that can come into the system. IT platforms should provide culturally tailored content and interfaces.. Firstly, translating content into multiple languages is a common practice. Then, incorporating culturally relevant examples and imagery helps. Finally, designing interfaces consistent with cultural norms avoids differences when starting.

Facilitating Intercultural Dialogue

IT platforms can allow intercultural dialogue and exchange. The platforms allow users from different cultures to connect, share knowledge, and learn from each other. It can help break down stereotypes and promote understanding of a global community (Zachariev, 2006).

Countering Information Bias

IT systems can be designed to counter information bias. This can be achieved by providing users with access to a wide range of perspectives and sources of information. Algorithms can be tuned to promote diversity in search results. Biased content should automatically get flagged. Users should be provided with tools to evaluate the credibility of information (Keshavarz, 2021).

Promoting critical thinking

IT can promote critical thinking (Pucer et al., 2014). It can build awareness and promote media literacy. Users can be provided with the tools and resources they need to evaluate the information they encounter online. Table 2 provides a list of tools that help users. These tools help users identify and challenge biased or inaccurate information. They can make informed decisions about the information they consume. They can cross-check facts before sharing them with others.

Table 2. Tools for promoting critical thinking and building media literacy

Name of tool	Description
Educational Materials	Online courses, tutorials, and guides provide critical thinking skills.
Fact-checking tools	Allow users to check the accuracy of claims and statements in news articles, social media posts, and other online content.
Bias checkers	Users will be able to identify potential biases in news articles, social media posts, and other online content.
News aggregators	Allow users to access a variety of news sources from different perspectives.
Social media monitoring tools	Allow users to track and analyze social media conversations to identify trends and spot misinformation.

Source: Author's creation

Addressing Linguistic Bias

Linguistic Representation: This topic is broad, but two popular representations exist. The Language Modeling Bias is an AI-based language technology that often favoring dominant languages, leaving underserved languages behind. Developers must address this bias by extending language models to diverse linguistic contexts (Wolfe & Caliskan, 2022). Similarly, Valuing Linguistic Differences is a simplistic understanding of diversity that hinders progress. Acknowledge profound linguistic differences and their impact on knowledge production.

Socio-Cultural Relevance: Systems should express socio-culturally relevant notions across communities, avoiding overemphasizing dominant powers.

Epistemic Justice: Recognizing Epistemic Injustice: Language technology biases can lead to epistemic injustice, where marginalized language communities' needs are overlooked. Developers must consider diverse perspectives.

Implicit Bias Reduction: This might exist in several forms, two of which are Demographic and Cultural Shifts and Inclusive Hiring Practices.

As AI becomes increasingly central to information systems, the risks and opportunities related to cultural bias grow more pronounced. AI not only inherits biases from data and design but also has the potential to continue stereotypes or promote intercultural understanding. The final section of this chapter explores how AI can be developed with cultural awareness and used as a tool for promoting cross-cultural dialogue and equity.

ARTIFICIAL INTELLIGENCE AND CULTURAL UNDERSTANDING

Artificial Intelligence (AI) is reshaping art and creativity and is expanding artistic audiences (Yadav et al., 2024). AI can play a vital role in cultural heritage protection (Zhan, 2025). Ensuring that AI systems are developed with cultural awareness and sensitivity becomes crucial as they are integral to daily lives. AI models can inadvertently perpetuate cultural stereotypes if trained on biased data or designed without input from diverse perspectives. Emerging neuro-AI tools raise cultural questions about cognition, agency, and identity (Thatta et al., 2024). Ethical considerations are essential when AI is applied in high-stakes domains like Education, Healthcare, and Employment. The cultural context significantly impacts outcomes in these domains. AI in leadership decisions must account for cultural variations in authority and hierarchy (V. K. Manda, Christy, et al., 2024). However, technology can also be used to bridge cultural divides. AI-powered language translation, cultural heritage preservation, and educational platforms can promote mutual understanding and appreciation across cultures. Collaboration between AI developers, cultural experts, and end-users helps create AI systems that are culturally inclusive and enriching.

This section deals with three questions:

1. How can AI be developed to be more culturally aware and avoid perpetuating cultural stereotypes?
2. What ethical considerations arise when AI is used in education, healthcare, or employment, where cultural context matters significantly?
3. How can IT be used to promote intercultural understanding through language translation tools, cultural heritage preservation initiatives, and AI-powered educational platforms?

Developing Culturally Aware AI

This theme deals with the following question:

How can AI be developed to be more culturally aware and avoid perpetuating cultural stereotypes?

Considering AI's increasing importance and presence in human lives, AI must be developed to be more culturally aware and avoid perpetuating cultural stereotypes. Figure 5 shows the key steps involved in developing culturally aware AI systems.

Figure 5. Steps involved in developing culturally aware AI systems

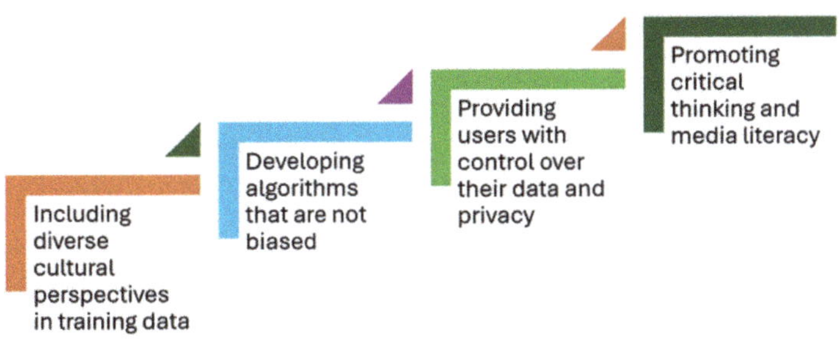

Author's creation

AI systems can generate fake news and other forms of misinformation. A whole branch of such systems, such as deepfakes, has evolved. People can identify and challenge cultural stereotypes when promoting critical thinking and media literacy. It leads to improved informed decisions about the information they consume.

Ethical Considerations Using AI in Cultural Context

This theme deals with the following question:

What ethical considerations arise when AI is used in education, healthcare, or employment, where cultural context matters significantly?

AI has increased use in education, healthcare, and employment. This increased use is now raising ethical considerations regarding the significance of cultural context. From a cultural perspective, the seven critical dimensions of ethics in AI are explained in Table 3.

Table 3. Ethical AI considerations from a cultural perspective

Dimension	Description
Bias & Fairness	Inherited biases and perpetuated stereotypes in training data cause unfair outcomes in AI systems.
Data Privacy & Ownership	Cultural norms and values vary regarding collecting and using personal data in large language models to train the AI system.
Transparency & Accountability	It requires transparency about how AI systems are designed and used and accountability for their decisions.
Cultural Sensitivity	Trust and acceptance of AI systems depend on how they respect diverse cultures' beliefs, values, and practices.
Cultural Homogenization	A one-size-fits-all approach for AI systems might ignore the unique cultural contexts of different communities.
Cultural Appropriation	AI systems may exploit cultural knowledge and practices without proper acknowledgment or compensation.
Cultural Diversity and Inclusion	AI systems should promote cultural diversity and inclusion, ensuring they are accessible and beneficial to individuals from all cultural backgrounds.

Source: Author's creation

Promoting Intercultural Understanding

This theme deals with the following question: *How can IT be used to promote intercultural understanding through language translation tools, cultural heritage preservation initiatives, and AI-powered educational platforms?*

Information Technology (IT) is a vital enabler of intercultural understanding. So, it can help promote intercultural understanding. Various technologies that are part of IT can facilitate cross-cultural dialogue and preserve and disseminate diverse cultural narratives. They help us appreciate the plurality of ways of knowing and being.

Language Translation Tools

Language translation tools can help to break down language barriers. They facilitate communication between people from different cultures. They make it easier for people to understand each other. Modern language translation tools have become technologically robust. Tools like Automated Speech Recognition (ASR), Natural Language Processing (NLP), and Machine Translation are becoming common. These tools enable more seamless communication across cultures. However, care should be taken that these tools are designed to account for the touches of cultural context, idioms, and figurative expressions. This way, the essence of cultural narratives is preserved during translation.

Cultural Heritage Preservation Initiatives

Information Technology can be used to preserve and promote cultural heritage. IT applications such as those that help in Digitizing Cultural Artifacts, Creating Online Archives, and Developing Educational Resources are some examples. These initiatives can help to raise awareness of different cultures and to foster a sense of global community. ASR and NLP were used to transcribe oral histories, digitize written records, and create interactive language learning applications. They help document different ways of speaking, like dialects, linguistic subtleties, and languages in danger of disappearing. This makes them available to more people who want to learn about them. AR and VR technologies can make cultural treasures accessible globally. They promote cross-cultural appreciation and understanding. Digital archives, virtual museums, and immersive experiences are some modes to achieve this.

AI-Powered Educational Platforms

AI-powered educational platforms help create personalized learning experiences (Barrera Castro et al., 2024). These learning experiences are created and customized according to the learner's needs. The platforms allow learners to access various cultural content and resources. They expose learners to different perspectives and worldviews. This way, they can promote intercultural understanding.

Several personalized AI tools exist - language-learning apps, chatbots, and virtual language tutors. These tools create immersive language-learning experiences. These platforms encourage younger generations to reconnect with their heritage languages (Hannaford et al., 2024). They bring different cultural perspectives, epistemologies (ways of learning), and teaching-learning methods (pedagogical approaches). They can challenge Western-centric biases. They help get a deeper understanding of different worldviews and knowledge systems.

Technology has a significant role in promoting intercultural understanding on a broader scale. It provides a platform for people from different cultures to connect and share their experiences. Social media, online forums, and other communication tools enabled by technology can help break down stereotypes and build bridges between people from diverse backgrounds. In this way, IT developers can create a more inclusive and interconnected world where people from all cultures can learn from and appreciate each other.

This exploration of AI's role in cultural understanding concludes the core thematic analysis of the chapter. These four dimensions—communication, access, bias, and AI—illustrate the deep interconnection between culture and technology through the lens of Cultural Epistemology. The following section outlines potential avenues for future research that can build upon and extend this framework.

FUTURE RESEARCH DIRECTIONS

This section aims to give some ideas for future researchers in this area. Future research can be pursued as a continuation or extension of this study in the following ways:

1. **Factors:** Discussing intersecting factors like gender, economic inequality, or institutional policies shapes digital access. Including these perspectives would help address cultural diversity more fully and make the arguments even more robust.
2. **Theoretical Connection:** Research can involve thorough studies that connect to various theories. For example, Hofstede's work on cultural dimensions, Hall's theory of high- and low-context communication, or more foundational literature on algorithmic fairness can be a topic of researcher interest. Similarly, the study can highlight existing frameworks or toolkits such as UNESCO's "Guiding Principles for AI and Ethics", WHO's Digital Inclusion Guidelines, or the World Economic Forum's insights on inclusive digital infrastructure.
3. **Case Study Approach:** Conducting cross-cultural case studies on AI systems is necessary. Some topics on this include facial recognition, hiring algorithms, and language models. The case study can critically examine how culturally specific biases emerge, are reinforced, and impact marginalized communities differently across global contexts.
4. **Real-world cases:** Real-world examples or case studies from different regions can be used to ground the theory. The case study approach can benefit both management practitioners and IT professionals. Case studies or practical examples

of culturally aware IT systems currently in use across various domains, such as education, healthcare, and public administration, can be discussed separately.

5. **User-Centered Design**: A thorough study on the role of user-centered design (UCD) methodologies can be conducted to assess how well the systems are in ensuring culturally responsive technology. This chapter touches on inclusive design principles. However, a more explicit discussion of how UCD processes can be done is provided. For example, the study can discuss co-design workshops, contextual inquiries, or participatory design with underrepresented groups. Such research can promote more profound cultural sensitivity and would be a valuable addition.
6. **Institutional Insights**: A detailed analysis of institutional or industry efforts to implement culturally responsive tech design (e.g., global design standards, corporate responsibility initiatives, or open-source community practices) can be conducted. Integrating such insights would provide a broader lens on how these issues are being addressed at scale.

Having reviewed the current landscape and proposed directions for future inquiry, the study concludes by synthesizing the key insights. It reiterates the importance of a culturally epistemological approach to IT and AI development.

CONCLUSION

This chapter discusses the relationship between Culture and Information Technology (IT) from the lens of Cultural Epistemology (CE). Four dimensions in which Culture and IT influence each other are first identified: Impact of Culture on Information Sharing and Communication, Digital Divide and Global Access to Information, Information Bias and Algorithmic Fairness, and Artificial Intelligence and Cultural Understanding. IT applications used in culturally sensitive contexts require additional ethical considerations. Culture and IT are deeply intertwined. Culture influences how people use and perceive IT and can shape cultural norms and values. This relationship has important implications for the design and use of IT systems. It is essential to consider the cultural context in which IT systems will be used and to ensure that they are designed to be culturally sensitive and inclusive. Understanding the relationship between culture and IT helps develop more effective policies, practices, and systems that promote the equitable use of IT and address the digital divide. IT systems thus built will be more culturally sensitive and inclusive. This helps to build a more just and equitable world.

REFERENCES

Adrian, A. (2012). An old narrative for a new culture: What cyber-culture can learn from indigenous culture. *International Journal of Liability and Scientific Enquiry*, *5*(3/4), 242. DOI: 10.1504/IJLSE.2012.051959

Ardichvili, A., Maurer, M., Li, W., Wentling, T., & Stuedemann, R. (2006). Cultural influences on knowledge sharing through online communities of practice. *Journal of Knowledge Management*, *10*(1), 94–107. DOI: 10.1108/13673270610650139

Ardielli, E. (2017). Usage of ICT in EU Culture Sector as a Tool of Cultural Participation. *International Journal of Entrepreneurial Knowledge*, *5*(2), 33–46. DOI: 10.1515/ijek-2017-0009

Babarykina, N., Demianenko, O., & Mahda, Y. (2022). Decentralization as a global trend of democracy development. *Cuestiones Políticas*, *40*(73), 454–467. DOI: 10.46398/cuestpol.4073.24

. Barrera Castro, G. P., Chiappe, A., Becerra Rodriguez, D. F., & Sepulveda, F. G. (2024). Harnessing AI for Education 4.0: Drivers of Personalized Learning. *Electronic Journal of E-Learning, 22*(5), 01–14. DOI: 10.34190/ejel.22.5.3467

Becker, S. J., Nemat, A. T., Lucas, S., Heinitz, R. M., Klevesath, M., & Charton, J. E. (2023). A Code of Digital Ethics: Laying the foundation for digital ethics in a science and technology company. *AI & Society*, *38*(6), 2629–2639. DOI: 10.1007/s00146-021-01376-w

Bigman, Y. E., Wilson, D., Arnestad, M. N., Waytz, A., & Gray, K. (2023). Algorithmic discrimination causes less moral outrage than human discrimination. *Journal of Experimental Psychology. General*, *152*(1), 4–27. DOI: 10.1037/xge0001250 PMID: 35758989

Bolander, E. C. (2024). Online Learning: Engaging Families Using Culturally Relevant Pedagogy. *Intervention in School and Clinic*, *59*(3), 199–206. DOI: 10.1177/10534512231156889

Broeder, P. (2021). Informed Communication in High Context and Low Context Cultures. *Journal of Education, Innovation, and Communication*, *3*(1). Advance online publication. DOI: 10.34097/jeicom-3-1-june21-1

. Broeder, P., & Lee, Y. (2016). Location Privacy Online: China, The Netherlands and South Korea. *International Journal of Security, Privacy and Trust Management, 5*(4), 01–10. DOI: 10.5121/ijsptm.2016.5401

Brown-Jeffy, S., & Cooper, J. E. (2011). Toward a Conceptual Framework of Culturally Relevant Pedagogy. *Teacher Education Quarterly*. https://files.eric.ed.gov/fulltext/EJ914924.pdf

Cao, Y., Chen, M., & Hershcovich, D. (2024). *Bridging Cultural Nuances in Dialogue Agents through Cultural Value Surveys* (Version 2). arXiv. DOI: 10.48550/ARXIV.2401.10352

Chaki, J., & Dey, N. (2021). Data Tagging in Medical Images: A Survey of the State-of-Art. *Current Medical Imaging Formerly Current Medical Imaging Reviews*, *16*(10), 1214–1228. DOI: 10.2174/1573405616666200218130043 PMID: 32108002

Chameroy, F., Salgado, S., De Barnier, V., & Chaney, D. (2024). In platform we trust: How interchangeability affects trust decisions in collaborative consumption. *Technological Forecasting and Social Change*, *198*, 122997. DOI: 10.1016/j.techfore.2023.122997

Christopoulos, A., Styliou, M., Ntalas, N., & Stylios, C. (2024). The Impact of Immersive Virtual Reality on Knowledge Acquisition and Adolescent Perceptions in Cultural Education. *Information (Basel)*, *15*(5), 261. DOI: 10.3390/info15050261

Ebrahimi, N., Singh, S. K. G., & Tabrizi, R. S. (2010). *Cultural Effect On Using New Technologies*. DOI: 10.5281/ZENODO.1056192

Fan, S., Barlas, P., Christoforou, E., Otterbacher, J., Sadiq, S., & Demartini, G. (2022). Socio-Economic Diversity in Human Annotations. *14th ACM Web Science Conference 2022*, 98–109. DOI: 10.1145/3501247.3531588

Gajendran, T., & Brewer, G. (2012). Cultural consciousness and the effective implementation of information and communication technology. *Construction Innovation*, *12*(2), 179–197. DOI: 10.1108/14714171211215930

Garber, D. (2004). Growing Virtual Communities. *International Review of Research in Open and Distance Learning*, *5*(2). Advance online publication. DOI: 10.19173/irrodl.v5i2.177

Gibbs, J. L., Gibson, C. B., Grushina, S. V., & Dunlop, P. D. (2021). Understanding orientations to participation: Overcoming status differences to foster engagement in global teams. *European Journal of Work and Organizational Psychology*, *30*(5), 653–671. DOI: 10.1080/1359432X.2020.1844796

Gildert, N., Pomfret, A., & Timmis, J. (2019). Comparing The Performance of Explicit and Implicit Communication in Simple Object Manipulation Tasks. *2019 IEEE 15th International Conference on Automation Science and Engineering (CASE)*, 1456–1462. DOI: 10.1109/COASE.2019.8843011

Gravili, G. (2016). Sharing Knowledge through Social Media: The Influence of National Cultures. *Proceedings of the 2016 International Conference on Artificial Intelligence: Technologies and Applications*. 2016 International Conference on Artificial Intelligence: Technologies and Applications, Bangkok, Thailand. DOI: 10.2991/icaita-16.2016.34

Grgić-Hlača, N., Zafar, M. B., Gummadi, K. P., & Weller, A. (2017). *On Fairness, Diversity and Randomness in Algorithmic Decision Making* (No. arXiv:1706.10208). arXiv. https://arxiv.org/abs/1706.10208

Gupta, R., & Pal, S. K. (2020). Learning and Developing With Each Other: Case of Collaborative Innovation Through ICT in India and Canada. In Vij Mali, N. (Ed.), (pp. 132–146). Advances in Electronic Government, Digital Divide, and Regional Development. IGI Global., DOI: 10.4018/978-1-5225-5412-7.ch006

Guţă, A. J. (2021). The role of social media platforms in increasing the efficiency of online communication activity. *MATEC Web of Conferences, 342*, 08011. DOI: 10.1051/matecconf/202134208011

Häkkilä, J., Wiberg, M., Eira, N. J., Seppänen, T., Juuso, I., Mäkikalli, M., & Wolf, K. (2020). Design Sensibilities—Designing for Cultural Sensitivity. *Proceedings of the 11th Nordic Conference on Human-Computer Interaction: Shaping Experiences, Shaping Society*, 1–3. DOI: 10.1145/3419249.3420100

Hannaford, E. D., Schlegel, V., Lewis, R., Ramsden, S., Bunn, J., Moore, J., Alexander, M., Barker, H., Batista-Navarro, R., Hughes, L., & Nenadic, G. (2024). Our Heritage, Our Stories: Developing AI tools to link and support community-generated digital cultural heritage. *The Journal of Documentation*. Advance online publication. DOI: 10.1108/JD-03-2024-0057

Hyun, K. K., & Yoan, R. S. (2018). Variations in the Asian Collectivistic Working Culture in Intercultural Collaboration: A Case of a South Korean Company in Malaysia. *Social Sciences & Humanities*, *26*(3), 1633–1650.

Izotova, N. N. (2012). Ethno-cultural characteristics of Japanese communication style. *MGIMO Review of International Relations*, *6*(27), 179–182. DOI: 10.24833/2071-8160-2012-6-27-179-182

Joshua, R. (2011). *How Competition over Limited Natural Resources Impacts the Formation of Culture* [Undergraduate Capstone Project / Master of Liberal Studies, University College, University of Denver]. https://digitalcommons.du.edu/ucol_gs/8/

Kampermann, A., Opdenakker, R., Heijden, B. V. D., & Bücker, J. (2021). Intercultural Competencies for Fostering Technology-Mediated Collaboration in Developing Countries. *Sustainability, 13*(14), 7790. DOI: 10.3390/su13147790

Keshavarz, H. (2021). Evaluating credibility of social media information: Current challenges, research directions and practical criteria. *Information Discovery and Delivery, 49*(4), 269–279. DOI: 10.1108/IDD-03-2020-0033

Khaliq, L. N., & Manda, V. K. (2023). Customer Experience in the Web 3.0 Era: The Meeting of Blockchain and the Metaverse. In Majeed, M., Ofori, K. S., Amoako, G. K., Alolo, A.-R., & Awini, G. (Eds.), (pp. 258–274). Advances in Marketing, Customer Relationship Management, and E-Services. IGI Global., DOI: 10.4018/978-1-6684-7649-9.ch015

Khosrow-Pour, D. B. A. M. (Ed.). (2018). *Enhancing Art, Culture, and Design With Technological Integration:* IGI Global. DOI: 10.4018/978-1-5225-5023-5

Korkut, P. K., Dolmaci, M., & Karaca, B. (2018). A Study on Communication Breakdowns: Sources of Misunderstanding in a Cross-Cultural Setting. *Eurasian Journal of Educational Research, 18*(78), 1–20. DOI: 10.14689/ejer.2018.78.7

Kumari, P., & Nirban, V. S. (2018). Intercultural communication competencies in the Indian Information Technology industry. *International Journal of Cross Cultural Management, 18*(3), 327–347. DOI: 10.1177/1470595818812996

Ladislav, Š. (2009). ICT influence on the formation of cultural knowledge. *Vysoká Škola Ekonomická v Praze*. https://core.ac.uk/outputs/30108111/?source=oai

Lazar, M., Miron-Spektor, E., Agarwal, R., Erez, M., Goldfarb, B., & Chen, G. (2020). Entrepreneurial Team Formation. *The Academy of Management Annals, 14*(1), 29–59. DOI: 10.5465/annals.2017.0131

Li, Y., Rho, E. H. R., & Kobsa, A. (2022). Cultural differences in the effects of contextual factors and privacy concerns on users' privacy decision on social networking sites. *Behaviour & Information Technology, 41*(3), 655–677. DOI: 10.1080/0144929X.2020.1831608

Litts, B. K., Searle, K. A., Brayboy, B. M. J., & Kafai, Y. B. (2021). Computing for all?: Examining critical biases in computational tools for learning. *British Journal of Educational Technology, 52*(2), 842–857. DOI: 10.1111/bjet.13059

Liu, J. (2010). Culture and Knowledge Transfe: Theoretical Considerations. *Journal of Service Science and Management, 03*(01), 159–164. DOI: 10.4236/jssm.2010.31020

Mahaffey, C. J., Heim, E., & Wu, C. (2011). *Asian American Social Interaction at the University of Illinois at Urbana Champaign*. University of Illinois. http://hdl.handle.net/2142/34683

Mampilly, A. J., & Manda, V. K. (2023). Driving Change: Securing Success through Digital Transformation. In *Modern Management Challenges: A Case Compendium* (pp. 196–205). Excel India Publishers. http://dx.doi.org/DOI: 10.13140/RG.2.2.15789.05605

Manda, M., Yadav, A., & Khaliq, L. N. (2024). Role of Payment Systems in Financial Well-Being: An Examination from India. In *Emerging Perspectives on Financial Well-Being* (p. 24). IGI Global., https://www.igi-global.com/gateway/chapter/347786

Manda, V. K., Abukari, A. M., Gupta, V., & Bharathi, M. J. (2024). Revolutionizing Finance With Decentralized Finance (DeFi). In Darwish, D. (Ed.), (pp. 127–150). Advances in Web Technologies and Engineering. IGI Global., DOI: 10.4018/979-8-3693-1532-3.ch006

Manda, V. K., Christy, V., & Jitta, M. R. (2024). Ethical AI and Decision-Making in Management Leadership. In Bhattacharya, P., Hassan, A., Liu, H., & Bhushan, B. (Eds.), (pp. 197–226). Advances in Human and Social Aspects of Technology. IGI Global., DOI: 10.4018/979-8-3693-4147-6.ch009

Martin, K. (2019). Ethical Implications and Accountability of Algorithms. *Journal of Business Ethics*, *160*(4), 835–850. DOI: 10.1007/s10551-018-3921-3

McKay, E. (2025). *Human-Computer Interaction Closes the Digital Divide: A multicultural, intergenerational ICT case study*. https://dl.acm.org/doi/pdf/10.5555/1151681.1151685

Ntoutsi, E., Fafalios, P., Gadiraju, U., Iosifidis, V., Nejdl, W., Vidal, M., Ruggieri, S., Turini, F., Papadopoulos, S., Krasanakis, E., Kompatsiaris, I., Kinder-Kurlanda, K., Wagner, C., Karimi, F., Fernandez, M., Alani, H., Berendt, B., Kruegel, T., Heinze, C., & Staab, S. (2020). Bias in data-driven artificial intelligence systems—An introductory survey. *Wiley Interdisciplinary Reviews. Data Mining and Knowledge Discovery*, *10*(3), e1356. DOI: 10.1002/widm.1356

Papadima-Sophocleous, S. (2003). A Web of Intercultural Communication. *Informatologia*, *3*(36), 166–241.

Passey, D. (2015). Inclusive Technologies and Learning: Research, Practice and Policy. *Social Inclusion (Lisboa)*, *3*(6), 1–6. DOI: 10.17645/si.v3i6.543

Pucer, P., Trobec, I., & Žvanut, B. (2014). An information communication technology based approach for the acquisition of critical thinking skills. *Nurse Education Today, 34*(6), 964–970. DOI: 10.1016/j.nedt.2014.01.011 PMID: 24581890

Reddick, C. G., Enriquez, R., Harris, R. J., & Sharma, B. (2020). Determinants of broadband access and affordability: An analysis of a community survey on the digital divide. *Cities (London, England), 106*, 102904. DOI: 10.1016/j.cities.2020.102904 PMID: 32921864

Rubino, F., Agostino, D., & Spallazzo, D. (2025). Strategizing blockchain adoption in public cultural services: A comprehensive scoping review. *International Journal of Public Sector Management, 38*(1), 77–97. DOI: 10.1108/IJPSM-12-2023-0383

Sabedini, M. (2017, October 28). Social Media, their Platform, Respectively the Users, Ethics, Advantages and Disadvantages. *2017 UBT International Conference*. University for Business and Technology International Conference, Durres, Albania. DOI: 10.33107/ubt-ic.2017.212

Schmeichel, M. (2012). Good Teaching? An examination of culturally relevant pedagogy as an equity practice. *Journal of Curriculum Studies, 44*(2), 211–231. DOI: 10.1080/00220272.2011.591434

Shao, Y., & Crook, C. (2015). The Potential of a Mobile Group Blog to Support Cultural Learning Among Overseas Students. *Journal of Studies in International Education, 19*(5), 399–422. DOI: 10.1177/1028315315574101

Stypinska, J. (2023). AI ageism: A critical roadmap for studying age discrimination and exclusion in digitalized societies. *AI & Society, 38*(2), 665–677. DOI: 10.1007/s00146-022-01553-5 PMID: 36212226

Sulistyono, I. B., & Purwasito, A. Warto, & Pitana, T. (2023). *Epistemology of Cultural Study*. DOI: 10.5281/ZENODO.7747729

Svensson, P. (2016). *Big Digital Humanities: Imagining a Meeting Place for the Humanities and the Digital*. University of Michigan Press., DOI: 10.2307/j.ctv65sx0t

Thatta, S., Manda, V. K., & Athota, V. S. (2024). Research and Applications of Neurotechnologies for Leadership. In Saluja, S., Kukreja, J., & Sharma, S. (Eds.), (pp. 36–62). Advances in Business Strategy and Competitive Advantage. IGI Global., DOI: 10.4018/979-8-3693-1785-3.ch004

Tsipi, L., Vouyioukas, D., Loumos, G., Kargas, A., & Varoutas, D. (2023). Digital Repository as a Service (D-RaaS): Enhancing Access and Preservation of Cultural Heritage Artifacts. *Heritage, 6*(10), 6881–6900. DOI: 10.3390/heritage6100359

Wang, X., Zeng, M. L., Gao, J., & Zhao, K. (2024). *Intelligent Computing for Cultural Heritage: Global Achievements and China's Innovations* (1st ed.). Routledge., DOI: 10.4324/9781032707211

Weber, D. M., & Kauffman, R. J. (2011). What drives global ICT adoption? Analysis and research directions. *Electronic Commerce Research and Applications, 10*(6), 683–701. DOI: 10.1016/j.elerap.2011.01.001

Wesner, S. (2022). Cultural values as policy directives: Techniques, technologies and symbolic work. *International Journal of Cultural Policy, 28*(7), 858–874. DOI: 10.1080/10286632.2022.2137153

Winschiers-Theophilus, H., Blake, E., Maasz, D., Muashekele, C., Gallert, P., Stanley, C., & Koruhama, A. K. (2021). Embracing Indigenous Knowledge Systems in ICT-Enabled Education. In Lubin, I. A. (Ed.), *ICT and International Learning Ecologies* (1st ed., pp. 132–157). Routledge., DOI: 10.4324/9780429345463-9

Wolfe, R., & Caliskan, A. (2022). American == White in Multimodal Language-and-Image AI. *Proceedings of the 2022 AAAI/ACM Conference on AI, Ethics, and Society*, 800–812. DOI: 10.1145/3514094.3534136

Xudoyqulov, X., & Abdurazzoqov, F. (2022). *Speech culture development technologies among youth.* DOI: 10.5281/ZENODO.7271962

Yadav, M., Kumar, M., Sahoo, A., & Rathnasiri, M. S. H. (2024). AI and the Evolution of Artistic Expression: Impacts on Society and Culture. In Kilinç, U. (Ed.), (pp. 15–36). Advances in Media, Entertainment, and the Arts. IGI Global., DOI: 10.4018/979-8-3693-3916-9.ch002

Yam, K. C., Tan, T., Jackson, J. C., Shariff, A., & Gray, K. (2023). Cultural Differences in People's Reactions and Applications of Robots, Algorithms, and Artificial Intelligence. *Management and Organization Review, 19*(5), 859–875. DOI: 10.1017/mor.2023.21

Zachariev, Z. (2006). Éducation, Dialogue Interculturel et Société de L'information. *International Review of Education, 52*(5), 425–442. DOI: 10.1007/s11159-006-9002-3

Zhan, X. (2025). Artificial Intelligence in the Cultural Industry: Applications and Challenges. *Innovation in Science and Technology, 4*(1), 62–68. DOI: 10.56397/IST.2025.01.06

Zhang, X., & Zhou, M. (2023). Information and digital technology-assisted interventions to improve intercultural competence: A meta-analytical review. *Computers & Education, 194*, 104697. DOI: 10.1016/j.compedu.2022.104697

KEY TERMS AND DEFINITIONS

Algorithmic Fairness: Algorithmic fairness is the practice of ensuring that algorithms and artificial intelligence (AI) systems are fair and unbiased and that they do not discriminate against individuals or groups based on their protected characteristics, such as race, gender, age, religion, or other sensitive attributes.

Artificial Intelligence: AI is a branch of computer science that focuses on creating intelligent machines that can perform tasks that typically require human intelligence, such as perception, reasoning, learning, decision-making, and natural language processing.

Cultural Bias: The tendency to prefer or favor one culture over another. It can affect individuals, groups, and institutions, leading to unfair or inaccurate judgments and outcomes.

Cultural Norms: Cultural norms are unwritten rules and expectations that govern (and guide) the behavior of individuals within a particular cultural group. These are shared beliefs about what is acceptable, desirable, or appropriate in each context.

Dataset: A dataset is a collection of data organized and structured in a specific way for analyzing, processing, drawing conclusions, or making decisions.

Digital Divide: The digital divide is the gap between individuals, groups, or communities with access to modern information and communication technologies and those without.

Immersive Experience: An immersive experience is a type of experience that fully engages in the senses and mind of an individual. It creates a feeling of presence within a virtual or simulated environment. Immersive experiences can be created using a variety of technologies, such as virtual reality (VR), augmented reality (AR), mixed reality (MR), and 360-degree video.

Indigenous Knowledge Systems: Indigenous Knowledge Systems (IKS) refer to the traditional unique knowledge, beliefs, practices, and worldviews developed, maintained, and transmitted by indigenous peoples over generations.

IT Platforms: IT platforms are the underlying technological infrastructure and software that supports the development, deployment, and operation of digital applications, services, and systems.

Chapter 2
Algorithmic Decision-Making and the Ambiguity of Transparency in EU Digital Regulation

Maria Romana Allegri
https://orcid.org/0000-0001-8535-5131
Sapienza University of Rome, Italy

ABSTRACT

This paper explores the concept of algorithmic transparency, which aligns closely with algorithmic explainability. After highlighting the challenges of translating technological rules from natural to machine language and presenting attempts to define some measurable indicators of transparency, it addresses a few inconsistencies in the recent EU regulations on data protection, digital services and AI systems. Furthermore, it examines notable rulings from the highest Italian courts concerning the use of algorithms in public administration decision-making, pointing out existing uncertainties. It concludes that it would be necessary to produce measurable standards at a European level to be applied to the use of artificial intelligence systems by public and private entities, to effectively evaluate their transparency and fairness.

1. INTRODUCTION

Under Article 3(1) of the EU Regulation on Artificial Intelligence (AI Act)[1] «An AI system is a machine-based system designed to operate with varying levels of

DOI: 10.4018/979-8-3693-8729-0.ch002

autonomy and that may exhibit adaptiveness after deployment and that, for explicit or implicit objectives, infers, from the input it receives, how to generate outputs such as predictions, content, recommendations, or decisions that can influence physical or virtual environments». Today, artificial intelligence systems – regardless of the various and not always univocal definitions of AI provided at different times and levels – have become an essential part of daily life, so much so that we can talk about "social machines" (Simoncini, 2023, p. 4; Cristianini and Scantamburlo, 2020) to indicate those systems that arise from the structural interaction of human and technical elements.

Indeed, we live in an ecosystem based on human-machine interaction. Algorithms define and limit the epistemic horizons of individuals since they shape and determine how we come into contact with the world, our opportunities for knowledge and our possibilities of choice. They have come to the point of interfering with our moral freedom and the very construction of our self-identity (De Vivo, 2024, p. 54). The growing pervasiveness of AI systems in all areas of human life raises doubts about the effectiveness of Article 22 para. 1 of the General Data Protection Regulation (GDPR)[2], which states that «the data subject shall have the right not to be subject to a decision based solely on automated processing, including profiling, which produces legal effects concerning him or her or similarly significantly affects him or her». Indeed, automated data processing has now become a common standard. This may also lead to the disruptive idea that algorithmic decisions, for the sole fact that they are produced by technologically advanced machines and simplify the lives of human beings, are assumed uncritically as the natural order of things (Sterpa et al., 2024, p. 1152).

Beyond the specific problems related to single applications of AI technologies, a fundamental risk underlying the entire system seemingly consists of the loss of centrality of the anthropological element, given that we are faced with a (total or partial) delegation properly human cognitive functions to technological systems. Technology is no longer just a tool used to carry out human decisions more efficiently but is the acting subject that autonomously processes data and makes decisions suitable of impacting the legal sphere of individuals (Simoncini, 2023, p. 8). No new technology, until now, had ever become autonomous with human beings and capable of shaping an environment that develops, grows and innovates autonomously, going beyond the human physical and cognitive limits (Sterpa et al., 2023, p. 1124). In other words, technology has turned from a tool to an acting subject: machines are not only supposed to carry out what humans have decided but are also asked to decide independently (Simoncini 2019, p. 69) and, additionally, the knowledge on the basis of which human agents make their decisions is increasingly provided by automated technological systems, capable of influencing (or even determining) individual and collective decisions through various forms of nudging (Simoncini 2019, p. 70).

Additionally, the ability of AI systems to influence human behaviour and shape consciousness puts into crisis the very foundation of constitutional law, based on the centrality of the person and the guarantee of individual rights, if not of law *tout court* understood as a technique for controlling human behaviour, which is performed through linguistic statements. Indeed, in the field of AI hard law competes with other regulatory systems, sometimes synergistically and some others being overcome by other forms of regulation: the overall risk is that traditional legal methods and instruments will be gradually condemned to ineffectiveness (Mobilio, 2020, p. 404). The effectiveness of the law, therefore, is linked to its ability to not limit itself to a reactive approach – that is, merely reacting to the undesired consequences resulting from the spread of AI – but to qualify as a proactive tool, capable of perceiving trends and risks and anticipate the resolution of problems that may arise very shortly, thorough interaction and integration among different regulatory systems (Mobilio, 2020, p. 405).

Somehow algorithms play an ordering function similar to that of law: reacting to the chaos of infinite possibilities, they both order and rationalize the multiplicity of elements, making them available for the human mind, that is, they both take on complexity and produce simplicity, except that the algorithm does it with enormously superior speed and processing capacity compared to legal tools. On the other hand, unlike algorithms, the law is a flexible ordering tool: it leaves room for innovation, discretion and interpretation, it allows human behaviours and choices that contrast with it, it is axiologically oriented (i.e. it also takes moral, ethical and value factors into account) and it is legitimized by popular will in democratic systems (Sterpa et al., 2023, p. 1124). Legal rules are formulated in general and abstract terms so that they can be applied to a variety of different situations, have binding effects over time and may also extend to new situations not contemplated at the time of the original regulatory production. Therefore, law is expressed in an intrinsically malleable and some ways ambiguous language and can only be applied by resorting to interpretation through human judgment. Instead, technical rules are formulated in precise and formalized language which uses well-defined categories and a pre-established methodological choice (Lombardi, 2020, p. 10). Since only the law can guarantee that any ordering activity is oriented towards the preservation of the dignity and freedom of the person and social pluralism, the primacy of the legal order (therefore of the democratic-liberal values of constitutionalism) must be somehow defended (Sterpa et al., 2023, p. 1128). However, transposing legal rules into code, so that they can be applied automatically, requires their transcription into the formal language of machines, which involves a reduction of their degree of generality and abstractness and, consequently, their flexibility and adaptability (Lombardi, 2020, p. 11).

Now, it's a fact that any technological system originally derives from a creative human act and that, therefore, the human activity of designing the functioning of algorithms is at the origin of AI systems. However, the inferential and statistical-probabilistic dynamics through which technological systems "learn" from data are such that the "logic" underlying the decisions taken by machines is not comprehensible to humans, not even to the programmers of the algorithms themselves: it cannot be explained before algorithmic decisions have been taken nor can it be reconstructed *a posteriori* by retracing the chain of processes backwards. This happens because most algorithmic decisions, unlike those that inspire human actions, are based on the calculation of probabilities and not on logical-deductive processes in which the judgment is a consequence of the evaluation of assumptions in terms that are not only utilitarian but also ethical. It follows that simply disclosing information regarding the various phases of the algorithmic processes does not necessarily establish transparency.

The question of algorithmic transparency (AT) as a possible antidote to "technocracy" or "algocracy" (Chomanski, 2022; Danaher, 2016; Marrone, 2022; Zambonelli, 2023) constantly recurs in recent legislation and jurisprudence regarding data processing, digital services, automated decision-making and artificial intelligence. There is no doubt that individuals must be placed in a position to understand the logic underlying algorithmic decisions, to be able to protect their rights. It is not clear, however, how this can be achieved in practice. The objective of this paper is precisely to highlight the ambiguity intrinsically connected to the concept of algorithmic transparency and the vagueness of the related regulatory requirements, due to the difficulty of translating algorithmic processes into common language and the absence of precise parameters to measure the adequacy of their disclosure.

For this paper, we may define the notion of explainability as the ability to provide an "interface" between the algorithmic decision-maker and its human recipients that is at the same time both an accurate proxy of the decision-maker and comprehensible to humans. Similarly, the concept of interpretability is meant as the ability to explain or provide meaning in understandable terms to a human. Both notions refer to the idea of translating information from one language (the algorithmic one) to another (the human one) and both are aimed at making algorithms comprehensible. Nevertheless, attempts to arrive at unambiguous parameters to achieve AT have so far been found to be substantially inconsistent and fragmented. Many questions remain still unanswered, such as the exact meaning of interpretable, transparent and explainable models, comprehensible explanations and best ways to provide them, and reasons why explanations are necessary (Guidotti et al., 2019, pp. 3-5).

To sum up, starting from an attempt to define the elusive concept of algorithmic transparency, this paper moves on to examine the ambiguous and unclear solutions adopted in EU regulation. Indeed, as it emerges from the relevant jurisprudence of the Italian highest courts examined in par. 4, the concrete application of the principle of algorithmic transparency proves to be highly problematic, due to an irremediable (and untranslatable) difference between human language and the "language" of the algorithm.

2. DEFINITIONAL ISSUES: WHAT IS MEANT BY ALGORITHMIC TRANSPARENCY?

Algorithms are a systematic procedure that produces the answer to a question or the solution of a problem in a finite number of steps[3]. So, the nature of algorithms is procedural and consists of a sequence of input, transformation and output. Rule-based algorithms follow a simple "if → then" logic: their application as every instruction must be predefined, and they are usually not capable of dealing with new information or unforeseen problems. Machine learning systems (MLS) are instead designed for improving through experience: they can test the strength of correlations between many variables and outcomes to create predicting models that can estimate the likelihood of future behaviours or events. MLS can be supervised (when the computer is presented with example inputs and desired outputs) or unsupervised (when the computer is not given any labels and freely discovers data patterns). Any type of algorithm, however, as well as the underlying data that they draw on, are always shaped by their human creators. As such, they are far from neutral pieces of technology but reflect the (un)conscious preferences priorities and prejudices of those who build them and can perpetuate biases, inequalities and even corruption (Kossow et al., 2021, pp. 5-7).

Given that the concept of AT is related to the lack of information on the inner workings of algorithms, it can be defined as the «disclosure of information about algorithms to enable monitoring, checking, criticism, or intervention by interested parties» (Diakopoulos and Koliska, 2017, p. 811). However, a distinction has to be made between AT as *action* (i. e. the act of provisioning information about an algorithm to its users or interested third parties, which depends on the attitude of algorithm's owners or developers in economical, ethical and technical terms) and AT as *perception* (the impact of the information about an algorithm on its recipients, which is influenced by the methodology used for its explanation and the characteristics and experience of its recipients), that may lead to a discrepancy between the actual and perceived understanding of the algorithm (Bitzer et al., 2023, pp. 304-306). Scholars differ in the relevance attributed to the various shaping factors of AT as action and as perception, which does not contribute to an agreed definition of AT (Bitzer et al., 2023, pp. 313-316). Some even maintain that AT possesses an intrinsic manipulative potential, because it may induce users of the algorithmic systems to trust the explained norms, thus reducing critical awareness (Wang, 2022; Wang, 2023), although some others dispute the idea of an automatic connection between

AT and manipulation of vulnerable users, believing that this only occurs in cases where AT has been intentionally designed for preventing users from understanding some decision-making rules (Klenk, 2023).

In 2017 the Association for Computing Machinery (ATM) defined seven principles to be respected to ensure that organizations use personal data and algorithms fairly[4] (Watson and Nations, 2019, p. 495; Kossow et al., 2021, pp. 14-15): 1) awareness (ensuring that individuals recognize what personal data organizations collect, analyze, use, and share and the extent to which they automate decisions; 2) access and redress (investigating and correcting erroneous decisions); 3) accountability (ensuring that companies and the people who develop and use algorithms are responsible for their actions); 4) explanation (communicating the algorithm's logic in human terms); 5) data provenance (knowing the data sources used and their trustworthiness); 6) auditability (recording the processes followed in developing and using algorithms so they can be reviewed); 7) validation and testing (ensuring that automated systems perform as intended). The concept of AT is somehow dependent on all these principles, but mainly on the application of principles No. 1, 4 and 5.

Another guideline containing nine non-exhaustive principles aimed at enhancing AT by design is that published by the German think tank Bertelsmann Stiftung in 2019, entitled *Algo Rules*[5]. Such principles are formulated in quite general terms. For instance, statement No. 6 (Ensure intelligibility) simply explains that «information about the data and models on which the system is based, its architecture and potential effects must be published in easily understood terms» and that «it is important to check whether an objective can be achieved without a significant loss in quality through the use of a less complex algorithmic system that involves an easier to understand mode of operation». Furthermore, rule No. 7 – relating to the need to always be able to maintain human control over every phase of the algorithmic process, abstaining from using machine learning systems whose complexity prevents this – appears a bit naive, considering the speed with which the technological progress reaches ever-new frontiers, unimaginable until recently.

Some progress was made by Bertelsmann Stiftung in 2020, through the development of a document entitled *From Principles to Practice. An Interdisciplinary Framework to Operationalise AI Ethics*[6], that represents a noticeable attempt to reach the definition of observable indicators that may facilitate the respect of ethical standards (transparency, accountability, privacy, justice, reliability, environmental sustainability) in the design and use of algorithms. Concerning the issue of transparency, for example, the document provides some indicators linked to the origin of the data sets expressed in question form, the answers to which would allow to assess the adequacy of the disclosure, the reliability of the algorithmic models used, the interpretability and understandability of the algorithmic models for the target groups of recipients.

The most interesting aspect of such documents is that of identifying an operational method for the practical implementation of algorithmic fairness. Indeed, what is missing in the discourse on algorithmic explainability, inextricably linked to its fairness, is a shared ethical-normative substrate for evaluating whether an algorithm is "correct" in its procedure, explainable and, therefore, "legitimate". Unbiased algorithms are conceivable only on condition that interpretative categories avail themselves of shared values (Sterpa et al., 2023, p. 1128).

The intrinsic algorithmic opacity and the rising prominence of AI have drawn attention to the concept of explainable AT (xAT) as an antidote to the so-called "black box problem" (Bitzer et al., 2023, p. 297). Indeed, implementing AT in practice may prove highly problematic concerning black box systems, i. e. sophisticated machine-learning models to predict individual information, that may also be sensitive, by processing the digital traces that people leave behind them while performing everyday activities (Guidotti et al., 2019, p. 1). The internal processes used by black box models and the various weighted factors remain unknown and no human – not even the programmers and administrators of the machine or algorithm – knows or understands how the output was reached. In other words, «a black box predictor is a data mining and machine-learning obscure model, whose internals are either unknown to the observer or they are known but uninterpretable by humans» (Guidotti et al., 2019, p. 5). This may happen not only because the inner workings of algorithms are kept secret to protect intellectual property but also because deep learning algorithms create a huge number of non-linear relationships between inputs and outputs whose complexity makes it difficult for a person to understand which features and interactions lead to specific outputs[7]. Given that black box models are undesirable both because they may produce unidentifiable and unexplainable biased outputs and logic errors and make it difficult to determine who should be held accountable when outputs are flawed or outright dangerous, there is a general agreement on the need for the implementation of the principle stated in Article 14 para. 2(g) of the General Data Protection Regulation regarding the right of the data subject to obtain at least «meaningful information about the logic involved, as well as the significance and the envisaged consequences of such processing for the data subject». This presupposes that we know what constitutes a meaningful explanation, although research on this question in the context of AI systems has been so far limited, apart from some attempts (Lu et al., 2019).

Indeed, algorithmic opacity can either be intentional (when aimed at preventing malevolent actors from gaming the system or protecting intellectual property rights or trade secrets) or unintentional (when due to the technical complexity of the systems) (Şimşek, 2021, p. 21-25). It can also be relational – i. e. depending on the degree to which different categories of individuals have access to data processing activities that are relevant to them – or technology-based, when it concerns the extent to which experts having full access to the system can understand its inner functioning (Grochowski et al., 2021, p. 48). When addressing the issue of transparency, one has to make clear which type of opacity is to be amended: "exoteric" transparency is reached when a system is not relationally opaque, as it lets people interacting with it have access to the information that is meaningful to them, whereas "esoteric" transparency is achieved when a system is not technologically opaque, being understandable to experts, having full access to its internal functioning (Grochowski et al., 2021, p. 54). Computer scientists have developed various explanatory techniques and models of esoteric transparency, which are intended for technological experts (Grochowski et al., 2021, p. 55): for example, esoteric transparency may be achieved through reverse engineering methodologies (i. e. understanding the overall logic behind black box systems by deconstructing and analyzing them, which is sort of a *post hoc* interpretation) or transparent box design (i. e. designing algorithmic models so that their functioning rules are *a priori* disclosed and made comprehensible at least for those who possess a certain degree of technical skills)[8]. This type of disclosure, however, is not relevant for many "common" users of the systems, who should rather be granted exoteric transparency, which means receiving an explanation suitable to make them aware of how the system's decisions affect or may affect their situation. The ultimate level of exoteric transparency will thus depend on a wide range of communicative and dialectical aspects, some of which need to be addressed *ex-ante* (before the subject undergoes data processing) and some others *ex-post* (when a decision is already taken), so that information is tailored to the recipients' characteristics (Grochowski et al., 2021, pp. 55-57).

3. REGULATORY ISSUES AND UNSOLVED PROBLEMS

All this considered, how can the principle of algorithmic transparency, which is a recurring *mantra* in European Union legislation in the digital sector, be implemented in practice? The following are examples of the tendency to use vague wording in legislative acts without any indication of their possible practical implementation.

Transparency is indeed mentioned, together with lawfulness and fairness, in Article 5(a) of the GDPR as a fundamental principle of data processing. Further references to transparency, especially in its exoteric meaning, can be found in Article 12, regarding information and communication to be given to the data subject «in a concise, transparent, intelligible and easily accessible form, using clear and plain language» (para 1), in Articles 13 para. 2(f), 15, para. 1(h), and 22, referred to the right of the data subject, in case of automated data processing, to obtain «meaningful information about the logic involved, as well as the significance and the envisaged consequences of such processing», and in Article 14 para. 2(g), which burdens the data controller with such obligation in cases where data are not obtained directly from the data subject. Notably, the scope of application of Articles 15 and 22 GDPR is different from that of Articles 13 and 14: while the latter refers to a moment preceding the start of data processing, Article 15 refers to a subsequent phase, when the processing is already underway, thus legitimizing the request for further information, which may deal, for example, with the new algorithmically generated data or the functioning of the algorithms used in the process (Colapietro, 2023, p. 160).

Moreover, while Article 22 para. 3 provides the right of the data subject to human involvement in the algorithmic decision-making, para. 1 and 2 place narrow limitations to the use of algorithms, aimed to avoid the dominance of the machine over the human mind (De Minico, 2021, p. 28): the data subject shall have the right not to be subject to a decision based solely on automated processing, unless necessary in the context of a contract between the data subject and a data controller, or authorised by UE or Member State law laying down suitable measures to safeguard the data subject's rights and freedoms and legitimate interests, or is based on the data subject's explicit consent. Nevertheless, the right to obtain an *ex-post* explanation of the algorithmic decision is not explicitly enshrined in any of the GDPR provisions, being solely relegated in Recital No. 71, which also mentions the right of the data subject to challenge such decision: «the lack of motivation prevents the data subject from checking how the public power has used the algorithms that affect him. Should this case occur, we would have an updated version of the arbitrary and unmotivated oppression of individual rights and liberties of which the history of modern democracies delivers many examples» (De Minico, 2021, p. 31). The GDPR, therefore, does not establish any rights or specific obligations regarding the minimum explainability requirements of algorithmic technologies (Spiller, 2021, p. 425). Apart from this, the scopes of the relevant provisions of GDPR are limited since they are focused on the rights of the data subject (and on the related duties of the data controller), without envisaging any obligation of algorithmic disclosure to other individuals or the public (Şimşek, 2021, p. 33).

More recently the Digital Services Act (DSA)[9] has made a step forward in terms of transparency requirements, establishing some transparency obligations also towards public authorities, independent auditors, vetted researchers, and the public, although mainly in the form of *ex-post* explanations (transparency reporting, referred to in Articles 15 and 24). Article 27 DSA, however, postulates *ex-ante* transparency for recommender systems (Leerssen, 2020), whose «main parameters» and any options to modify or influence them must be set out «in plain and intelligible language». Such parameters should include at least the most significant criteria in determining the information suggested to the recipient of the service, and the reasons for their relative importance. Still, no explanations are provided about what exactly these parameters and criteria should consist of, not even in Recital No. 70, which basically repeats the wording of Article 27. Another important element in terms of transparency by design can be considered the prohibition, expressed in Article 25, of online interfaces that deceive or manipulate the recipients of the service, compromising their ability to make free and informed decisions (dark patterns). However, as explained by Recital No. 67, this does not have to do so much with algorithmic transparency, but rather with the design of the user interface, which must not be designed to be deceptive or excessively nudging. Finally, under Article 26 para. 1(c) and Recital No. 68, digital platforms presenting advertising must ensure, among other requirements, that meaningful information about the main parameters used to determine the recipient(s) of the advertisement and about how to change them is directly and easily accessible from the advertisement itself. But, again, the nature and type of these parameters are not made explicit, apart from the fact that they must provide «meaningful explanations of the logic used» expressed «in a clear and unambiguous manner».

However, the most delicate issue is the frequent normative recourse to the principle of transparency of AI systems in the EU regulation on Artificial Intelligence (AI Act), which is supposed to ensure that users are put in the condition to understand and adequately make use of such systems having acquired a general knowledge of their functioning. Notably, the concept of risk related to the implementation and use of AI systems, and that of the graduation of such risks, is the cornerstone around which the European regulation on artificial intelligence revolves (Casonato and Marchetti, 2021). Considering the need for intellectual and industrial property protection, the regulation provides that the algorithm's owner is not bound to disclose it entirely, but to limit the disclosure to the information necessary to the deployer for knowing and understanding the reasons (i. e. the "logical" procedure used) underlying the algorithmic decision itself (Colapietro, 2023, p. 172). In the case of high-risk AI systems, some transparency obligations are foreseen also towards public authorities and independent third parties involved in conformity assessment procedures.

Indeed, according to Recital No. 72, «high-risk AI systems should be designed in a manner to enable deployers to understand how the AI system works, evaluate its functionality, and comprehend its strengths and limitations». Deployers of such systems should receive information about «the characteristics, capabilities and limitations of performance of the AI system. These would cover information on possible known and foreseeable circumstances related to the use of the high-risk AI system, including deployer action that may influence system behaviour and performance, under which the AI system can lead to risks to health, safety, and fundamental rights, on the changes that have been pre-determined and assessed for conformity by the provider and on the relevant human oversight measures, including the measures to facilitate the interpretation of the outputs of the AI system by the deployers». Moreover, «deployers should, *inter alia*, be in a better position to make the correct choice of the system that they intend to use in light of the obligations applicable to them, be educated about the intended and precluded uses, and use the AI system correctly and as appropriate». To this end, the instructions of use should contain illustrative examples where appropriate, so to enhance legibility and accessibility of the information. Finally, providers of AI systems «should ensure that all documentation, including the instructions for use, contains meaningful, comprehensive, accessible and understandable information, taking into account the needs and foreseeable knowledge of the target deployers», which also implies that instructions are «made available in a language which can be easily understood by target deployers». Now, apart from the fact that Recitals are not legally binding norms but solely interpretative guidelines, not even the exhaustive wording of Recital No. 72 adequately clarifies how technical parameters can be fully explained in natural language and which degree of algorithmic comprehensibility is needed.

Coming to the legal provisions of the AI Act, in terms of transparency by design Article 13 sets out provisions referring to high-risk AI systems, that «shall be designed and developed in such a way to ensure that their operation is sufficiently transparent to enable deployers to interpret the system's output and use it appropriately» (i. e. the principle of esoteric transparency envisaged in para. 1) and «shall be accompanied by instructions for use in an appropriate digital format or otherwise that include concise, complete, correct and clear information that is relevant, accessible and comprehensible to deployers» (i. e. exoteric transparency in para. 2). Concerning the content of these instructions, para. 3 specifies a list of indicators which currently represents the most precise and exhaustive effort to formalize operational rules of algorithmic transparency: interestingly, some of the enlisted items aim to provide an overall insight into the inner workings of the algorithmic systems and some others are intended to provide the persons or groups of persons on which the system is intended to be used with a certain degree on "insight view" on it (Şimşek, 2021, p. 43). However, apart from the fact that this "decalogue" refers only to high-risk systems, even this list cannot escape ambiguity when it states in para. 2 that information must be «comprehensible to deployers» or when, in the last bullet point, it establishes that instructions shall contain «a description of the mechanisms included within the AI system that allows deployers to properly collect, store and interpret the logs»: what level of accuracy should this description have and in what language should it be expressed to be comprehensible? However, under Article 96, the Commission shall develop guidelines on the practical implementation of the Regulation, including those concerning transparency obligations: the effective regulation of transparency requirements is therefore postponed to a later time, without a specific deadline.

The above-mentioned concerns are particularly relevant in the case of the riskiest AI systems, namely those intended to directly interact with natural persons, those generating audiovisual deepfakes and those based on emotion recognition or biometric categorization. Article 50 solely requires that «the concerned natural persons are informed that they are interacting with an AI system, unless this is obvious from the point of view of a natural person who is reasonably well-informed, observant and circumspect, taking into account the circumstances and the context of use». Article 50 continues stating that «deployers of an emotion recognition system or a biometric categorization system shall inform of the operation of the system the natural persons exposed thereto» and that «deployers of an AI system that generates or manipulates image, audio or video content constituting a deep fake shall disclose that the content has been artificially generated or manipulated». Apart from the fact that the concept of "obvious" remains undefined from a regulatory point of view, what seems most worrying is that Article 50 limits the obligation to the mere disclosure of the exposition to certain types of AI systems, without considering the need to make their operating methods clear in some way. These concerns are particularly relevant when we think of the risks in terms of manipulation of consciousness that can derive from the development of technological brain-computer interface (BCI) systems. The Parliamentary Assembly of the Council of Europe explicitly addressed BCI systems as capable of creating «a fully symbiotic connection between the human brain and digital computing systems», with the consequent risk of a «profound violation of individual privacy and dignity, with the potential to subvert free will and violate the ultimate refuge of human freedom – the mind»[10]. The European Parliament, in its Resolution of 20 October 2020 relating to the ethical aspects of artificial intelligence[11], also underlined, among other issues, the need to avoid any use of AI «that might lead to inadmissible direct or indirect coercion, threaten to undermine psychological autonomy and mental health or lead to unjustified surveillance, deception or inadmissible manipulation» (para. 29). Subsequently, in the Resolution of 3 May 2022 on artificial intelligence in a digital age[12], the EP highlighted the risks of this type of manipulation for the stability of democratic systems.

The AI Act (Articles 51-56) pays particular attention to the so-called GPAIMS (general purpose AI models), i.e. systems (such as ChatGPT, for instance) having multiple intended and unintended purposes, among which image/speech recognition, audio/video generation, pattern detection, question answering, translation, etc., which raise a lot of hot questions because of their large size, opacity and potential to develop unexpected capabilities beyond those intended by their producers (EPRS, 2023; Trigueiro et al., 2024). An attempt to better specify GPAIMS transparency parameters is contained in Annex IX, dedicated to "transparency information", but even such a detailed list fails to unravel the issue of the clear comprehensibility of data processing and algorithmic decision-making procedures.

A similar ambiguity is reproduced in Article 14, which states that high-risk AI systems must be effectively overseen by a natural person when in use (*human in the loop* or HITL), thus reinforcing the provision already contained in Article 22 GDPR, para. 3 (Marchetti, 2021, p. 377). The natural person in charge of oversight must be enabled to properly understand the relevant capacities and limitations of the high-risk AI system, to remain aware of its possible tendency of automatically relying or over-relying on its output produced ("automation bias"), to correctly interpret its output employing the available interpretation tools and methods: one wonders exactly what type of explanation should these human supervisors receive to be put in the condition to efficiently perform these tasks[13]. In this regard, Recital No. 73 specifies that it is essential "that high-risk AI systems include mechanisms to guide and inform a natural person to whom human oversight has been assigned to make informed decisions if, when and how to intervene to avoid negative consequences or risks, or stop the system if it does not perform as intended"; nevertheless, the content of such information remains undetermined.

From this review of EU legislation, a certain subordination of legal rules to technical ones emerges. Instead, «mandatory law should take responsibility for the "first move". This expression calls upon the supranational decision-maker to defend its necessary independence from codes while dictating the main guidelines that discipline them» (De Minico, 2023, pp. 21-22). Indeed, although the presence of indeterminate and interpretable concepts in EU regulation implies flexibility in their application and, consequently, room for maneuver in favour of the Member States, the related risk is that this indeterminacy generates uncertainty, opening up the possibility of differentiated applications of the discipline and consequent increase in judicial litigation (Casonato and Marchetti, 2021, p. 422).

4. SOME SUGGESTIONS (AND MANY AMBIGUITIES) FROM THE ITALIAN CASE LAW

It has been argued that full algorithmic transparency is unachievable, not only due to the impossibility of explaining the reasons behind algorithmic decisions but also because, more simply, translating software into natural language inevitably involves a high degree of imprecision and approximation, not to mention the fact that some languages– especially English – seem more suitable than others to be used both in algorithm programming and in the translation of algorithmic language. Furthermore, unlike natural languages, which evolve automatically over time as a result of the use that speakers make of them, algorithmic languages are rather rigid and can only be modified through intentional processes of programming, development and updating (Caravaggion and Orofino, 2023, pp. 181-183).

So, what level of comprehensibility of the algorithm can be considered adequate to integrate the requirement of transparency? This issue has been addressed by some recent rulings of Italian courts (commented by Cappellazzo, 2022; Corrado, 2023; Grimaldi, 2022; Lo Sapio, 2021; Papa, 2022; Simoncini, 2023; Valsecchi, 2020), although with substantially unsatisfactory results. All these decisions deal with the use of algorithmic technologies by public entities that are bound to justify citizens the correctness and legality of their decision-making processes regardless of the tools used for this purpose. None of them, however, managed to resolve the problem of the intrinsic ambiguity of the concept of algorithmic transparency.

To start with, not even the display of the source code of the algorithm, i.e. the software developed with the programming language – as prescribed, for example, by TAR Lazio[14], section III bis, in judgement No. 3769 of 2017 (Grimaldi, 2022, pp. 81-82; Simoncini, 2023, pp. 27-30), concerning the use of algorithms to form the ranking of teachers destined to different places of employment – would solve the problem of algorithmic transparency, not only because computer language is obscure to non-experts, but also because it would not necessarily be suitable to explain the reasons behind a decision made by machine-learning algorithms, designed to "learn" from data and automatically evolve accordingly. In any case, Italian legislation and jurisprudence seem to look favourably on the idea of public administrations using exclusively or mainly open-source software (Corrado, 2023, pp. 210-214).

Secondly, natural language and specifically legal language may prove unsuitable to effectively translate algorithmic rules. According to Section IV of the Council of State[15] (judgement No. 2270 of 8 April 2019) (Grimaldi, 2022, pp. 87-88; Marchetti, 2021, pp. 374-376), the principle of algorithmic transparency implies the full knowability of a rule expressed in a language other than the legal one. Therefore, it is necessary that the "technical formula", which represents the algorithm, is accompanied by explanations translating it into the "legal rule" underlying it, to make it readable and understandable, both for citizens and for the judge. In particular, the knowability of the algorithm must be guaranteed in all aspects: namely its authors, the procedure used for its development, the decisional mechanism, including priorities assigned in the evaluation and decision-making procedure and the data selected as relevant. Only in this way, in fact, will the judge be put in the condition to examine the logic and reasonableness of the rules governing the algorithm and producing the final administrative decision. The same principles were expressed by the Council of State, section VI, in judgement No. 8472 of 13 December 2019 (Lo Sapio, 2021), according to which the minimum guarantee for any hypothesis of use of algorithms in public decision-making must consist of: a) full *ex-ante* knowability of the system used and the criteria applied; b) responsibility for the decision attributed to the public body that took it, which must be able to carry out the necessary assessment of the logic and legitimacy of the choice and the outcomes entrusted to the algorithm. Particularly, algorithmic knowability – specified with the same wording as in judgement No. 2270/2019 – is demanded to verify that algorithmic criteria, assumptions and outcomes comply with requirements and purposes pre-established by law or by the administration itself so that the rules according to which the algorithm is set are also reviewable.

Here, transparency takes on a "reinforced" dimension with variable geometry, which means it's intended as modulated differently depending on the recipient: it will be broader for the public official adopting the decision and for the judge in charge of the judicial review, more limited for the person on whom the effects of the decision reverberate. However, the failure of this reasoning is that it explicitly moves from the erroneous assumption that technical rules governing algorithms are the same as general administrative rules, produced by humans and simply applied by machines. Conversely, today artificial intelligence systems do not limit themselves to applying pre-established rules but can produce new rules based on the experience acquired through data processing. Indeed, the concept of transparency as a "right to explanation" is imposed by the algorithmic "linguistic opacity", which can hide the risk not only of non-comprehensibility but also of non-correspondence between legal prescriptions and rules "transposed" into machine language, considering that the transition from legal to computer language implies the task of choosing one of the 2500 programming languages available today (Lo Sapio, 2021, p. 247).

It is, therefore, necessary to clarify what exactly needs to be explained and how explainability can be conveyed in the legal status of algorithms, as well as how to resolve the dilemma between explainability and the black box, which characterizes machine learning and deep learning systems. Precisely in this regard, the Council of State (judgement No. 7891 of 25 November 2021) (Cappellazzo, 2022) has made a distinction between the notion of algorithm – a definite sequence of instructions, well defined and unambiguous, so that they can be executed mechanically and such as to produce a certain result – and that of artificial intelligence, where machine learning mechanisms create a system that is not limited only to applying software rules and pre-set parameters (as "traditional" algorithms do) but, on the contrary, constantly elaborates new inference criteria for data processing and makes efficient decisions based on these elaborations, using automatic learning processes.

More or less in line with the Council of State, the Italian Court of Cassation[16] (1st civil section, order No. 28358 of 10 October 2023) deemed the transparency and knowability requirements of the algorithm to be satisfied (in this case, a reputational rating algorithm) when the user is put in a position to understand the "executive scheme of the algorithm", i.e. the sequence of steps necessary to achieve a result, starting from certain data: these steps must be elementary, unambiguous, of a finite number, operable in a finite time and with a single result and must be explained to the user "in an unambiguous and detailed manner" using the terms of the common language and certainly not the mathematical language, which deployers are not required to understand. But the problem remains: how can mathematical procedures be described in common language unambiguously and with a sufficient level of detail? And, in the event of disputes, what indicators would allow an assessment of the adequacy of this explanation? Indeed, precisely in this case, the Court of first instance had considered that the level of algorithmic transparency was not satisfied since its description was limited to the impact of the data taken into consideration, without explaining the methods by which the final result was reached (i.e., the "specific weight" of the single factors considered and the interaction mechanisms among them).

This last objection, although not shared by the Court of Cassation, is indicative of an insoluble problem: the explanation of the algorithm "in words" inevitably produces a certain degree of obscurity and ambiguity. Perhaps a remedy, however partial, could consist of the legislative definition of indicators and parameters as widely shared as possible, according to which the level of algorithmic transparency could be measured. Indeed, when public bodies make use of AI systems in decision making, they end up entrusting a significant role in administrative organization and administrative action (often silently and almost surreptitiously) to systems engineers and software designers. Therefore, the use of AI in the adoption of administrative decisions must be governed by legal rules providing both for the possibility of delegating decision-making power to software systems (formal legality) and for the provisions regulating the exercise of such delegation (substantial legality) (Otranto, 2021, p. 199; Rangone, 2022, p. 487).

5. CONCLUSIONS

Despite all the above-mentioned definitional and applicational difficulties, it is unquestionable that we are all part of a "techno-anthropological" or even "anthropo-technocratic" ecosystem, in which we are simultaneously designers-producers of technological systems and passive and often unaware recipients of tech regulation. Therefore, resolving the issue of transparency and comprehensibility of algorithmic and AI software is an unavoidable priority if we do want to keep the human person at the centre of decision-making processes. This means that the entire regulation on algorithms and AI should not be understood as intended for the technology itself, but for the human beings who produce and make use of technology, so that the person is put at the centre of digital innovation and a new "technological humanism" can be achieved, respecting the Kantian imperative of the person as an end and not as a means (Colapietro, 2023, p. 172).

In principle, this assumption is unquestionable. However, many problems remain unsolved in this regard.

Firstly, a precondition for legitimizing the use of AI in public decisions is the reasonable expectation – not yet proven – that "intelligent" machines can differentiate without discriminating (Papa, 2022, p. 107). Conversely, the presumed neutrality of AI systems may conceal discriminatory practices caused by a flawed and prejudicial dataset, thus hindering the identification, challenge and sanction of discrimination perpetrated by this technology.

Secondly, the opacity of the algorithmic processes results in a serious deficiency in terms of knowledge of the motivation underlying the decision and, consequently, of its legitimacy (Fasan, 2022, p. 186; Rangone, 2021, pp. 482-483). It has been effectively noted that «the presence of technology in the public decision-making process demands the reshaping of the "mixed" decision-making process, due to the coexistence of human and mechanical will … In other words, if the public authority uses algorithms, this delegation of operation to the machine should not become an easy alibi for a reintroduction of authoritative obscurity» (De Minico, 2023, pp. 31-32). Therefore, transparency is crucial for the effectiveness of procedural guarantees: the right to contest a decision affecting a person cannot be fully exercised without a broad and clear motivation on how the automated decision has been adopted and the reasons why the public authority has accepted the machine's outcome or has decided to deviate from it.

Thirdly, unfortunately, the recent European regulation on AI does not distinguish between public and private entities, but only between providers and deployers of AI services, although public decision-makers should be asked to provide a much higher level of transparency than private ones to grant all citizens equal and reasonable treatment (Papa, 2022, p. 110). In particular, EU regulation does not oblige public bodies to disclose the fact that they avail themselves of algorithms in decision-making, unless it involves the processing of personal data: this does not ensure prior knowledge of how the algorithm has been "trained" to decide (Papa, 2022, pp. 110-112). Nevertheless, if we want to combine the efficiency of public decision-making processes with respect to the rule of law, the political decision becomes crucial in favouring the application of understandable technologies (white boxes) over those opaque (black boxes) (Grimaldi, 2022, p. 92). Transparency of public decision-making processes, to be extended to the artificial logics underlying them, constitutes an indispensable bulwark for the legitimacy of the decision guided by algorithms, since it allows citizens to discern the contribution of technique from the evaluation of politics, thus compensating the democratic deficit occurring when decision-making is delegated, even only partially, to automated machines (Papa, 2022, p. 110).

To this respect, the *Model Rules on Impact Assessment of Algorithmic Decision-Making Systems Used by Public Administration*, set up in 2022 by the European Law Institute (ELI) based in Wien[17], constitute an interesting attempt to supplement

European legislation on AI in the specific context of public administration that would provide solid safeguards to improve citizens' confidence in this technology. The central idea underlying the Model Rules, meant to serve as inspiration to national legislators, is an impact assessment procedure, aimed at evaluating the seriousness of the risks of different AI systems. Rule No. 6, in particular, prescribes the content of the impact assessment report, which contains various items applicable to algorithmic transparency, e. g. the nature and technical characteristics of the system, the selection of training, validation and testing data, the context in which the system is used, the system's interrelation with other digital systems, an assessment of the performance, effectiveness and efficiency of the system for reaching the intended public objectives, an assessment of the specific and systemic impact of the system on fundamental or other individual rights or interests, as well as on democracy, societal and environmental well-being. Another valuable experience is that of the UK government, which launched the *Algorithmic Transparency Recording Standard Hub*[18]: to help public sector organizations provide clear information about the algorithmic tools they use, and why they're using them, a standardized way of recording and sharing information has been designed, consisting of a template to be filled with the specific information requested[19], accompanied by guidelines to support public sector teams completing algorithmic transparency reports[20]. It would perhaps be appropriate to effectively implement the prescriptions of the AI Act, to develop something similar at the EU level, intended not only for public entities but also for private ones. Perhaps some operational solution could come shortly from the European Centre for Algorithmic Transparency (ECAT), inaugurated on 18 April 2023, which brings together researchers and experts in the fields of computer science, data analysis, intelligence artificial sciences and legal and social sciences, to analyze the functioning of the software of the large platforms and search engines.

REFERENCES

Bitzer, T., Wiener, M., & Cram, W. A. (2023). Algorithmic transparency: Concepts, antecedents, and consequences. A review and research framework. *Communications of the Association for Information Systems, 52*, 293–331.

Cappellazzo, N. (2022, March 23). Algoritmi, automazione e meccanismi di intelligenza artificiale: la classificazione proposta dal Consiglio di Stato. Retrieved from https://www.federalismi.it/nv14/articolo-documento.cfm?Artid=46943

Caravaggion, G., & Orofino, M. (2023). Lingua e costituzione: L'irrompere dei linguaggi algoritmici. *Rivista AIC, 4*, 155–184.

Casonato, C., & Marchetti, B. (2021). Prime osservazioni sulla proposta di regolamento dell'Unione Europea in materia di intelligenza artificiale. *BioLaw Journal, 3*, 415–437.

Chomanski, B. (2022). Legitimacy and automated decisions: The moral limits of algocracy. *Ethics and Information Technology, 3*, 1–9.

Colapietro, C. (2023). Gli algoritmi tra trasparenza e protezione dei dati personali. *Federalismi.it, 5*, 151-174.

Corrado, A. (2023). La trasparenza necessaria per infondere fiducia in una amministrazione algoritmica e antropocentrica. *Federalismi.it, 5*, 175-215.

Council of Europe. Parliamentary Assembly. (2020, October 22). The brain-computer interface: new rights or new threats to fundamental freedoms? Resolution No. 2344. Retrieved from https://pace.coe.int/en/files/28810

Cristianini, N., & Scartamburlo, T. (2020). On social machines for algorithmic regulation. *AI & Society, 35*, 645–662.

Danaher, J. (2016). The threat of algocracy: Reality, resistance and accommodation. *Philosophy & Technology, 3*, 245–268.

De Minico, G. (2021). Fundamental rights, European digital regulation and algorithmic challenge. *Medialaws, 1*, 9–38.

De Vivo, I. (2024). Il sé allo specchio dell'algoritmo. Libertà epistemica e identità individuale. In Sterpa, A. (Ed.), *L'ordine giuridico dell'algoritmo* (pp. 51–69). Editoriale Scientifica.

Diakopoulos, N., & Koliska, M. (2017). Algorithmic transparency in the news media. *Digital Journalism (Abingdon, England), 5*(7), 809–828.

European Parliament. (2020). Framework of ethical aspects of artificial intelligence, robotics and related technologies. Resolution P9_TA 0275. Retrieved from https://www.europarl.europa.eu/doceo/document/TA-9-2020-0275_EN.html

European Parliament. (2022). Artificial intelligence in a digital age. Resolution P9_TA 0140. Retrieved from https://www.europarl.europa.eu/doceo/document/TA-9-2022-0140_EN.html

European Parliament Research Service (EPRS). (2023, March). General-purpose artificial intelligence. PE 745.708. Retrieved from https://www.europarl.europa.eu/RegData/etudes/ATAG/2023/745708/EPRS_ATA(2023)745708_EN.pdf

Fasan, M. (2022). I principi costituzionali nella disciplina dell'Intelligenza Artificiale. Nuove prospettive interpretative. *DPCE Online*, *1*, 181–199.

Grimaldi, L. (2022). Costituzionalismo "post-umano" alla prova della decisione pubblica algoritmica. *Federalismi.it, 34*, 75-95.

Guidotti, R., Monreale, A., Ruggieri, S., Turini, F., Giannotti, F., & Pedreschi, D. (2019). A survey of methods for explaining black box models. *ACM Computing Surveys*, *5*, 1–42.

Klenk, M. (2023). Algorithmic transparency and manipulation. *Philosophy & Technology*, *4*, 1–20.

Kossow, N., Windwehr, S., & Jenkins, M. (2021). Algorithmic transparency and accountability. Retrieved from https://knowledgehub.transparency.org/assets/uploads/kproducts/Algorithmic-Transparency_2021.pdf

Leerssen, P. (2020). The soap box as a black box: Regulating transparency in social media recommender systems. *European Journal of Law and Technology*, *2*, 1–51.

Lo Sapio, G. (2021). La trasparenza sul banco di prova dei modelli algoritmici. *Federalismi.it, 11*, 239-252.

Lombardi, E. M. (2020). Norma e algoritmo: alcune considerazioni sul nuovo ordine tecnologico. *Giustizia Civile.com, 7*, 1-16.

Lu, J., Dokyun, L., Tae Wan, K., & Danks, D. (2019). Good explanation for algorithmic transparency. Retrieved from https://papers.ssrn.com/sol3/papers.cfm?abstract_id=3503603

Marchetti, B. (2021). La garanzia dello human in the loop alla prova della decisione amministrativa algoritmica. *BioLaw Journal*, *2*, 367–385.

Marrone, P. (2022). Democrazia epistemica, epistocrazia, algocrazia: Alcuni problemi. *Paradigmi*, *2*, 307–324.

Mobilio, G. (2020). L'intelligenza artificiale e i rischi di una "disruption" della regolamentazione giuridica. *BioLaw Journal*, *2*, 401–424.

Otranto, P. (2021). Riflessioni in tema di decisioni amministrative, intelligenza artificiale e legalità. *Federalismi.it, 7*, 187-204.

Papa, A. (2022). Intelligenza Artificiale e decisioni pubbliche tra tecnica, politica e tutela dei diritti. *Federalismi.it, 22*, 101-114.

Rangone, N. (2022). Intelligenza artificiale e pubbliche amministrazioni: Affrontare i numerosi rischi per trarne tutti i vantaggi. *BioLaw Journal*, *2*, 473–488.

Simoncini, A. (2019). L'algoritmo incostituzionale: Intelligenza artificiale e il futuro delle libertà. *BioLaw Journal*, *1*, 63–89.

Simoncini, A. (2023). Il linguaggio dell'intelligenza artificiale e la tutela costituzionale dei diritti. *Rivista AIC*, *2*, 1–19.

Şimşek, C. (2021, May). Algorithmic transparency in the EU. Retrieved from https://www.sciencespo.fr/public/sites/sciencespo.fr.public/files/SIMSEK%20Can.pdf

Spiller, E. (2021). Il diritto di comprendere, il dovere di spiegare. Explainability e intelligenza artificiale costituzionalmente orientata. *BioLaw Journal*, *2*, 419–432.

Sterpa, A., De Vivo, I., & Capasso, C. (2023). L'ordine giuridico dell'algoritmo: La funzione regolatrice del diritto e la funzione ordinatrice dell'algoritmo. *NAD*, *2*, 1120–1155.

Trigueiro, I., Molina, D., Poyatos, J., Del Ser, J., & Herrera, F. (2024). General purpose artificial intelligence systems (GPAIS): Properties, definition, taxonomy, societal implications and responsible governance. *Information Fusion*, *103*, 1–16.

Valsecchi, A. (2020, September 14). Algoritmo, discrezionalità amministrativa e discrezionalità del giudice. Retrieved from https://www.iusinitinere.it/algoritmo-discrezionalita-amministrativa-e-discrezionalita-del-giudice-nota-a-cons-stato-sez-vi-sent-del-4-febbraio-2020-n-881-30630

Wang, H. (2022). Transparency as manipulation? Uncovering the disciplinary power of algorithmic transparency. *Philosophy & Technology*, *3*, 1–25.

Wang, H. (2023). Why should we care about the manipulative power of algorithmic transparency? *Philosophy & Technology*, *1*, 1–6.

Watson, H. J., & Nations, C. (2019). Addressing the growing need for algorithmic transparency. *Communications of the Association for Information Systems*, *1*(45), 488–510.

Zambonelli, F. (2023). *Algocrazia. Il governo degli algoritmi e dell'intelligenza artificiale*. Scienza Express.

ENDNOTES

1 Regulation (EU) 2024/1689 of 13 June 2024, in force since 1 August 2024.

2 Regulation (EU) 2016/679 of 27 April 2016, in force since 25 May 2018.

3 This definition is taken from https://www.britannica.com/science/algorithm

4 https://www.acm.org/binaries/content/assets/public-policy/2017_joint_statement_algorithms.pdf.

5 https://www.bertelsmann-stiftung.de/en/publications/publication/did/algorules-en

6 https://www.bertelsmann-stiftung.de/en/publications/publication/did/from-principles-to-practice-wie-wir-ki-ethik-messbar-machen-koennen

7 https://www.techopedia.com/definition/34940/black-box-ai

8 Some of these technological solutions are explained by Guidotti et al., 2018.

9 Regulation (EU) 2022/2065 of 19 October 2022 on a Single Market for Digital Services.

10 Resolution No. 2344 of 22 October 2020: https://pace.coe.int/en/files/28810

11 https://www.europarl.europa.eu/doceo/document/TA-9-2020-0275_EN.html

12 https://www.europarl.europa.eu/doceo/document/TA-9-2022-0140_EN.html

13 Moreover, in the case of algorithms used by public bodies in decision-making, one wonders (Marchetti, 2021, p. 377) if the human supervisor should be a public official, who should be provided with adequate technical skills, or the developer of the algorithm, who is a person external to the public administration.

14 TAR Lazio is the administrative tribunal of the first instance of the Italian region named Lazio. Regional TARs are responsible for judging appeals brought

against administrative acts by private individuals who consider themselves harmed (in a manner not in conformity with the legal system) in their legitimate interest. TAR's decisions can be appealed before the Council of State.

15 The Italian Council of State is both a legal-administrative consultative body and the highest Italian administrative court (the second and ultimate level of decision after TARs). It ensures the legality of public administration, i. e. that the acts of public bodies are coherent with primary laws and administrative procedures.

16 The Supreme Court of Cassation is the highest court in Italy. It ensures the correct application of law in the inferior and appeal courts and resolves disputes as to which lower court (penal, civil, administrative, military) has jurisdiction to hear a given case. It doesn't re-examine the facts of a case but only interprets the relevant law and can reject or confirm a judgement from a lower court.

17 https://www.europeanlawinstitute.eu/projects-publications/publications/eli-model-rules-on-impact-assessment-of-algorithmic-decision-making-systems-usey-by-public-adminstration/

18 https://www.gov.uk/government/collections/algorithmic-transparency-recording-standard-hub

19 https://www.gov.uk/government/publications/algorithmic-transparency-template/algorithmic-transparency-template

20 https://www.gov.uk/government/publications/guidance-for-organisations-using-the-algorithmic-transparency-recording-standard

Chapter 3

The Integration and Impact of AI in the Greek Media Ecosystem: Challenges, Opportunities, and the Digital Divide

Andreas M. Panagopoulos
https://orcid.org/0009-0001-1513-9481
Aristotle University of Thessaloniki, Greece

Sotirios Triantafyllou
https://orcid.org/0009-0003-6618-3532
University of Peloponnese, Greece

Panagiotis Kapos
https://orcid.org/0000-0002-5148-248X
Panteion University, Greece

ABSTRACT

The integration of Artificial Intelligence (AI) into journalism constitutes a transformative development in the evolution of news reporting and the broader media landscape. This chapter explores the emergence of AI in the Greek media system through qualitative analyses of journalists and CEOs, complemented by an examination of five case studies. The findings suggest that the majority of media organizations in Greece adopt a cautious "wait-and-see" approach before fully incorporating AI tools into their workflows. Nevertheless, they are taking steps in order to prepare for the integration of AI into their value chain. Journalists are experimenting with AI tools, using them primarily as Intelligent Assistants (IA) while in certain cases,

DOI: 10.4018/979-8-3693-8729-0.ch003

AI is employed for specific tasks. These varied approaches are shaped by concerns about AI's limitations, such as linguistic challenges, ethical considerations, intellectual property issues, and potential job displacement. Moreover, the digital divide and a lack of technical skills emerge as significant barriers to the adoption of AI.

INTRODUCTION

Media organizations are experiencing ongoing transformation, increasingly driven by advancements in AI technologies (Deuze & Witschge, 2018). AI has rapidly transitioned into a critical component of contemporary journalism (Nurelmadina et al., 2021). Early applications focused on basic algorithms assigned automated routine tasks, such as generating financial summaries or reporting sports results. Over time, the scope of AI's involvement has significantly broadened to include advanced data analysis, personalized content delivery, support for investigative reporting and even influencing editorial strategies (Bahroun et al., 2023). This disruption is characterized by the growing integration of Artificial Intelligence (AI), which has begun to closely link journalists with generative systems and synthetic technologies.

Background

Initially, AI was introduced as an experimental tool. Advancements in AI are steadily enhancing its functions, with algorithmic processes now performing many fundamental journalistic tasks, transforming journalistic practices (Diakopoulos, 2019). This mirrors the larger trend of digital transformation, altering news gathering, reporting, and engagement. The production of news is undergoing rapid transformation (Stray, 2021), driven primarily by advancements in technology, particularly AI (Rouhiainen, 2018). AI supports journalists by automating repetitive tasks, leveraging data for uncovering stories, and presenting narratives in new and engaging ways. Using intelligent computer programs, AI generates news articles, visual content, analyzes big data in real time, and provides concise summaries. This capacity enables faster, more captivating news delivery (Broussard et al., 2019). The growing reliance on AI underscores not only technological progress but shifts in news consumer behavior and expectations (Gollmitzer, 2023).

As newsrooms consider AI adoption to improve sourcing, organization, and distribution, a gap unfolds between resource-rich organizations and those with fewer resources (Newman et al., 2020). AI can automate content creation, personalize feeds, and process large real-time data sets (Pavlik, 2023). Yet, AI's impact is often limited by resistance to change, knowledge gaps, regulatory and ethical challenges, and technological constraints (Makwambeni et al., 2023; Munoriyarwa et al., 2023).

Power Dynamics and Considerations

Power dynamics among stakeholders, the use of AI enforcement tools, and compliance with legal and ethical guidelines require careful attention (Broussard et al., 2019). AI tools and algorithms may reflect the values and biases of their creators; technology companies often lack incentives for transparency regarding biases (Bird et al., 2016). Bias sources in AI can emerge from various stages of the machine learning process, from data collection and algorithm design to user engagement.

The contemporary media environment brings both opportunities and challenges. A major concern is digital inequality—disparities in access to and effective use of digital technologies. As AI transforms newsrooms, understanding adoption and responses in less affluent countries is crucial (Kothari & Cruikshank, 2022). This divide affects both media professionals' information access and the quality and reach of news (Jamil, 2023). The digital era has set new standards for journalism, shifting how news is sourced, verified, and circulated online. Evolving norms are key to news credibility, with journalism leaders shaping these standards. Media self-regulation, often implemented via internal newsroom guidelines, is vital and functions independently from government control (Ruggiero et al., 2022).

Examining the increasing institutionalization of journalism and the digital divide offers insight into AI's influence within media. The present "AI-fication" of journalism is not simply about technological upgrades; it signals deep intelligence integration into organizational structures—a leap for which many media organizations may still be unprepared. This makes it urgent to address both the opportunities and challenges linked to AI, particularly digital inequalities limiting its reach. Ensuring that AI innovations in journalism remain accessible, inclusive, and ethical is paramount.

Recent research from the Reuters Institute (Cherubini & Sharma, 2023) sheds light on industry views regarding AI. When asked about the effects of generative AI and workflow automation in newsroom employment in the coming decade, 74% of the respondents stated that AI would improve efficiency without fundamentally changing the essence of journalism. In contrast, 21% expected AI to radically alter workflows and processes, significantly impacting all roles within the newsroom. Only 2% felt that AI would have no effect on newswork, reflecting a widespread anticipation of AI's role in the industry. Another report by Nic Newman, titled "Journalism, Media, and Technology Trends and Predictions 2024," indicates that this year will witness the complete integration of AI into newsroom operations. Among the global media outlets surveyed, 16% reported having appointed someone responsible for AI-related tasks, with an additional 24% planning to do so soon. However, challenges such as technological disparities and the digital divide remain, especially in regions like Latin America, which trails behind the U.S. and parts of Europe regarding AI adoption in the media sector (Pennacchio, 2024). Notably,

Greece was not included in this survey, although other European nations, including Germany, Spain, France, Austria, Finland, Norway, Denmark, the Netherlands, Poland, Hungary, and Slovakia, were represented.

Before examining the Greek media landscape and, in particular, the impact of AI on it, a literature analysis provides the necessary context of the digital age. The expected impact of AI on the media value chain and the new business models that have emerged from the evolving techno-economic and communication paradigm are explored.

MEDIA ORGANIZATIONS AND THE IMPACT OF AI: BUSINESS MODELS AND VALUE CHAIN

Media organizations, with information and content as core drivers, are at the center of digital change, facing intensified challenges with the spread of AI tools. While maintaining traditional and analog aspects, they are compelled to devise new strategies for digital transition, adapting established production, communication, and business models.

Media organizations are active across conception, production, dissemination, and distribution of goods and services, using information and content as primary inputs. Their function extends beyond transmission—they enable thematic diversity, help prevent misinformation, and ensure pluralism crucial for democracy.

The new techno-economic and communication paradigm catalyzes profound change in media industries.

Transitioning to this paradigm involves:

- Digitization of content and an altered information life cycle;
- Multiple forms of convergence spanning technology, media, information, and communication, as well as editorial and access spaces (Leandros, 2010, 2011)
- Economic dematerialization—shifting from owning goods to accessing services, and the rise of the attention economy (Esposito, 2010).

This evolution incorporates features like increased interactivity, communication capacity, inclusivity, remote networking, mass customization, and new content providers and distribution platforms (Kapos, 2023).

Digital tools now transform consumers from passive receivers into prosumers—simultaneously producers and consumers of content (Toffler, 1980).

All these changes reshape media production by altering business models, the value and supply chain, labor relations, organizational forms, and consumption patterns (Castells, 2000).

INNOVATIVE AND REFORMED PRODUCTION AND BUSINESS MODELS IN MEDIA ORGANIZATIONS

New business models in media must reflect consolidated realities internationally and in Greece. For example:

- Revenues from print advertising and sales have shrunk dramatically;
- Traditional print models have suffered, digital advertising has not compensated;
- Public trust in traditional media is declining among younger generations and is changing consumption patterns;
- ICT innovations have radically altered how news is created, distributed, and consumed (Leandros, 2010, 2011).

To survive and thrive, media organizations must move beyond linear, static, and mass-focused models of the analogue economy, embracing non-linear and dynamic models.

The "digital disruption" led by the diffusion of AI accelerates structural change in business models and industry value chains, creating new value propositions and redefining sectors.

The fourth industrial revolution brings robotics, AI, big data, IoT, and more. Sectors driven by information, including media, are experiencing accelerated digital transformation at all levels, particularly in production, distribution, and communication (Kapos, 2023).

Business models often center on the "value chain"—the added value at each step until the end user (Osterwalder & Pigneur, 2010; Thompson, 2010). Performance of value creation and coordination of resources to realize and deliver it are central to most definitions.

The approach of Osterwalder & Pigneur (2010) highlights business models as descriptions of creating and capturing value by connecting technology and strategy. Digitalization, internet, ICT, and AI tools drive the creation of new value, transforming workflows, relationships, and value distribution.

Key digital-era business models (Cook & Sikkunen, 2012; Kapos, 2023; Leandros, 2010, 2011):

- **Subscription models**: Predictable upfront revenue, relying on unique or niche news content
- **Paywall models**: "Premium" digital content for fee-based access.
- **Micropayments**: Small payments for individual articles, akin to pay-per-view or streaming.
- **Open-access (Free/Freemium)**: Basic content free, premium content comes with a charge.
- **Expanded revenue sources**: E-commerce, custom publishing, branded book products, and cross-promotions by leveraging journalistic/editorial assets

Media organizations now operate in a hybrid analog/digital world, requiring integration of networks, channels, audiences, and the broader news ecosystem (Anderson, 2016). Business models today must embed value networks at their core, moving beyond linear, process-driven structures to dynamic, interconnected frameworks.

The new landscape enables platforms to revolutionize industries—personalized recommendations by Amazon, Google, Netflix, and algorithmically curated social media feeds define user experience (Murschetz, 2024; Napoli & Roepnack, 2018; Smith & Telang, 2016).

Media business model sustainability rests on resilience to changes in consumption patterns, competition, work conditions, and the effects of innovation like AI along the value chain (European Commission, 2022).

THE VALUE CHAIN OF MEDIA ORGANIZATIONS IN THE AGE OF ARTIFICIAL INTELLIGENCE

Porter's "value chain" groups activities into primary (production, sales, distribution, service) and support categories, each adding competitive advantage (Porter, 1985). Media industries now see every link in the value chain reshaped by AI.

AI is defined by computer systems performing tasks formerly requiring human intelligence. The evolution of generative AI and machine learning challenges human creativity, as AI now mimics activities once uniquely human (Pavlik, 2023).

AI's impact covers all media business value chain levels—from input (information, text, sound, image), to production, promotion, and content consumption.

In content production and operations, AI streamlines and automates:

- New content formats, automated scripts and articles, translation, editing, and proofreading;
- Layout/design, image creation, summaries, AI-driven fact-checking, big data collection, and voice-to-text technologies;

- Automatedaggregation and curation.

In marketing and sales, AI facilitates:

- Massive data analysis, personalized recommendations, customer service bots, campaign targeting, ad personalization, paywall automation, reader engagement, visitor trend analysis, and sentiment forecasting.

In outbound logistics and supply chain, AI transforms:

- Stock management, distribution coordination, and demand forecasting.

These uses of AI fundamentally advance efficiency, creativity, responsiveness, and economic sustainability across every segment of the modern media organization's value chain.

AI in Media: Applied Dimensions

The integration of Artificial Intelligence (AI) into newsrooms has become a widespread phenomenon across global media organizations. Several outlets began adopting AI technologies as early as 2014 or even earlier, employing a variety of tools. Major media organizations abroad began integrating early applications of Artificial Intelligence (AI) into content production as early as 2014 – such as using AI for generating financial reports (Associated Press, n.d.; Miller, 2015) and sports news (Eule, 2013; McFarland, 2014) through the Wordsmith tool developed by Automated Insights (Automated Insights, n.d.) – these systems primarily relied on transforming structured data into text using rule-based AI technology, which had existed for decades (Pressman, 2017).

The Associated Press (AP) was among the pioneering organizations that incorporated AI into its operations, beginning in 2014. The AP has reported that AI implementation has resulted in a 20% increase in journalists' available time, along with enhanced accuracy and an expanded volume of news production (Triantafyllou, 2020; Weber, 2014).

Similarly, The Washington Post introduced AI in 2016 with the development of its proprietary algorithm, Heliograf. Initially, Heliograf was deployed to cover significant events, in August 2016, and in November 2016 (Albeanu, 2016a; Albeanu, 2016b). The Post recently launched a generative AI tool, "Ask the Post AI," in November 2024. This experimental tool provides users with answers to their queries, leveraging the journalistic expertise of The Washington Post (Washington Post, 2024).

The New York Times (NYT) also makes extensive use of AI technologies and has appointed an AI Officer.It has more recently applied these technologies to its homepage modules, website, and mobile application (Yang, 2024). Notably, the NYT won Pulitzer Prizes using AI-based tools. (Deck, 2024).

In addition to these cases, other media organizations have embraced AI. In Europe, prominent media outlets, such as the BBC, Sky News, The Telegraph, Daily Mail, and The Economist utilize various AI tools. Furthermore, Norway's National News Agency and Sweden's Mitt Media have adopted AI technologies.

A particularly innovative application of AI is observed in China's Xinhua News Agency, which introduced AI-powered news presenters in April 2018. These virtual presenters, modelled on real journalists, are capable of delivering news continuously, 24 hours a day, in both English and Chinese (Triantafyllou, 2020; Baraniuk, 2018).

Generative Artificial Intelligence (Generative AI) gained widespread recognition with the launch of ChatGPT 3.5 for free preview on November 30, 2022, though OpenAI had been publishing research on generative models trained on text, images, and audio since 2016 and had released GPT-2 as early as 2019 (Hines, 2023). Generative AI, a subset of Artificial Intelligence, is distinct in its ability not only to analyze existing data, but also to create new data based on inputs or prompts provided by users (Airan et al., 2024).

The Case of Greece

The Greek media landscape is marked by significant media concentration, fragmentation of digital brands, and significant low level of trust in news (Kalogeropoulos 2024, RSF 2024). Political polarization and concerns about the influence from politicians and influential business figures characterize the media market.

Digital media in Greece remain committed to advertising supported models first developed in the 1990s and early 2000s. They have been slow to introduce online subscription models compared to elsewhere in Europe, in part because of their fear of losing audiences and influence. *Kathimerini*, a leading legacy newspaper, introduced a soft paywall last year, being one of the very few Greek newspapers to have done so.

Despite being classified as a Global North country, Greece's digital technology sector reflects characteristics typical of the Global South. A pronounced digital divide exists between the Global North and Global South, significantly affecting Greece's media landscape.

Our recent survey (Panagopoulos, et.al., 2024) of Greek journalists indicates that AI has not yet been fully integrated into Greek newsrooms. Journalists currently use AI on an experimental basis for personal purposes, lacking established guidelines or formal training in its application. Instead, self-directed learning is common, and there is no cohesive strategy for AI adoption in the Greek media landscape.

Nonetheless, Greek media outlets are gradually incorporating AI technologies into their daily operations. For instance, protothema.gr launched an AI-supported English version in July 2024 (Protothema, 2024a) and implemented AI-driven comment moderation (Protothema, 2024b). Around the same time, the digital edition of the Sunday newspaper *To Vima* introduced AI-cloned voices for the op-eds of notable columnists, including Yannis Pretenteris (Tovima, 2024). Additionally, a new AI-only news website, robonews.gr, was launched in May 2024. Previous research also found that Gazetta.gr, a leading sports infotainment website, employs specific AI tools for video production and social media content.

Taking the above into account, this chapter seeks to provide a thorough and well-rounded analysis of how AI is reshaping the Greek media landscape, with a focus on the challenges and opportunities it presents for journalism and media organizations.

More specifically, by focusing on qualitative analysis and case studies of the aforementioned media outlets, we will investigate how leading media organizations are preparing for and utilizing AI within the journalistic field.

Research Methodology Analysis

This chapter aims to explore, through qualitative research, the usage, challenges, and risks associated with AI in the Greek media ecosystem.

The research methodology adopted for this study is based on an interdisciplinary approach, integrating concepts and practices from media economics, cultural management, communication, and journalism. The methodology is qualitative in nature and involves the following key components:

Case Studies

A key methodological approach adopted in this investigation is the case study method, used to examine the impact of Artificial Intelligence technologies on Greek media organizations. This methodological tool serves as a valid method of analysis in various sciences and is considered particularly useful and effective, especially when the subject under investigation is broad, complex, and lacks an extensive theoretical foundation (Dul & Hak, 2007), as the case study of Greece. In a previous study (Triantafyllou, Panagopoulos &Kapos, 2025) we found 5 media organizations which announced the use of AI systems in their workflow. Thus, our research focused on AI-generated content produced by Protothema.gr, Gazzetta.gr, Tovima.gr, Huffpost.gr, and Robonews.gr. Subsequently, between August and

November 2024, we carried out five semi-structured interviews with the respective managerial personnel of these outlets.

Semi-Structured In-Depth Interviews

Purposive sampling was utilized to recruit CEOs and journalists representing various media sectors, including print, television, radio, and digital platforms, as well as spanning different age groups. This approach was employed to ensure the inclusion of a diverse range of experiences and perspectives concerning the adoption of AI in journalism. Purposive sampling is particularly advantageous for selecting information-rich cases that offer comprehensive and nuanced insights into the phenomenon under study (Paton, 2015)

We conducted 14 semi-structured interviews with CEOs and News managers – journalists (Table 1). The aim was to gain deeper insights into the applied practices related to the use of Artificial Intelligence tools, as well as to explore their perceptions, concerns, needs, expectations, and plans regarding the development of Artificial Intelligence tools in their media and, more broadly, in journalism and the production of journalistic content. More specifically we interviewed 4 CEOs from distinguished Media Groups in Greece. Two of them are managing Media organizations in TV, Print and Digital and the rest 2 run Digital only Media Groups. Also, we conducted 10 different semi structured interviews with news managers in Print, Broadcasting and Digital news outlets during the period 01/09/2024- 30/09/2024.

All interviews were transcribed, anonymized and coded by the researchers.

Table 1. Sample of interviews

Medium	N	Role
TV, Print, Digital	2	CEO
Digital (only)	2	CEO
TV, Radio, Digital	2	News Manager
Print, Digital	3	News Manager
Digital (only)	3	News Manager
TV, Print, Digital	2	News Manager
Total	14.	

The interviews were conducted through Zoom, each lasting approximately 15-20 minutes. They were recorded and transcribed using AI technology, with researchers overseeing the process and making verbatim adjustments as necessary. NVivo14 software was employed to systematically code, organize, and analyze the qualitative data collected from journalists. The analysis was performed through an iterative process of close reading and thematic coding, directed by the study's research questions.

We will highlight the 'Digital Divide' in AI adoption within Greece compared to other EU countries, illustrating how this divide manifests across various dimensions, including unequal access to new media technologies. Additionally, we will identify different patterns of AI usage across media organizations and mediums. A significant focus of our research will be on the implications of AI, including potential job losses, cost reductions, and ethical and copyright concerns.

The Five Case Studies of Greece

With the introduction of ChatGPT 3.5, both Greek journalists and the general public were able to access and experiment with the tool. Interest in ChatGPT in Greece started to grow in early December 2022, as reflected in search activity on Google. Data from Google Trends indicates that during the week of November 27 to December 3, 2022, only two searches related to ChatGPT were recorded in Greece (Panagopoulos, Triantafyllou, Kapos, 2024).

Our research focused on five Greek media.

Protothema.gr

SimilarWeb data[1] identifies protothema.gr as the leading digital news outlet in the country, with 4.1 million unique users and 67 million visits per month. However, according to the Reuters Institute's *Digital News Report* (Kalogeropoulos, 2024), it ranks fourth in weekly usage, capturing 20% of the market share, and achieves a 10% share for usage three days per week. Additionally, *Protothema* publishes a weekly newspaper.

The media is collaborating with Cloudevo.ai and partners with Google in case studies. In November 2024, the English edition of Protothema implemented an AI Semantic Search system, developed in collaboration with Google, FT Strategies, and Cloudevo. This innovative feature enables readers to engage interactively with news content (Protothema c, 2024). "Offering AI-powered responses, it helps readers better navigate and comprehend the news by providing quick and accurate information, whilst the questions are typed by the users in a conversational style (ask me anything in the search bar) and not as keywords", explains the CTO of the outlet. However, the Greek language constitutes a barrier at the moment for the leading Greek digital edition. Despite the obstacles, the AI model will be fed with more Greek content and, hopefully, it will work soon for the Greek audience, too.

Months earlier (July 2024), the English version was enhanced using AI for translation from Greek content. The tool automatically translates and transfers the most popular articles from the Greek version into the English version based on readability and with minimal human intervention (Protothema,2024a). It is noteworthy that, on January 4, 5 stories were published, on March 4, 23 stories, on June 4, 35 stories and on November 4, almost double, 69 stories. The news stories production increased, using AI translation. However, the Greek language still limits its potential, since the new stories production in the Greek edition is more than 300 per day. Using the English edition, "Protothema is expanding its audience and aspires to have more traffic and revenue"adds the CTO of the publication. At the moment, the English edition requires humans being kept informed.

Protothema relies on the comment section for audience engagement and traffic increase. Due to the polarized political environment, the perception of the audience and opposition political parties that it has a pro-government stance has as a result that the comments section is gaining a lot of attention and at the same time hate speech. They collaborated with a US-based company for AI comment moderation. The AI algorithm autonomously processes user comments, determining their approval or rejection without direct human involvement. For cases in which the algorithm cannot reliably assess the content, human moderators intervene to make the final decision. The algorithm is programmed for continuous learning, enhancing its performance and accuracy progressively. The published comments increased by more than 1000% per month (Protothema, 2024b). "In the comments section, they were working with 8 journalists, but after the AI implementation, they have only 2 journalists", as the CTO points out. Currently, there may be a 75% reallocation of human resources; however, the prolonged integration of AI in news production has the potential to result in workforce reductions.

Gazzetta.gr

The digital news outlet gazzetta.gr is a brand of Liquid Media Group, which claims that, in total, it has 16M Unique Users and 180M pageviews per month[2]. Gazzetta, a digital platform specializing in sports and infotainment, ranks third among news websites in the country and holds the top position for sports content in the digital domain. According to SimilarWeb data, the site attracts 3.5 million unique users and records 20.4 million visits per month.

Liquid Media is promoting itself on the company's home page as "the science behind the digital content". Gazzetta.gr, emphasizes the integration of AI tools in news production. To enhance reader engagement, the outlet leverages its prominent sports journalists by cloning their voices, allowing users to listen to stories narrated in the journalist's voice. Journalists sign separate agreements that grant the company the rights to utilize their voice recordings only for the specific purpose – voice blogs or short podcasts as they call them. This content is disseminated as video reels in social media edited with AI tools. "The feedback we have so far from the audience is exciting, despite the concerns we've had before launching it" mentions the News Manager. AI tools are also employed at gazzetta.gr for generating images and creative content. As an example, News director highlights a project led by the art director, who has a passion for comics. The director created a comic that was transformed into a video, narrating the story of a sports fan tragically killed by rival hooligans. In video production, gazzetta.gr utilizes AI tools for automated subtitling in Greek, translation, and editing tasks. "Missing technical skills from the journalists are not issue any more, because of AI tools". Additionally, the outlet is experimenting with AI-driven journalist avatars, exploring new ways to integrate these technologies into their content creation processes. It employs AI tools for transcription tasks as well as specialized tools for research, and summarizes large volumes of unstructured data in documents, streamlining their workflow and enhancing efficiency. "We must adapt to the rapidly evolving and transformative nature of technology, taking into consideration the existing ethics concerns" states the News director. He emphasizes that these tools are viewed as assistants rather than replacements for professionals in the media production sector. Furthermore, these tools serve as valuable complements to journalistic work, particularly by alleviating repetitive and monotonous tasks, allowing journalists to focus on more creative and impactful aspects of their roles. Prompting skills and a combination of different tools for the same task will provide better results. However, text production is the responsibility of the journalists.

tovima.gr

The digital edition of the century-old, prominent weekly newspaper To Vima serves as a flagship brand within the Alter Ego Media Group. The Group's portfolio includes an additional newspaper (TA NEA), the leading television station MEGA, six digital brands, several magazines, and a radio station. In a significant acquisition, the Group purchased the rights to the domain enet.gr and the brand name of the historic left-wing newspaper Eleftherotypia—along with its Sunday edition Kyriakatiki Eleftherotypia—for €8.1 million. This newspaper ceased operations in 2014 following bankruptcy.

The digital edition tovima.gr is ranked 22nd among digital news outlets and, according to SimilarWeb data, garners 2.07 million unique users and 5.05 million visits per month. The Group's digital flagship, in.gr, ranks 3rd in the Digital News Report 2024 (Kalogeropoulos, 2024) and attracts 2.14 million unique visitors alongside 9.5 million visits per month, according to SimilarWeb.

Notably, tovima.gr has utilized AI to clone the voice of its prominent columnist and journalist Yannis Pretenteris, enabling audiences to listen to his articles daily. The digital edition continues to experiment with advanced AI tools, while the Alter

Ego Media Group is actively developing a comprehensive strategy for AI adoption across its operations. According to the manager, "AI tools are particularly valuable for supplementing journalists' technical skills that may currently be lacking". The publication is testing more tools, however, there are concerns for transparency of the AI tools owners and providers.

HuffingtonPost.gr

The Greek edition of Huffington Post, operating under a licensing agreement, holds a modest share of the Greek audience. According to SimilarWeb data, it attracts 0.62 million unique visitors and records 1.8 million visits per month. Originally expanding from the U.S. to Europe during the 2010s, *Huffington Post* sought to achieve scale by leveraging its brand reputation, content, and expertise developed in the United States to establish a presence in international markets (Nichols et al., 2016). In Greece, other U.S.-based media also operate under licensing agreements, including *cnn.gr*, managed by the DPG Media Group, and *Vice*, operated until 2023 by the Antenna Media Group. CNN Greece has 1.70 million users and 9.42 million visits per month.

The Greek edition of Huffington Post maintains close communication with its U.S. headquarters, receiving guidance and content support.

As part of this collaboration, they exchange ideas and experiments with AI tools. One such tool, used for generating SEO-optimized headlines, has proven successful in the U.S. edition. However, "in the Greek context, the AI-proposed headlines are often unsuitable, with around 50% being deemed "laughable", underlines the News manager.

A News Manager of the Greek edition notes that "the complexities of the Greek language pose a significant challenge for the effective implementation of AI text-based tools. However, they are fine for other tasks, such as for ideas, search, etc.". Due to these linguistic barriers and the fact that AI algorithms are typically trained on larger, non-Greek datasets, Greek journalists currently face little risk of job displacement or replacement by AI in the near future. While AI can assist with brainstorming content ideas and conducting research, human supervision remains essential to address algorithmic errors and biases.

Robonews.gr

Robonews is a beta-stage Greek website that exclusively generates content –both text and images– using AI. Its founder, Sotiris Poulopoulos (robonews, n.d.), a renowned publisher and sports journalist, is actively experimenting with AI, attempting to create stories and visuals. Over the course of a year, he has been testing and refining the generated texts by adjusting his prompts to achieve optimal results. Stories that meet his quality standards are published on the platform.

Robonews will be ready for public access in a few months. Although his experimentation does not involve reinforcement learning techniques, "the results have often been remarkably effective", he claims.

Greek newsrooms currently refrain from utilizing AI for tasks requiring ethical decision-making, maintaining relationships with sources, conducting face-to-face interviews, or producing content based on primary sources. Similarly, AI is not employed for fact-checking or investigative journalism. Significant barriers to broader AI adoption include linguistic challenges and a lack of adequate training in AI technologies.

Despite these limitations, contemporary newsrooms are evolving in order to rely not solely on traditional roles, such as editors and reporters, but increasingly on a diverse range of specialized professionals to meet the demands of modern journalism.

The integration of AI into journalistic workflows remains limited in Greek newsrooms, confirming our previous research (Panagopoulos et al., 2024) with its application primarily confined to specific departments such as translation, comment moderation, and video production. These tools are predominantly used to address gaps in technical expertise among journalists. The adoption of AI technology is still in its infancy, with respondents recognizing its potential as the future of media, but emphasizing the need for significant advancements in both human and technological capabilities to fully realize its benefits.

A key priority for media organizations should be to enhance employees' competencies in using AI effectively. Raising awareness of how AI can function as a supportive tool or a content creator is essential. Interviewees in the Case studies for this research highlighted noticeable disparities in IT and AI skill levels across staff members within media organizations, irrespective of age.

Respondents generally rated their newsrooms' understanding of AI technologies as moderate, largely due to the recent establishment of AI-focused teams. Concrete outcomes from these teams are expected to emerge over time.

Currently, the use of AI in media work is concentrated in the following areas:

- Transcription
- Text-to-speech conversion
- Video content creation
- Translation into English
- Text editing and moderation

Efforts to address these challenges and maximize AI's potential will require investments in training, infrastructure, and ethical frameworks tailored to the unique needs of the Greek media industry.

The findings align with those of previous studies, highlighting the pivotal role of AI tools in shaping public information. Since the media significantly influence public perception, the integration of AI into journalism and information dissemination carries profound implications. While there is widespread recognition that AI is an enduring and rapidly evolving factor in the media landscape (Stahl et al., 2023), critical questions persist regarding its broader role within journalism.

AI tools for journalists are advancing at a remarkable pace, offering functions, such as content writing, visual creation, and audio generation for text articles. These capabilities enable newsrooms and journalists to streamline workflows, save time, and reduce operational costs (Dhiman, 2023). Increasingly, AI is also being employed to produce news articles directly distributed to the public, assuming tasks traditionally carried out by journalists. These applications are primarily tested on service-oriented information that requires minimal analytical depth or subjective commentary. Nonetheless, there have also been ongoing experiments with AI systems generating more complex, analytically rich news content (Arguedas & Simon, 2023). However, language might be a barrier.

The CEO's View

Though technological innovation moves at lightning speed, Artificial Intelligence (AI) remains in a fledgling phase. News organizations are understandably grappling with what AI means to both operations and long-term sustainability.

Digital transformation strategy is absent in the Greek media according to a quantitative survey, conducted on 148 professionals from 86 media organizations investigating the ongoing digital transformation landscape in four European countries: Greece, France, Portugal, and Cyprus (Mandenaki et al, 2024). Greek respondents confirmed the absence by 53.5%. Respondents from Cyprus by 50%, from Portugal by 48% and from France by 12%.

At present, in our research, only one Greek media organization employs an AI expert, reflecting broader challenges in adopting AI in the industry. Several factors contribute to this slow uptake:

1. AI remains a relatively novel tool in many media outlets and its full potential is not yet well understood.
2. Many organizations lack sufficient human resources to implement AI solutions effectively.

3. Financial constraints limit the development of AI tools tailored to specific organizational needs.
4. Ethical concerns persist over the accuracy and reliability of AI-generated content in newsrooms.
5. The Greek language poses a significant barrier, as many AI tools are not trained to accommodate this linguistic market.

To explore the integration of AI in Greek media, we conducted four semi-structured interviews with CEOs – two from organizations spanning TV, print, and digital media, and two managing digital-only media groups.

According to their insight, the primary constraint is a lack of understanding regarding the availability and applicability of AI tools. This knowledge gap, combined with the technical skills of IT staff and journalists, fosters uncertainty about transitioning to AI-driven workflows. As one CEO of a Media Group underlined "there are myriad tools and systems promising (revenue) heavens, but there are no enough data to support their cases. Moreover, my executives can't support such a transformation'

The CEOs expressed concern over AI's propensity for errors and its reinforcement of human biases. Moreover, the fact that most AI tools are developed and controlled by private entities raises issues related to financial dependencies and ownership, which, in turn, influence trust in generative AI technologies. To preserve the integrity of journalism in the AI era, there is a critical need to understand who controls these tools and their implications for the media industry. In this context, many CEOs who lack familiarity with AI technology prefer a cautious approach, waiting until these tools are perceived as safe and reliable before integrating them into their organizations." I said 'wow' when AI started, but now I'm waiting'' mentions one CEO from a Digital only Media group, a point view close to Gartner's Hype Cycle (Khandabattu, 2025)

Greek digital media publishers, having relied heavily on platforms like Google and social media to attract audiences over the past decade, remain skeptical about dependency on new platforms.

Concerns about evaluating the quality and reliability of AI tools were a recurring theme. As one Media Group CEO remarked, "We can't embrace an unknown tool in our systems and workflows without proper evaluation and in-depth research."

Another CEO highlighted apprehensions about IT teams' ability to assess risks, including the potential for AI tools to misuse data or compromise confidentiality. "There are many issues undefined regarding data security and more over we are concerned about copyright issues", mentions especially for Print one CEO from a Media Group.

A recurring concern among publishers and journalists is the issue of intellectual property rights, particularly as it pertains to AI-generated content. This issue is closely related to the need for sustainable revenue streams, with many viewing intellectual property as a potential new source of income. However, one interviewee expressed a willingness to navigate through intellectual property challenges if intellectual property offered an opportunity to innovate with AI and generate revenue through alternative applications. This perspective highlights the tension between traditional revenue models and the pursuit of novel opportunities enabled by emerging technologies.

Additionally, the rapid pace of technological advancements poses challenges. "We are thinking of the need to train journalists and other staff in AI basics at least because we believe that the Group will benefit shortly", a CEO from a Digital Media noticed.

Another CEO from a Media Group observed, "If we invest in a tool today –even if it's affordable– it will require significant effort, time, and resources to train our staff. Yet, the next day, a more advanced tool might render everything obsolete."

CEOs' Discussion

We find a "wait-and-see" approach, with some organizations hesitating to invest until a significant breakthrough occurs. However, this perspective risks neglecting the value of early training and expertise in AI architecture, which could serve as a foundational asset for future implementations.

Despite these reservations, discussions between CEOs, news managers and tech-savvy personnel are ongoing, focusing on identifying tools that can be integrated into daily workflows. Notably, two media groups have begun documenting and categorizing repetitive, structured tasks –not limited to journalism– to streamline operations. Examples include Accounting and BI, analyzing large datasets from TV ratings and integrating this data with advertising placement systems, automating video archiving, generating reports from web analytics, and personalizing content distribution and diffusion.

These priorities reflect a growing recognition of AI's potential to enhance efficiency and improve the technical aspects of media operations.

However, we find nuances in the adoption of AI among CEOs from Only Digital Groups and CEOs from Print, TV, and Digital. AI's augmentation in Digital newsrooms seems to have fewer obstacles than the others. Greek media organizations in broadcasting, print and Digital are more cautious, possibly due to the investment needed and lack of personnel to integrate AI in their newsrooms. However, they are exploring opportunities at other sections, such as Accounting, Data Analysis, Advertisement. We find, in other words, a divide on priorities among Digital

Media Groups and Media Groups that expand into many publication formats. The perspectives and strategic orientations of Chief Executive Officers (CEOs) play a pivotal role in organizational decision-making. Given their position at the apex of corporate governance, CEOs exert significant influence on the direction, priorities, and long-term vision of the company. Their viewpoints shape not only operational choices but also the broader strategic trajectory of the organization.

The News Managers' View

Mapping the use of AI in Greek newsrooms reveals a nuanced pattern of adoption, showcasing both its potential benefits and limitations. AI tools are particularly advantageous for enhancing efficiency and managing data, especially for smaller, independent media organizations that often operate with limited resources and a workforce lacking advanced technical skills. Insights from interviews with ten news managers from leading Greek media outlets, who have experimented with, but not adopted, various AI applications, underscores the current boundaries of these tools.

Unanimously, Journalists said that they are using AI for personal but not official use. 6 to ten use free versions of the tools they use, 2 to ten they have subscriptions to ChatGpt, Gemini and Perplexity and 2 to ten to transcriptions and subtitling tools. "We need and our newsrooms need training. Our Upper management must prepare us for a possible transformation" they underline.

By automating routine and repetitive tasks, AI enables journalists to focus more on complex and significant topics. However, Greek newsrooms are experimenting in most cases and not adopting AI tools in their workflow. It is a sporadic and personal use by tech-savvy and curious managers and journalists. This aligns with our previous research (Panagopoulos et.al, 2024). This trend highlights that Greece is a Global South country, despite the official ranking in Global North. According to a 2024 survey involving 314 media leaders from 56 countries –in which Greece is not included–, the primary use of AI in the media industry remains focused on back-end automation tasks, such as transcription and copyediting (56%), followed by recommender systems (37%), content creation with human supervision (28%), and commercial applications (27%) (Newman, 2024).

Despite its potential, AI adoption in newsrooms remains cautious and relatively moderate. Key barriers include language limitations and technological challenges. One respondent emphasized the importance of addressing talent gaps and knowledge-sharing, stating, "We must ensure that those proficient in video editing can leverage AI tools effectively and without hesitation to share this expertise with the broader team." This comment highlights concerns regarding skills development as well as fears of job displacement.

While AI offers significant benefits, such as saving time, automating menial tasks, and assisting with initial drafts of articles, skepticism persists regarding its reliability without human supervision. "Video subtitling is fast and easy with AI tools, but there are always mistakes and we have to check and edit before publishing'' admits a News Manager who is responsible for a Digital only Media. However, he adds that "I need no one with special skills to adjust the subtitles". Journalists express concerns that errors and bias from AI tools could harm the credibility of their outlets. "In my opinion there is a need for a Greek LLM which will be trained by Greek literature to avoid 'language raping' and be safer on that issue", underlines a News Manager for Print. Ethical considerations and the need for transparency underscore a broader tension between leveraging technological advancements and maintaining journalistic integrity. "We must understand as a profession that we have to regain the trust of the audience. There is no other way. We must always have a human in the loop and never publish AI-produced content without verification. And we have to say to the audience when we used AI to produce content," says a News Manager from Broadcast and Digital outlet. Although AI promises streamlined operations, improved content personalization, and enhanced data management, it also raises concerns about its role in perpetuating fake news. Simultaneous usage of synthetic media (deepfakes) from the audience and for information from media and journalists might cause at least confusion in the audience. "The EU AI Act has no rules- actually allows- for User Generated Content of deepfakes. This might confuse the audience and increase distrust of the media", notes a News Manager from a Digital Media.

The potential pitfalls of the adoption of AI are significant. As one respondent cautioned, "Without appropriate policies, AI could result in reputational damage, legal issues, and erosion of audience trust."

For example, when CNET, a popular U.S. tech site, quietly published dozens of AI-generated feature articles as part of an experiment, it faced backlash and accusations of betraying journalistic principles (Farhi, 2023).

Text creation remains a contentious issue for journalists, often described as the "holy grail" of their profession. Respondents unanimously rejected the idea of fully delegating text creation to AI. " AI can be trained and write pieces using my writing style, but no one will accept it", says a News Manager from a Print and Digital media. However, they acknowledged using AI tools for less critical tasks, such as generating listicles, creating content ideas for vacation planning, or rephrasing existing articles. "AI as IA – Intelligent Assistant- will be wonderful", admits a News Manager from Broadcast and Digital Media. More common applications include transcribing, voice reproduction, text-to-speech conversion, subtitling, and video editing for social media, though these have not been fully integrated into newsroom workflows. Many newsrooms are cautiously experimenting with these tools to prepare for eventual adoption, but hesitate to formally make a request for them due to fears of failure and potential conflicts within their teams. "In Breaking News, I use AI tools as search engines for a short historic recap, and it is always useful. But, usually I avoid sharing the usage of AI with the team due to fears of possible misuse by them," underlines a TV, Print and Digital News Manager.

The growing reliance on generative AI in journalism has intensified discussions about the need for transparent communication regarding its use, particularly given the ethical implications. Audiences increasingly demand disclosure when AI contributes to news content creation, although Greek journalists fear that labelling such content might further erode trust –already at the lowest levels in Europe. "I fear that my audience will reject my content if it is labeled as AI-assisted. They might not understand that I was involved in producing it" states a News Manager from Print and Digital. There is an ongoing debate about which AI-generated content should be labelled. Research suggests that audiences are generally more accepting of content created by human journalists, especially on serious topics like politics, while being relatively more tolerant of AI involvement in less critical subjects such as sports (Newman et al., 2024; Fletcher & Nielsen, 2024).

International newsrooms, particularly in Europe, have begun implementing guidelines for AI use, ensuring that humans remain involved in the production process (de Lima Santos et al., 2024). In contrast, Greek newsrooms have yet to adopt similar measures. However, the Panhellenic Federation of Journalists' Unions (PFJU) published a Code of Ethics for AI use in journalism. This initiative aims to prevent misuse by publishers and safeguard journalists' roles and integrity in an increasingly AI-driven media landscape (Poesy, 2024).

News Managers' Discussion

News Managers are experimenting and using AI tools, despite the stance of their Upper Management. They are not using them officially in everyday's routine and workflow of the Newsroom, but in many cases, they publish content which is assisted by AI tools. This hybrid approach allows them to be at the forefront of AI use and at the same time to have a cautious stance until they feel safe to implement specific ones in the Newsroom's workflow. It seems that video subtitling and transcription will be 'must tools' for all of the Greek Newsrooms soon. When it is for Digital-only Mediums, it looks like the News Managers are more positive about adopting AI tools. We found a similar approach from the CEOs of Digital Groups. Ethical considerations regarding language, transparency and trust are raised by all the participants, but especially from the News Managers who run Print outlets. Data security issues are not mentioned by the News managers, even though there is a possibility of 'leaking documents' when AI tools are used to edit them. If AI is used as an IA (Intelligent Assistant), Journalists will integrate many tools in their workflows. Training is requested by them in order to have quality results when using AI tools. Finally, article writing remains the 'Holy Grail' for the profession.

CONCLUSIONS – CHALLENGES AND PERSPECTIVES

Undoubtedly, productive AI creates structural challenges and perspectives that can be found in every aspect of the value and supply chain of media organizations, from content inputs to its distribution, sales, and communication, in the workflows of journalists, content producers, business models and strategies of media industries.

Just as the "invention of printing as the result of a far-reaching irreversible process" (Eisenstein, 1979, p. 57) changed written culture in general, the same could be said of the advent of AI, which irreversibly leads the media industries into a new era, full of challenges and prospects.

Considering the above, it is evident that the changes in the value chain are expected to have a significant impact on media organizations, creating new challenges and perspectives in areas such as:

- Business models such as subscriptions, personalized access to content, micropayments
- Automation in Accounting, BI, Marketing, Business Data Analysis
- Automations in Archiving, pre- and post-production
- News gathering, Research and Big data analysis
- Multimedia content creation and editing
- Data journalism and visualization
- Development of deepfakes and fake news detection systems
- Distribution strategies: From mass production content to personalized services and products
- Social Media engagement
- Intellectual property, copyright, content ownership
- Limits of use, bias and credibility
- Labor market in media companies

Greek journalism operates within a politically polarized environment characterized by widespread public distrust, economic and financial instability, media fragmentation, and ownership concentration. Despite being part of the Global North, the Greek media landscape faces significant digital disparities, a lack of training in modern journalistic practices, and a critical shortage of AI experts. These challenges contribute to a growing gap between Greece and other European countries in leveraging AI for journalism. Only a handful of media organizations possess the technological capabilities, infrastructure, or skilled personnel required to advance in this area.

Our case studies highlight contrasting examples of AI adoption in Greek media. A retired, self-taught journalist demonstrated a pioneering spirit by launching a website entirely powered by AI. Another media organization with a tech-forward approach has integrated multiple AI tools to enhance audience engagement and production. Meanwhile, a third company is expanding its audience and potential revenue streams through an English-language edition supported by automated translation. However, the majority of Greek newsrooms, especially smaller ones, lack the human resources and strategic vision to incorporate AI into their workflows. AI usage remains sporadic and experimental, with little evidence of an AI-driven culture.

Currently, Greek media outlets rely predominantly on external AI solutions, as there is no possibility for in-house development. Moreover, the limitations of existing AI models, particularly their insufficient training in the Greek language, hinder rapid adoption. Effective journalism requires time and resources, and AI has the potential to provide tools for creative and investigative work, while addressing skill gaps among journalists. However, only a few media organizations are actively exploring how AI could be integrated into their operations.

The success of AI implementation depends on a collaborative approach rather than a top-down strategy. Journalists and other personnel, and possibly media executives, must receive comprehensive training to view AI as a tool that enhances, rather than replaces, their work. Developing and effectively utilizing AI in newsroom contexts requires acquiring new skills while building on foundational data literacy competencies, which are pivotal to data journalism.

Specialized roles in machine learning, generative AI, and data analysis are increasingly essential in modern newsrooms. Although editors and reporters will continue to be central figures, they must also acquire enough expertise to critically evaluate AI outputs, such as those from AI models, and collaborate with technical specialists. However, ethical considerations, source management, and face-to-face interviews remain in areas where AI is not applicable, underscoring the need for a balanced integration of technology into journalistic practices.

Technology has been omnipresent throughout the history of the media, primarily as a means of dissemination, a catalyst for accelerating the circulation of content, and a driver of changes in its form. Never before, however, has it served as a vehicle for such effortless and immediate production of the content itself (text, image, sound, etc.) as is now the case with the advent of Artificial Intelligence and the tools that are continually being developed.

Media enterprises, as information and content-intensive industries, have historically focused on human resources, since content and creativity in general originate from the human factor. It is precisely these resources that are now subject to the influence of Artificial Intelligence.

In the Greek media landscape, as is evident from the present chapter, Artificial Intelligence is taking its first steps primarily as a tool currently functioning as an assistant in various areas of the value chain (e.g., the gathering, production, and dissemination of journalistic content), rather than in the actual authorship of textual content. However, digital-born media seem more adaptive to AI's augmentation, probably due to their familiarity with technology.

In any case, over time, with the anticipated bridging of the digital divide and the prospect of deeper integration of Artificial Intelligence into the value network of the Greek media sector, its impact remains to be seen—both on the sustainability of existing business models and, more broadly, on the development of new business and production models for a sustainable news and media ecosystem.

REFERENCES

Airan, A., Patil, V., Narayan, V., & Kappor, A. (2024, July 2). Examining the Impact of Generative Artificial Intelligence on the Video Creation Industry. *aapti institute*. https://aapti.in/blog/examining-the-impact-of-generative-artificial-intelligence-on-the-video-creation-industry/, Retrieved 25.07.2024.

Albeanu, C. (2016a, Feb., 24). Around two thirds of Independent's UK readers access it only from mobile. https://www.journalism.co.uk/news/around-two-thirds-of-independent-readers-in-the-uk-access-the-title-only-from-their-mobile-devices/s2/a613542/ Retrieved 10.08.2024.

Albeanu, C. (2016b, Aug., 17). The Washington Post rolls out automation for the Olympic Games, gearing up for US election coverage. https://www.journalism.co.uk/news/the-washington-post-rolls-out-automation-for-the-rio-olympic-games-gearing-up-for-us-election-coverage/s2/a665769/ Retrieved 10.08.2024.

Anderson, C. W. (2016). News Ecosystems. In Witschge, T., Anderson, C. W., Domingo, D., & Hermida, A. (Eds.), *The Sage Handbook of Digital Journalism*. SAGE.

Arguedas, A. R., & Simon, F. M. (2023). *Automating Democracy: Generative AI, Journalism, and the Future of Democracy* (p. 21). Balliol Interdisciplinary Institute, University of Oxford. Associated Press (n.d). ARTIFICIAL INTELLIGENCE: Leveraging AI to advance the power of facts. *Associated Press*. https://www.ap.org/solutions/artificial-intelligence/ Retrieved 10.08.2024.

Automated Insights (n.d) Wordsmith. *Automated Insights*.https://automatedinsights.com/wordsmith/, Retrieved 07.07.2024.

Bahroun, Z., Anane, C., Ahmed, V., & Zacca, A. (2023). Transforming Education: A Comprehensive Review of Generative Artificial Intelligence in Educational Settings through Bibliometric and Content Analysis. [Switzerland]. *Sustainability*, *15*, 12983. Advance online publication. DOI: 10.3390/su151712983

Baraniuk, C. (2018, Nov.8). *China's Xinhua agency unveils AI news presenter*.https://www.bbc.com/news/technology-46136504 Retrieved 15.10.2024

Bird, S., Barocas, S., Crawford, K., Diaz, F., & Wallach, H. (2016). Exploring or exploiting? Social *and ethical implications of autonomous experimentation in AI*. Workshop on Fairness, Accountability, and Transparency in Machine Learning. https://papers.ssrn.com/sol3/papers.cfm?abstract_id=2846909

Broussard, M., Diakopoulos, N., Guzman, A. L., Abebe, R., Dupagne, M., & Chuan, C. H. (2019). Artificial Intelligence and Journalism. *Journalism & Mass Communication Quarterly, 96*(3), 673–695. DOI: 10.1177/1077699019859901/ASSET/1077699019859901.FP.PNG_V03

Castells, M. (2000). *The Rise of the Network Society*. Wiley Blackwell.

Cherubini, F., & Sharma, R. (2023, December 5). Changing Newsrooms 2023: Media leaders struggle to embrace diversity in full and remain cautious on AI disruption. *Reuters Institute* https://reutersinstitute.politics.ox.ac.uk/changing-newsrooms-2023-media-leaders-struggle-embrace-diversity-full-and-remain-cautious-ai#header--2, Retrieved 10/08/2024.

de-Lima-Santos, M. F., Yeung, W. N., & Dodds, T. (2024). *Guiding the Way: A Comprehensive Examination of AI Guidelines in Global Media*. arXiv:2405.04706. https://doi.org//arXiv.2405.04706DOI: 10.48550

Deck, A. (2024, May, 9). For the first time, two Pulitzer winners disclosed using AI in their reporting. https://www.niemanlab.org/2024/05/for-the-first-time-two-pulitzer-winners-disclosed-using-ai-in-their-reporting/ Retrieved 11.10.2024.

Deuze, M., & Witschge, T. (2018). Beyond journalism: Theorizing the transformation of journalism. *Journalism, 19*(2), 165–181. DOI: 10.1177/1464884916688550 PMID: 29417952

Dhiman, D. Bharat, Does Artificial Intelligence Help Journalists: A Boon or Bane? (March 24, 2023). Available at *SSRN*: https://ssrn.com/abstract=4401194

Diakopoulos, N. (2019). *Automating the news: how algorithms are rewriting the media*. Harvard University Press.

Dul, J., & Hak, T. (2007). *Case study methodology in business research*. Elsevier.

Eisenstein, E. (1979). *The Printing Press as an Agent of Change* (Vol. I and II). Cambridge University Press.

Esposito, J. (2010). Publishing After the Apocalypse. *Logos, 21*(3-4), 13–19.

Eule, A. (2013, Aug.31). Big Data and Yahoo's Quest for Mass Personalization. *BARRON'S*. https://www.barrons.com/articles/SB50001424052748704719204579028752466630862. Retrieved 07.08.2024.

Farhi, P. (2023, Jun, 17). A news site used AI to write articles. It was a journalistic disaster. *The Washington Post*.https://www.washingtonpost.com/media/2023/01/17/cnet-ai-articles-journalism-corrections/ Retrieved 12.10.2024.

Fletcher, R., & Nielsen, R. K. (2024). *AI and the Future of News: What does the public in six countries think of generative AI in news?* Reuters Institute for the Study of Journalism.

Gollmitzer, M. (2023). Journalism Ethics with Foucault: Casually Employed Journalists' Constructions of Professional Integrity. *Journalism, 24*, 1015–1033. DOI: 10.1177/14648849211036301

Hines, Kr. (2023, June 4). History Of ChatGPT: A Timeline OfThe Meteoric Rise Of Generative AI Chatbots. *Search Engine Journal*. https://www.searchenginejournal.com/history-of-chatgpt-timeline/488370/, Retrieved 07.07.2024.

Jamil, S. (2023). Evolving newsrooms and the second level of digital divide: Implications for journalistic practice in Pakistan. *Journalism Practice, 17*(9), 1864–1881. DOI: 10.1080/17512786.2022.2026244

Kalogeropoulos, A. (17 June 2024). Digital Report 2024, Greece. Reuters Institute for the Study of Journalism. https://reutersinstitute.politics.ox.ac.uk/digital-news-report/2024/greece retrieved 15/11/2024.

Kapos, P. (2023). *The Ecosystem of the book publishing industry in the 21st century: Digital transition and the Greek case (2009-2019)*. I. Sideris Publications. (in Greek)

Khandabattu, H. (July 8 2025). The 2025 Hype Cycle for Artificial Intelligence Goes Beyond GenAI. *Gartner*. https://www.gartner.com/en/articles/hype-cycle-for-artificial-intelligence Retrieved 10.07.2025.

Kothari, A., & Cruikshank, S. A. (2022). Artificial intelligence and journalism: An agenda for journalism research in Africa. *African Journalism Studies, 43*(1), 17–33. DOI: 10.1080/23743670.2021.1999840

Leandros, N. (2010). Media Concentration and Systemic Failures in Greece. *International Journal of Communication, 4*, 886–905.

Leandros, N. (2011). The International Media System in Transition. A Human Development Perspective. Retrieved from: https://www.newmedia21.eu/analizi/the-international-media-system-in-transition-a-human-development-perspective/

Makwambeni, Bl., Trust, M., & Bulani, J. G. (2023). Between Utopia and Dystopia: investigating journalistic perceptions of AI deployment in community media newsrooms in South Africa. In Dralega, C. A. (Ed.), *Digitisation, AI and algorithms in African journalism and media contexts* (pp. 17–32). Emerald Publishing Limited., DOI: 10.1108/978-1-80455-135-620231002

Mandenaki, K., Sotirakou, C., Karabella, A., & Mourlas, C. (2024). *Digital Transformation in Media.* IQ Media Hub, NKUA. Research Presentation at the 18th PFJU Conference 18-20 October 2024, Samothraki.

McFarland, M. (2014, September 23). The algorithm that can make fantasy football even more fun. *The Washington Post.* https://www.washingtonpost.com/news/innovations/wp/2014/09/23/the-algorithm-that-can-make-fantasy-football-even-more-fun. Retrieved 10.08.2024.

Miller, R. (2015). AP's 'robot journalists' are writing their own stories now. *The Verge*. https://www.theverge.com/2015/1/29/7939067/ap-journalism-automation-robots-financial-reporting, Retrieved, 10.08.2024.

Munoriyarwa, A., Chiumbu, S., & Motsaathebe, G. (2023). Artificial intelligence practices in everyday news production: The case of South Africa's mainstream newsrooms. *Journalism Practice*, *17*(7), 1374–1392. DOI: 10.1080/17512786.2021.1984976

Murschetz, P. (2024). Business Models in the Media Industry: A Synoptic Overview. *Handbook of Media and Communication Economics*. https://doi.org/DOI: 10.1007/978-3-658-39909-2_19425

Napoli, P. M., & Roepnack, A. (2018). Big data and media management. In Albarran, A. B., Mierzejewska, B. I., & Jung, J. (Eds.), *Handbook of media management and economics* (2nd ed., pp. 410–423). Routledge.

Newman, N. (2024, January 9). Journalism, media, and technology trends and predictions 2024. *Reuters Institute*https://reutersinstitute.politics.ox.ac.uk/journalism-media-and-technology-trends-and-predictions-2024#header--0, Retrieved 10/08/2024.

Newman, N., Fletcher, R., Robertson, C. T., Ross Arguedas, A. A., & Nielsen, R. K. (2024). *Reuters Institute Digital News Report 2024*. Reuters Institute for the Study of Journalism.

Newman, N., Fletcher, R., Schulz, A., Simge, A., & Nielsen, R. K. (2020). Journalism, media, and technology trends and predictions 2020. *Reuters Institute Digital News Report*. https://www.digitalnewsreport.org/publications/2020/journalism-media-and-technology-trends-and-predictions-2020

Nichols, T., Shabir, N., & Nielsen, R. K. (2016). *Digital- Born News Media in Europe. Digital News Project 2016*. Reuters Institute for the Study of Journalism. University of Oxford., DOI: 10.60625/risj-d7k0-sk74

Nurelmadina, N., Hasan, M. K., Memon, I., Saeed, R. A., Ariffin, K. A. Z., Ali, E. S., Mokhtar, R. A., Islam, S., Hossain, E., & Hassan, M. A. (2021). A Systematic Review on Cognitive Radio in Low Power Wide Area Network for Industrial IoT Applications. [Switzerland]. *Sustainability*, *13*, 338. Advance online publication. DOI: 10.3390/su13010338

Osterwalder, A., & Pigneur, Y. (2010). *Business Model Generation. A Handbook for Visionaries, Game Changers, and Challengers*. John Wiley & Sons.

Panagopoulos, A. M., Triantafyllou, S., & Kapos, P. (2024, June 29-30). *AI usage in the Greek Newsrooms*. CCLabs Conference, (In Greek) https://www.youtube.com/watch?v=w3qbeP205Lo&t=4235s 1.05.07 – 1.20.50.

Pavlik J.V. (2023). Collaborating with ChatGPT: considering the implications of generative artificial intelligence for journalism and media education. *Journal Mass Communication Education* 78(1):84–93. https://doi.org/ to ref 2023 in articleDOI: 10.1177/10776958221149577Return

Poesy (2024). PFJU establishes a Committee for a Code of Ethics on AI. In Greek. https://www.poesy.gr/poesy-idryei-epitropi-gia-ti-systasi-k/ Retrieved 15.11,2024.

Porter, M. E. (1985). *Competitive Advantage. Creating and Sustaining Superior Performance*. Free Press.

Pressman, L. (2017, June, 14). The State of Artificial Intelligence in 2017. *Automated Insights*. https://automatedinsights.com/blog/the-state-of-artificial-intelligence/ , Retrieved 06.06.2024.

Protothema.gr. (2024a, July 4). The new AI-powered English version of Proto Thema with the support of Google. *Proto Thema.* https://en.protothema.gr/2024/07/04/the-new-ai-powered-english-version-of-proto-thema-with-the-support-of-google/ Retrieved 10.08.2024.

Protothema.gr. (2024b, July 22). How Artificial Intelligence is changing the comments on protothema.gr. *Proto Thema*https://en.protothema.gr/2024/07/22/how-artificial-intelligence-is-changing-the-comments-on-protothema-gr/ Retrieved 10.08.2024.

Protothema.gr. (2024c, November, 15). ProtoThema Introduces AI Semantic Search. https://en.protothema.gr/2024/11/15/protothema-introduces-ai-semantic-search/ Retrieved 20/11/2024.

Robonews (n.d.) https://robonews.gr/%cf%80%ce%bf%ce%b9%ce%bf%ce%b9-%ce%b5%ce%b9%ce%bc%ce%b1%cf%83%cf%84%ce%b5/ (In Greek) Assessed 10.11.2024.

RSF (2024). Index, Greece. https://rsf.org/en/country/greece, Retrieved 15/10/2024.

Ruggiero, C., Karadimitriou, A., Lo, W. H., Núñez-Mussa, E., Bomba, M., & Sallusti, S. (2022) The professionalisation of journalism: global trends and the challenges of training and job insecurity. In: Trappel J, Tomaz T (eds) *Success and failure in news media performance: comparative analysis in the media for democracy monitor 2021.* (pp 309–335) Nordicom, University of Gothenburg, Gothenburg, https://doi.org/DOI: 10.48335/9789188855589-15

Smith, M. D., & Telang, R. (2016). *Streaming, sharing, stealing: Big data and the future ofentertainment.* MIT Press.

Stahl, B. C., Antoniou, J., & Bhalla, N.. (2023). A systematic review of artificial intelligence impact assessments. *Artificial Intelligence Review, 56*, 12799–12831. DOI: 10.1007/s10462-023-10420-8 PMID: 37362899

Stray, J. (2021). Making artificial intelligence work for investigative journalism. *Algorithms, Automation, and News*, 97–118.

Thompson, J. B. (2010). *Merchants of culture. The Publishing Business in the Twenty First Century.* Polity Press.

Toffler, A. (1980). *The Third Wave.* Bantam.

tovima (2024, July 1). Τα άρθρα του Γιάννη Πρετεντέρη στο ΒΗΜΑ αποκτούν τη φωνή του μέσω τεχνολογίας AI. *To Vima*https://www.tovima.gr/2024/07/01/media/ta-arthra-tou-gianni-pretenteri-sto-vima-apoktoun-ti-foni-tou-meso-texnologias-ai/ retrieved 10.08.2024.

Triantafyllou, S. (2020). *Media and journalism in transition. The changing working conditions of journalists in the digital age.* Salonika: Disigma publications. In Greek.

Triantafyllou, S., Panagopoulos, A. M., & Kapos, P. (2025). AI Pioneers and Stragglers in Greece: Challenges, Gaps, and Opportunities for Journalists and Media. *Societies (Basel, Switzerland), 15*(8), 209. DOI: 10.3390/soc15080209

Washington Post (2024, Nov, 7). The Washington Post Launches “Ask The Post AI,” a New Search Experience. https://www.washingtonpost.com/pr/2024/11/07/washington-post-launches-ask-post-ai-new-search-experience/ Retrieved 27.11.2024.

Weber, H. (2014, Jun, 30). Associated Press backs Automated Insights to automate boring earnings reports. http://venturebeat.com/2014/06/30/associated-press-backs-automated-insights-to-automate-boring-earnings-reports/ Retrieved 10.09.2024.

Yang, Z. (2024, Oct.,9). How The New York Times incorporates editorial judgment in algorithms to curate its home page. https://www.niemanlab.org/2024/10/how-the-new-york-times-incorporates-editorial-judgment-in-algorithms-to-curate-its-home-page/ Retrieved 13.10.2024.

ENDNOTES

1 In this chapter, all references to SimilarWeb monthly data pertain to October 2024.

2 https://www.liquid.gr/

Chapter 4
Automated Fact-Checking in Combating Mis-/Disinformation:
An Investigation of Fact-Checkers' Perceptions in 18 Countries

Achilleas Karadimitriou
https://orcid.org/0000-0003-3670-968X
Panteion University of Social and Political Sciences, Greece

ABSTRACT

Contemporary platform society necessitates fact-checking initiatives to combat fake news proliferation and enhance the credibility of digital discourse. While automated fact-checking experiments show promise in improving verification efficiency, they also raise concerns regarding the diminishing authority of human fact-checkers. Based on the theories of "disruptive technologies" and "affordances", this study explores the perception and application of automated technologies in the routines of fact-checkers across multiple countries, using data from an online survey of thirty-two fact-checkers and interviews with seven editors-in-chief of fact-checking organisations affiliated with the International Fact-Checking Network. Findings indicate that, while fact-checkers recognise the utility of AI tools for specific tasks, they resist full automation of the fact-checking process, highlighting the necessity of human oversight. AI is viewed primarily as an aid in the fight against disinformation rather than a complete solution due to inherent limitations that human capabilities surpass.

DOI: 10.4018/979-8-3693-8729-0.ch004

INTRODUCTION

Since the 2010s, the increased use of social media platforms and the rise of platform society have facilitated the spread of false or distorted information, amid growing public distrust towards news media organisations and (socio-political) institutions. This crisis of trust has fostered social polarisation, driven by disinformation and misinformation trends, shaping the image of the modern risk society. This perception of social reality is reflected in the World Economic Forum's latest report ("The Global Risks Report 2024"), where misinformation and disinformation are identified as the top global risks for the next two years, exploited by socio-political actors seeking to deepen social and political divides.

In the face of these challenges, scholars perceive fact-checking either as a new element of modern journalism (Graves & Anderson, 2020) or as part of a global effort to raise public awareness about disinformation (Moreno-Gil Ramon & Rodríguez-Martínez, 2021, p. 252), within a creative endeavour to revive the journalistic tradition of truth-seeking (Graves, 2016). Fact-checking initiatives have emerged as a response to the credibility crisis facing journalism following the rise of social media platforms as the primary source of news for online audiences, alongside an increase in the dissemination of fake or falsified news. The concept of fact-checking encompasses various practices, from simple verification of claims made by public figures to an official process: identifying statements or news items (media monitoring) that are suitable for checking, verifying them with appropriate sources, and then issuing a reasoned verdict that either confirms or disconfirms the information being examined, based on a rating scale that indicates its level of accuracy (Graves & Amazeen, 2019).

During the first two decades of the third millennium, fact-checking initiatives have expanded significantly, adopting organisational structures as sub-entities within news organisations or grassroots movements, functioning as digital-born non-profit platforms (Graves and Cherubini, 2016). Despite these developments, research remains limited regarding the perspectives of news fact-checkers on their practices within a communication environment that is increasingly open to automation based on Artificial Intelligence (AI). Although the fact-checking process has defined the platform era of journalism, it has been described by contrasting empirical evidence concerning its effectiveness and implications (Carnahan & Bergan, 2021; Fridkin, Kenney & Wintersieck, 2015; Garrett & Weeks, 2013). Therefore, ongoing research efforts are essential to clarify developments in a constantly evolving communication landscape.

Despite empirical differences, it is widely recognised that fact-checkers face a complex task that requires the use of new digital technologies. However, there is debate over whether innovative technologies that enable the automation of key parts of the fact-checking process can serve the goal of enhancing and maintaining trustworthiness in the information ecosystem. Since the late 2010s, AI-based tools used in fake or falsified news detection and fact-checking systems have proven to be somewhat effective in addressing information disorder, with potential for more substantial contributions in the future (Školkay & Filin, 2019).

Research on fact-checking has faced challenges in achieving multidimensionality, as it highlights the diverse understanding of its aims. Still, little is known about fact-checkers' perceptions of the impact of their practices (Dias & Sippitt, 2020, p. 605). The present study aims to address one aspect of this gap by examining how fact-checkers perceive the importance of automating the fact-checking process and the challenges associated with automation across eighteen countries. Regarding the theoretical framework, we discuss the idea of "disruptive technologies" (such as artificial intelligence) to suggest a pathway towards a sustainable, disinformation-free public discourse. Furthermore, inspired by the research of Dierickx, Sirén-Heikel, and Lindén (2024), this study applies the theory of affordances to the utilisation of automated tools in fact-checking. It aims to explore how fact-checkers in different social contexts perceive and interact with applications that support automation in their professional work. The primary focus is on analysing the "imagined affordances" of these technologies, which relate to expectations not yet fully realised (Nagy & Neff, 2015). Nevertheless, this concept underscores how people utilise and interpret technologies, as well as how technologies influence users' visualisation (Nagy & Neff, 2023).

Based on an online questionnaire survey involving thirty-two fact-checkers and seven interviews with managing directors or editors-in-chief of fact-checking initiatives affiliated with the International Fact-Checking Network (IFCN), this study aims to provide insights into automated fact-checking practices. It posits that, in the future, a comprehensive and efficient fact-checking model will necessarily include elements that combine automated assistance tools with professionals and crowds, equipped with relevant soft skills and a higher level of media literacy. The current research supports the claims of previous studies, arguing that solutions to enhance the detection of fake news are essential from the perspective of the communication discipline, as research has mainly been focused on IT engineering, applied mathematics, and IT systems management perspectives (García-Marín, Elías & Soengas-Pérez, 2022, p. 128).

Although fact-checking is a relatively new process in several countries, it is gradually becoming more professionalised and organised. Consequently, research into how fact-checkers work to restore truth and how they perceive new technological tools can deepen the understanding of this innovative practice's epistemology, which is a dominant force in the communication field. This study aims to contribute to this understanding, especially since fact-checking still faces scepticism. Despite its short history, it has already sparked significant debates and controversy in the public sphere. This is exemplified by the intense public discussions that arose, particularly after the Meta Corporation head decided to end the third-party fact-checking programme on his platforms—initially in the USA—and shift to a Community Notes model, asserting his commitment to free expression (Kaplan, 2025).

Disruptive Technology as Innovation?

Bower and Christensen (1995) introduced the concept of disruptive innovation to describe the unsettling impacts of new technologies, mainly on market processes, shaping how disruptive technologies are understood. Their conceptualisation, common in economics and business fields, emphasises innovative technologies capable of triggering the deregulation of existing supply chains and business models, thus disrupting markets (Christensen, 1997). The need for a broader understanding of social changes caused by new technologies has led scholars to discuss technological disruption as "techno-social disruption", meaning technologies that influence the fundamental forces of techno-social disruption as "socially disruptive technologies" (Hopster, 2021, p. 1).

Disruptive technologies are perceived as disturbing or unbalancing the current order (Bongomin et al., 2020), altering traditional operational standards (Laukyte, 2020), or exhibiting radical computational dynamics that drive technological progress (Jekov, 2022). They have been associated with the fourth industrial revolution (Majumdar, Banerji & Chakrabarti, 2018, p. 1248), the platform era or the widespread trends of media digitalisation and polymediation (Omenugha, 2019, p. 4), offering significant opportunities to enhance human life but also posing threats (Majumdar, Banerji & Chakrabarti, 2018).

They are also viewed as driving forces that help societies transition to new socio-economic standards, paving the way for innovative methods of production and consumption, as well as new business cycles (Melnyk et al., 2019, p. 24). Disruption is regarded as occurring when new technology replaces mainstream technology within a relevant industry (Adner, 2002). Therefore, disruptive technologies are understood as both creative (the foundation and basis of innovations and new development cycles) and destructive within societies (undermining existing production processes) (Melnyk et al., 2019, p. 26). The speed of adaptation to disruptive technologies varies among countries, which are characterised by different dynamics in terms of resources, knowledge, and finances (Bongomin et al., 2020, p. 10).

Table 1. Key arguments comparing the "disruptive technologies" and "disruptive innovations" theory to the "sustaining technologies/innovations" theory.

General Views on "disruptive technologies" or "disruptive innovations" theory	Conceptualisation of "Sustaining Technologies/ Innovations"	Conceptualisation of "Disruptive Technologies/ Innovations"
• They refer to disruptive impacts of new technologies across different sectors (Bower & Christensen, 1995) • Challenge the existing norms in markets and industries (Bongomin et al., 2020; Christensen, 1997) • Alter the typical operational patterns (Laukyte, 2020) • Hold radical computational dynamics that promote technological evolution (Jekov et al., 2022) • They are perceived as creative (prerequisites/the basis of innovations and new cycles of development) and as destructive (undermining an existing productive process) (Melnyk et al., 2019: 26)	• They derive their success in the market by offering gradual improvement in the characteristics of products or services that the public has already experienced and evaluated (Christensen, 2014)	• They create an entirely new market by introducing a new product or service • They initially attract only a "fringe" market • They can quickly capture a significant market share by serving needs that did not previously exist (Christensen, 2014)

In the field of communication, the disruptiveness caused by digital media innovations has been highlighted, especially regarding "smart" technologies that shape how media content is produced, disseminated, and consumed (Picard, 2003). Digital disruption has also been described as driving the emergence of technological giants amid a convergence between new media technologies and the trend towards digitalisation (Omenugha, 2019, p. 2). As disruptive technologies, new media also affect media management by presenting challenges and opportunities for developing skills and tasks, prompting media managers to prioritise strategic changes based on their cognitive abilities to improve the competitiveness of media organisations (Omenugha, 2019, p. 3). Although the disruptiveness of new technologies can be associated with both positive and negative implications (Omenugha, 2019, p. 2), from a research perspective, they are often viewed positively regarding their influence across corporate, consumer, and professional sectors (Păvăloaia & Necula, 2023).

Digital technologies are recognised as key drivers of significant disruptive innovations in the media industry. Due to digitalisation, what matters is not only the properties of technologies but also how users navigate the digital space, a process known as "polymediation" (Gholampour Rad, 2017, p. 10). AI is ranked among the most important disruptive technologies (Girasa, 2020), alongside the Internet of Things, Blockchain, 5G, 3D printing, cloud technologies, robotics, renewable energy methods, and virtual and augmented reality. Scholars generally agree that it has influenced the most affected fields, including healthcare/medicine, business, the labour sector, sustainable agriculture, education, and governance (Păvăloaia & Necula, 2023, pp. 29-33). However, as a disruptive technology, AI is believed to be responsible for causing dramatic shifts in how digital users access and consume news, information, and entertainment within the digital ecosystem, amid increasing audience fragmentation and potential information or communication divides (Lin, 2019, p. 592).

Responding to digital disruption presents a challenge for both the audience and the management sector of the media industry. For example, a study examining the role of first-order dynamic capabilities in addressing digital disruption in the newspaper industry has shown how the adjustment of a media organisation's resources, processes, and values can lead to digital platform capabilities that influence the response to digital disruption (Karimi & Walter, 2015).

The Theory of Affordances: A Brief Overview of Key Perspectives

The concept of affordances clarifies the properties of different technologies by assessing how they are used in real-world settings. These affordances have consistently shaped technological communication environments throughout media development (Hutchby, 2014). Gibson, who focused on visual perception of the world, introduced the term "affordances" to describe action possibilities that are directly perceivable by living beings from objects within their environment, suggesting that what we perceive is not just objects, but the potential actions they offer (Gibson, 1979).

Detecting objects' affordances, that is, the meanings of things within an environment, is considered a prerequisite for perceiving the action possibilities offered by that environment. The way in which objects within an environment can be used is related to the living entity's position in that context. In this manner, a mutual relationship between organisms and the environment is emphasised, based on the idea that an organism and its environment complement each other through affordances, which refer to the properties an object has that communicate potential actions to a living creature. How affordances are perceived depends on the context and the state of a living creature or animal (Chemero & Turvey, 2007), existing between personal interpretation and the objective characteristics of an object (Schmidt, 2007).

Although this concept was introduced to clarify ideas in ecological psychology and has generated much controversy and conflicting interpretations, its influence on other fields—ranging from industrial design (Norman, 1988) to autonomous robotics (Murphy, 1999; Şahin et al., 2007)—has been significant. Since its inception, various reformulations of the concept have demonstrated a dynamic role in shaping our understanding of the perception-action relationship (Chong and Proctor, 2020), despite its initial ambiguity regarding the overall meaning it conveys (Chemero & Turvey, 2007).

Scholars defining the conceptual framework of affordances have proposed several types of this concept, reflecting different perspectives. Norman (1988), who rearticulated the theory in a refined way using the term "perceived affordances," emphasised the visible properties that could reveal an object's accepted use, differentiating his perspective from that of Gibson (physical affordances) (Nagy & Neff, 2023). His approach highlighted the action possibilities that people perceive when they discover a new tool, linking its use with a cognitive evaluation process of how it might be employed (Norman, 1999). From this perspective, affordances are first perceived and then acted upon, and these perceived possibilities for action influence people's attitudes towards (technological) objects. Gaver (1991) adopted a different view, identifying three types of affordances: the perceptible ones, which refer to the uses of objects based on what they seem to be able to do; the hidden ones, which refer to the non-obvious uses of objects, meaning that affordances are not perceived but alter utility; and false affordances, which describe uses of objects characterised by veneer in terms of actual usability.

Graves (2007) adopted a different perspective, describing affordances as triggers of action. Despite several deviations from the original theory, as Nagy and Neff (2023) emphasise, affordances are regarded as "dispositional" and "relational"; the former relates to their recognisable status when perceived, while the latter pertains to their dependence on the relationship between users' goals, the properties of technology, and the context of its use. The relational aspect of technology affordances is also seen as tangible when the user interacts with a new digital tool (Vyas et al., 2006).

Touching on the issue of affordance and perception through a user-centred perspective, Nagy and Neff (2015) introduced the concept of "imagined affordances," which refers to the idea that in digital technologies, it is crucial to consider how social actors appropriate them—specifically, what they imagine about the uses of these technologies—instead of focusing solely on their actual and perceived properties (Kammer, 2020, p. 342). Building on this concept, they aimed to describe a mechanism through which scholars reflect on "technological environments' material qualities that mediate affective experiences" (Nagy and Neff, 2015, p. 2).

When the concept of affordances was applied to the field of human-machine communication, new approaches emerged that allowed researchers to explore and understand how users and designers "imagine" the potential uses of interactive technologies. According to Rodríguez-Hidalgo (2020), the way scholars in the communication discipline have conceptualised affordances has led to new perspectives that highlight the relational aspects of technology use. He argued that they tend to describe them as a "multifaceted structure", encompassing a relational pattern between an object or technology, its attributes, and the user (Rodríguez-Hidalgo, 2020, p. 59).

This concept of affordances has been applied to users' experiences in digital environments, such as social media platforms (Majchrzak et al., 2013), chatbots (Beattie et al., 2020), and robots (Cheong, 2020). Researchers developing robots recognised the importance of this idea and used it in artificial systems, considering it helpful for creating autonomous agents. This development is illustrated by the growing research into affordances in robotics (Yamanobe et al., 2017).

As a result, affordances have taken on a new role in academic discussions amidst the ongoing digitalisation process. They serve as a widely used theoretical tool for conceptualising and interpreting the features of new media technologies (Kammer, 2020: 338). However, the term lacks a clear definition within communication and media studies (Nagy and Neff, 2015, p. 1). It is argued that Gibson's analysis of affordances closely parallels Hall's discussion of how representation evolves into meanings ("Encoding/Decoding" model), and combining these theories can assist scholars in interpreting the implications of audience engagement with new interactive technologies and platforms shaped by ideologies (Shaw, 2017). In this context, all interactive media technologies are described as dynamic objects that encourage or discourage certain types of use based on their design (encoded into their design), with users, when they possess them, driving various emergent uses (decoded uses) (Shaw, 2017, p. 597).

PERCEIVED AND IMAGINED AFFORDANCES

Communication discipline has contributed to the theory of affordances by emphasising the relational perspective of this concept. This is reflected in the main argument of communication scholars, who assert that affordances exist in the interaction between an individual's assessment of utility and the actual characteristics of a technology (Schrock, 2015, p. 1231). Surroundings include various affordances perceivable by humans based on their attention.

This study focuses on investigating perceived and imagined affordances. Perceived affordances refer to what users believe and expect from technology in relation to their tasks. These expectations and mental models can influence their ability to adapt to technological advancements or their willingness to overcome barriers that hinder the integration of digital tools into their daily routines. Users clearly identify perceived affordances, guide actions (Norman, 2013), and can express their impressions of technological interfaces, thereby shaping their behaviour. The study uses the theory of affordances through Norman's (1999) framework, emphasising the action possibilities recognised by human actors (such as fact-checkers) when they encounter a new tool.

Imagined affordances include mediation, materiality, and affect—dimensions explored in this study. The first relates to the capacity of technologies to influence people's perceptions of the world. The second covers the inherent physical and digital characteristics of technologies, which become evident through the interaction among the object, infrastructure, designer, and user. The third pertains to people's emotional responses to machines, reflecting how they perceive and use technologies (Nagy & Neff, 2023). These three dimensions make imagined affordances a useful framework for examining human-machine communication (Nagy & Neff, 2023).

The Origins and Evolution of Fact-Checking

The origins of the modern fact-checking process through platforms can be traced to the USA during the 2000s. The creation of FactCheck.org in 2003 by the Annenberg Public Policy Center at the University of Pennsylvania sparked the emergence of The Washington Post's Fact Checker initiative and PolitiFact.com venture a few years later (2007). The latter is a non-profit project developed by the Tampa Bay Times that now operates under the auspices of the Poynter Institute for Media Studies. Since its inception, fact-checking has evolved considerably, shifting from a process affecting only individuals, journalists, and political actors to a multidimensional workload involving various tasks (editorial work, social media management, in-depth research and data analysis, advocacy, and policy work) aimed at enhancing the overall information society (Juneja and Mitra, 2022, p. 25).

The emergence of initiatives dedicated to verifying previously published news or assessing the truth and accuracy of statements made by public figures seems to have been influenced by the 2016 U.S. presidential elections. After all, political issues are often characterised by their controversial nature, which makes verification procedures concerning the accuracy of information spread by the media a common focus (Nyhan & Reifler, 2012; Young et al., 2018). The strong public support for these projects ensured that fact-checking's role in public discourse remained lively and prominent during Donald Trump's presidency (Bauder, 2017; Mantzarlis, 2016). Since the 2010s, fact-checking initiatives have increased even within Europe. This rise can be linked to the escalation of disinformation and the growing public mistrust of the media.

Research has demonstrated that the external conditions influencing the development of fact-checking initiatives worldwide are mainly the level of democratic governance and Internet access. However, the latter alone is not enough. Conversely, when assessed independently, press freedom has been closely linked to the emergence of fact-checking. It has also been shown that fact-checking aims to support democracy-building goals rather than being merely a product of a healthy democratic culture (Amazeen, 2020, pp. 104-107).

Recent studies describe fact-checking as a multidimensional process that combines both human and technological systems. Effective implementation of modern "long-term advocacy-centric fact-checking," based on research, policy, and advocacy, requires a collaborative effort among various stakeholders, supported by technology and policies. It includes both visible and invisible aspects: the first involves editorial and investigative work by editors and fact-checkers to debunk misinformation (short-term, claims-based fact-checking), while the second pertains to investigators, researchers, and advocates whose role is vital in improving the accessibility and quality of information at the national level through advocacy, policy, and research efforts ("long-term advocacy-centric fact-checking", Juneja and Mitra, 2022, p. 5).

Fact-Checking Practice as a Value-Informed Process

Fact-checking activities worldwide share many standard features related to traditional journalism's investigative and verification processes, such as the disclosure of truth based on accuracy, integrity, and a comprehensive, multifaceted presentation of views. Through various professional routines, fact-checkers combine conventional journalistic skills, like information gathering and source verification, with critical thinking and evidence-based data analysis to deliver accurate and reliable narratives of reality. However, fact-checking practices have gained independence and a distinct identity, becoming a specialised journalism subgenre (Mena, 2019; Singer, 2021). This is due to a unique work protocol that begins from the conception of a topic and continues through to its final publication, guided by strict steps and ethical standards. Among these, non-partisanship, transparency in sources, organisational funding, and methodology, together with an open and honest corrections policy, stand out as core principles that steer and permeate the fact-checking process to achieve excellence (IFCN, 2025).

Therefore, fact-checking cannot be regarded as an arbitral process. It is closely connected to a set of values that apply during its implementation. The success and credibility of fact-checks depend on a format that allows understanding of the reasoning behind assessing the information presented to the public (Young et al., 2018). At the same time, source transparency is regarded as a significant journalistic value (Hellmueller, Vos & Poepsel, 2013).

Research from various countries indicates that the details of the information environment and the type of fact-checking organisation influence source transparency. Compared to internal editorial fact-checkers in traditional newsrooms, independent fact-checkers are more inclined to use transparent sources in their fact-checked publications by frequently linking to external documents or sources. Additionally, journalistic professionalism at the national level affects source transparency, while a low-trust attitude toward the media ecosystem is not associated with the level of source transparency (Humprecht, 2019).

Although the fact-checking initiative's policy aims to uphold all these values mentioned above, applicable to practices related to truth verification and restoration, research has revealed the questionable effectiveness of such initiatives in improving public knowledge (Walter et al., 2019).

Automation Processes in Fact-Checking Practices

Technological innovation has facilitated and improved fact-checking practices, enabling this global trend in transnational journalism to demonstrate its practical use. Advances in computer science, particularly in natural language processing (NLP), have allowed for automation in fact-checking routines, from identifying claims suitable for verification to presenting the results. Notably, strategies to counteract viral misinformation spread on social media platforms have been supported through collaborations with technology companies that manage these platforms. This approach is often referred to as ML-supported fact-checking or debunking on social media, as machine learning (ML) systems play a crucial role in exploring, verifying, and correcting factual claims. An outstanding example is the algorithmic system in Meta Company's third-party Fact-Checking Programme, which scholars describe as an "intermediary actor in the struggle over truth in the public debate on social media platforms," acting between the fact-checkers and the citizens being fact-checked (Schjøtt & Bengtsson, 2024, p. 2).

The need to automate this process arises because, despite numerous fact-checking organisations and applications, the response rate to restoring the accuracy of fake or falsified stories circulating online remains slow over time. Consequently, fact-checking algorithms have been developed to identify and interpret claims using supervised machine learning (Adair et al., 2017). However, they are limited by issues such as the lack of comprehensive methods for handling multiple languages beyond the dominant English, reduced precision and accuracy in their tasks, dependence on pre-existing datasets of facts, and a deficiency in technologies that can analyse evidence, not just the claims (Lazarski, Al-Khassaweneh & Howard, 2021, pp. 20-21).

A study examining the impact of generative AI technology on the fact-checking process argues that its effectiveness depends on the organisational structures within which it is integrated. AI-based practices are used in various aspects of the fact-checking mission, including editing, investigation, and advocacy, with aims such as quality assurance, trend analysis, and information literacy (Wolfe & Mitra, 2024). Another study discusses utilising AI technology during the pre-fact-checking phase to identify claims with high dissemination potential that are deemed worth monitoring and further examining. AI is also employed in analysing content within images, texts, or videos, as well as in claim detection tasks, to save time and boost efficiency. Aside from its significant contribution to analysing data sets, its role has also been crucial in cross-checking data or facilitating the debunking process by converting audiovisual material into text (Gutiérrez-Caneda & Vázquez-Herrero, 2024). During the investigation process, other AI applications include transcribing and translating content, creating lead paragraphs, synthesising documents, supporting research, and analysing trends (Wolfe & Mitra, 2024, pp. 1534-1535).

In the next stage of content editing, generative AI technology improves written reports, transforming "dense content" into an engaging format for digital audiences while correcting grammatical and factual mistakes. It is also seen as a tool for identifying inaccurate assumptions in fact-checked articles and for pre-writing them before human review (Wolfe & Mitra, 2024, pp. 1534-1535). In contrast, the dissemination of verified case results is an area within the fact-checking process where AI use has been relatively limited (Gutiérrez-Caneda & Vázquez-Herrero, 2024, p. 62). Nonetheless, in the post-fact-checking stage, AI tools such as plug-ins and chatbots offer digital users an accessible way to stay alert and vigilant against misleading content (Gutiérrez-Caneda & Vázquez-Herrero, 2024, p. 63).

Despite various uses, media professionals hold differing views regarding AI-based technological tools: some see them as helpful in enhancing journalism's value by assisting in fighting fake news, editing news according to editorial policies, and providing content personalisation (Ali & Hassoun, 2019), while others remain sceptical about the ability of Generative AI to produce meaningful and accurate journalistic narratives (Ogola, 2025; van Dalen, 2024; Gutiérrez-Caneda, Vázquez-Herrero & López-García, 2023). Amid this debate, automated fake news tracking systems have not been widely adopted by fact-checking and debunking organisations (Pavleska et al., 2020). For example, in the four Nordic fact-checking organisations (Denmark, Finland, Norway, and Sweden), where media regulation and media literacy promote resistance to disinformation (Andersson, 2023), the relationship between fact-checkers and AI-based technology is evolving. Fact-checkers see these technologies as tools to enhance, support, and complement their work practices. AI-based applications are mainly used to verify the authenticity of images and videos, in a context of both scepticism and openness (Dierickx, Sirén-Heikel, & Lindén, 2024, pp. 465-466).

Against a backdrop of artificial intelligence's rise, fact-checking initiatives are often caught in economic and technological dependencies on tech giants such as GAFAM. An example is Africa, where financial motives and dominance issues characterise the partnerships between the domestic fact-checking platform (Africa Check) and the tech giants that control the design and deployment of intelligent tools and have achieved supremacy in the online information field. The automation process remains incomplete, and AI-assisted innovative procedures worsen technological dependency (Sarr, 2024).

Overall, research has shown scepticism and distrust among fact-checkers regarding the use of AI and automation in fact-checking tasks. Their reasoning is based on the complexity of their work, which requires several human interventions aimed at constructive discussions and decision-making. Their concerns focus on AI's ability to detect falsehoods in political statements and the integrity of algorithms used by tech companies (such as Meta) to identify content on social media platforms worth fact-checking. Despite their scepticism, AI has already been adopted by some fact-checking organisations, especially in cases involving low-stakes decision-making. Trust can increase if AI tools become more transparent in their outputs, allowing for human oversight (Juneja & Mitra, 2022, p. 21).

Although using AI-powered tools for automated content creation in journalism and fact-checking brings several benefits to media organisations, such as increased efficiency, lower costs, and easier live news coverage (Subroto et al., 2024; Horska, 2020; Huang et al., 2023), ethical concerns also emerge in this area. A significant issue is the transparency of data sources, and the quality of the data utilised (Monti, 2019: 11-12), along with the risk of biased algorithms or datasets that reflect prejudices, potentially resulting in biased or distorted news stories. Additionally, there are concerns about the trustworthiness of AI-generated information, especially when reports involve complex political events or human-interest stories, as well as difficulties in ensuring accountability when AI systems make errors (Hassim & Mohamad Nasir, 2025: 49-50) or in assigning responsibility for the content produced (Monti, 2019: 8-9). Ethical considerations also encompass the transparency of data used by content creators, relating to the audience's ability to interact with and distinguish between content authored by humans and that generated by robots (Ali & Hassoun, 2019: 44). These challenges, which do not encompass all issues raised by AI-supported technologies, demand careful examination and proactive responses (Reynolds and Nolan, 2025).

Users' trust in AI systems greatly relies on the quality of data, which, according to scholars, must adhere to three key principles: prevention, detection, and correction, to ensure reliable machine learning applications (Ehrlinger et al., 2019). Dierickx et al. (2024), based on the idea that accuracy and efficiency in machine learning tasks depend on high-quality data, suggest a data quality assessment framework that aims to link technical concerns with ethical responsibilities, thereby promoting data and AI literacy among journalists.

MEDIA PROFESSIONALS' CHALLENGES OF ADAPTABILITY IN THE PLATFORM ERA

While the integration of technology in journalism has long been pursued, it has increased in recent years due to rising competition in the media industry. In this competitive environment, journalism has become increasingly driven by technology, with journalists assuming technologist roles or collaborating with technology experts within reputable news organisations. Additionally, given the ongoing technological advancements, the ability of media organisations to adapt is a vital factor that shapes their innovation processes in a digital era where journalism prioritises direct audience participation and engagement as the core feature of the new digital landscape. A typical example of this is the digital initiative AJ+ launched by Al Jazeera, which demonstrates that investing in innovative digital projects does not necessarily lead to a radical change in journalistic practices, organisational culture, or the media organisation's business model (Zayani, 2020).

Modern journalism relies heavily on interconnected infrastructural sectors, including subsea fibre cables, content delivery networks, and data centres, which significantly influence how news organisations operate. This ecosystem is primarily influenced by dominant players, including Google, Meta, and Amazon, which significantly impact data management and dissemination due to their substantial ownership stakes (Sjøvaag et al., 2024). The rise of digital platforms has placed the journalism profession in a fragile position, leading to a decline in ethical and professional standards (Gür Omay, 2023). Consequently, journalistic practices have been modified to align with the principles of the platform society. The evolution of platform-oriented journalism is reflected in the emergence of journalistic knowledge platforms, a new generation of platforms that revolutionise the operation of newsrooms through practices aimed at enhancing the quality and mitigating the cost of news production (Gallofré Ocaña & Opdahl, 2020).

These long-term systematic challenges differ from other short-term operational ones, as they relate to the need for renewed journalism training and education that address the specificities of global media reform. The automation of journalism practices requires innovative, technology-driven training models for journalists, alongside improvements to national education systems (Wu and Wang, 2022). By developing various skills, including knowledge of new media technologies (such as content curation, understanding data and algorithms, managing social media platforms, awareness of updated intellectual property laws, and entrepreneurship), journalists are better equipped to face labour market challenges and adapt to disruptions caused by digital transformation. Past research on training print journalists for audiovisual storytelling and smartphone reporting has highlighted that this shift requires not only skill development but also a deep understanding of ethical and legal considerations, alongside a necessary change in mindset (Salzmann, Guribye, & Gynnild, 2021).

Considering the above challenges, scholars have discussed several approaches to the governance of algorithms and AI, ranging from focusing on ethical principles and their operationalisation (Floridi et al., 2018) to emphasising integrated methods and instruments for managing the accountable use of AI-driven tools (Gianni et al., 2022). An insightful study by Helberger and Diakopoulos (2022) examines the complexities of AI governance within media and journalism, particularly in the context of the European AI Act. The authors highlight that a responsible approach to AI deployment must encompass not only its use but also its design and broader contexts, which often overlook the perspectives of both journalists and consumers. They emphasise that policymakers face significant technological, economic, and societal complexities that challenge their ability to create effective regulatory strategies for AI and digital technologies.

RESEARCH QUESTIONS

Building on the theoretical framework analysed above, the main question this study addresses is how fact-checkers view automated fact-checking —such as using artificial intelligence technologies in their processes— as a promising step in combating disinformation/misinformation. The research aims to critically contribute to the debate about the challenges automated tools pose to fact-checkers' authority. The key questions raised by the study can be summarised as follows:

- What are the main “imagined affordances” of AI-based technologies relevant to fact-checking? How do fact-checkers envisage automation contributing to their practices, based on their interactions with emerging technologies?
- What action possibilities are indicated by the fact-checkers (“perceived affordances”, according to Norman, 1999) during their interactions with AI-based tools? How do fact-checkers cognitively assess the ways emerging automated technologies might be utilised?
- To what extent and in what ways are affordances related to technologies enabling the automation of professional practices applied to work routines adopted by fact-checkers?

RESEARCH DESIGN, METHODOLOGY AND SAMPLING

This study explores perceived and imagined relationships with automated processes through a case analysis of fact-checking initiatives in eighteen countries. It is based on online questionnaires completed by 32 fact-checkers and seven interviews with managing directors or editors-in-chief of fact-checking initiatives registered with the International Fact-Checking Network (IFCN). Data collection took place from October 2024 to January 2025. Both quantitative and qualitative methods have been employed to examine how fact-checkers across different media contexts perceive and interact with automated practices in their work. The study aims to explore and interpret their "perceived" and "imagined affordances" regarding AI-based technological tools ("perceived affordances" as a dynamic interaction between perceptions of usefulness/serviceability and the technology's qualities, and "imagined affordances" as a dynamic relationship between users and technology). All fact-checking initiatives in the sample, as members of the IFCN, adhere to ethical standards, including transparency in their working methods, resource allocation, and sustainability strategies.

The study employed a snowball non-probability sampling method for the questionnaire survey. Specifically, an exponential, non-discriminative snowball technique was used, based on the list of signatories on the IFCN website. Participants were recruited via emails—initial research subjects who then referred others—until a sufficient number of participants was assembled for the research.

The respondents' views on integrating AI-based tools into the fact-checking process were examined across nine variables using multiple questions: some were based on an 11-point scale (zero to ten), while others used a 5-point scale (strongly disagree, disagree, neutral, agree, strongly agree). Before analysing the data, the 11-point scale was converted into a 3-point scale (1. Action of no or limited importance, 2. Action of moderate importance, 3. Action of great to outstanding importance), a coding pattern better suited to the research aims. The survey addressed topics such as the role of fact-checkers in the public domain, their core values, and their perceptions of how automation and AI-based tools—an evolving field—impact professional routines and etiquette in fact-checking. It also included two open-ended questions, through which respondents offered insights regarding the potential contribution of automated tools to fact-checking implementation. The study employed thematic analysis to examine responses to these open-ended questions alongside results from closed-ended questions, enabling the researcher to identify common perceptions and patterns concerning the value of integrating automated processes into fact-checking routines.

The semi-structured interviews consisted of 12 open-ended questions that examined participants' personal experiences and opinions on the significance of fact-checking and its transition towards automation. The discussions centred on key themes that complemented those raised in the questionnaire survey, such as working strategies, challenges in responsibility tasks, and interactions with automated digital tools. We employed thematic content analysis of the interview data to address the research question. The study adhered to the six-phase framework outlined by Braun and Clarke (2006).

Main Characteristics of the Research Sample

The research sample for the survey comprises a diverse group of fact-checkers with different levels of familiarity with digital technology tools. They use AI applications for information verification mainly occasionally (34%), followed by those who do so minimally (in the minority of cases under investigation, 25%) or never (19%) (Figure 1). When it comes to using big data, the trends are more varied, with participants reporting experiences as those who utilise these tools extensively (25%), occasionally (31%), to a limited extent (22%), or minimally (19%).

Figure 1. Frequency of use of digital technology-related tools by survey participants

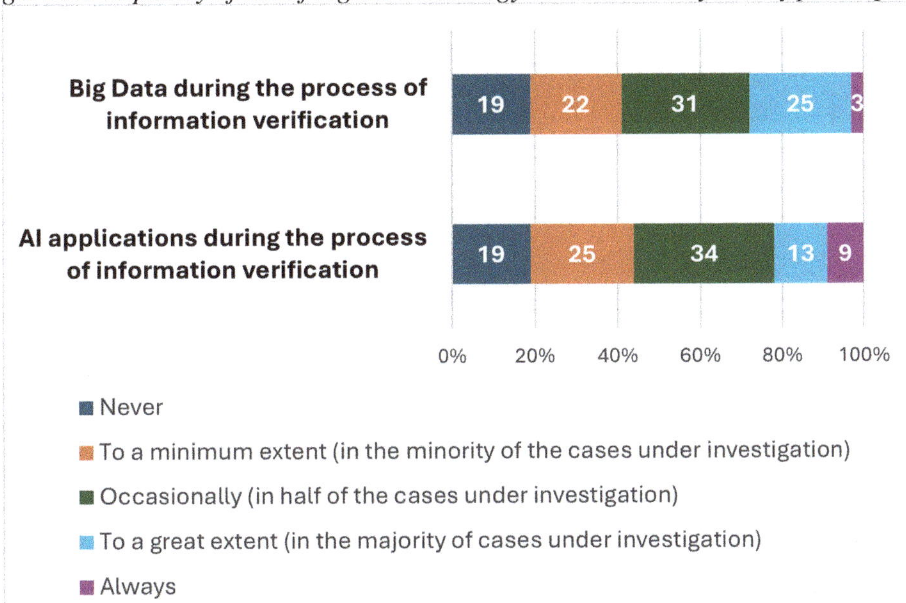

(Source: Author's own research)

Most survey respondents work full-time on fact-checking (69%) and have a background in journalism (56%) (Table 2). In terms of demographic characteristics, the research sample has more men than women (69% and 31%, respectively). Its members are split across various age groups, with the largest groups being 25-34 (38%) and 35-44 (38%). One-quarter of the participants hold a university degree in Media and Communication, and another quarter have an MA/MSc degree in the same field. Moreover, most work as fact-checkers in non-profit independent organisations (44%), followed by those in mainstream media organisations (19%) or organisations working with universities or research centres (19%). The most common source of income maintaining the financial stability of these organisations is grants from corporate entities (e.g., Google and Meta, 69%), followed by grants from philanthropic organisations (34%) and scientific or research institutions (22%).

Table 2. Main features of the survey's research sample (%)

Type of Feature	Percentage (%)
Gender	
Men	**69**
Women	**31**
Age Category	
Up to 24 years old	**3**
25-34 years old	**38**
35-44 years old	**38**
45-54 years old	**18**
55-64 years old	**3**
65+	**0**
Educational Level and Scientific Discipline	
Up to a high school degree	**6**
University Degree in journalism/media /communication	**25**
University Degrees (far from journalism/media/communication discipline)	**17**
MA/MSc in journalism/media /communication	**25**
MA/MSc University Degree (far from journalism/media/ communication discipline)	**9**
PhD in journalism/media/communication	**3**
PhD (far from journalism/media/communication discipline)	**6**
Other	**9**
Employment Status	
Full-time fact-checker	**69**
Part-time fact-checker	**22**
Other	**9**

continued on following page

Table 2. Continued

Type of Feature	Percentage (%)
Knowledge and skills background	
Journalistic background/experience	**56**
Academic research background/experience	**25**
Both journalistic and academic research background/experience	**16**
Other	**3**
The type of organisations fact-checkers are affiliated with	
Journalistic background/experience	**56**
Academic research background/experience	**25**
Both journalistic and academic research background/experience	**16**
Other	**3**

Source: Authors' own research.

The participants in the semi-structured interviews, conducted in conjunction with data from the questionnaire survey, come from five countries: Germany, France, Poland, Armenia, and Greece (Table 3). They represent media markets of varying sizes and dynamics, with distinct traditions concerning the key factors that have shaped the evolution of the communication field and fact-checking initiatives (Table 3).

Table 3. Key features in fact-checking – The cases of Germany, France, Poland, Armenia, and Greece

Country	Key features in fact-checking
Germany	Germany's fact-checking landscape features prominent organisations, such as dpa and Correctiv, as well as public broadcaster initiatives like ARD-Faktenfinder, many of which are certified by the IFCN. These entities focus on critical issues, including climate change, immigration, and crises such as the war in Ukraine and the COVID-19 pandemic. The government actively promotes media literacy and supports independent media, while platforms must adhere to transparency regulations (García-Gordillo, Rivas-de-Roca, & de-Lima-Santos, 2025).
France	Fact-checking initiatives in France involve several key organisations, including *AFP Fact Check*, which addresses misinformation globally, and the *CrossCheck* project, which focuses on combating online falsehoods, especially during elections. Moreover, regional hubs have been established under the *European Digital Media Observatory* (EDMO) to enhance local strategies against disinformation. French public service media organisations also participate in collaborative efforts, such as the European Broadcasting Union (EBU)'s *Spotlight initiative*, with *France Télévisions* developing tools for verifying visual content.
Poland	In Poland, key fact-checking organisations such as the *Demagog Association* focus on evaluating political statements and promoting media literacy through educational initiatives. Additionally, projects such as the "Guardians of Truth" aim to combat disinformation and educate voters. Initiatives in Poland are also related to the *European Digital Media Observatory* (EDMO), a platform designed for collaboration among fact-checkers and researchers in the fight against disinformation across Europe.
Armenia	Fact-checking in Armenia is primarily led by organisations like Hetq.am, which has formed a dedicated team to verify official statements and combat misinformation. Efforts to improve journalistic standards are supported by projects such as the European Media Facility Armenia, funded by the EU and DW Akademie, which aims to enhance media professionalism and credibility amid ongoing challenges of disinformation. A coordinated response involving civil society and the private sector is considered crucial, as emphasised in a report by Freedom House.
Greece	In Greece, fact-checking is conducted by platforms such as Ellinika Hoaxes, Check4facts Science and MediaWatch, which are supported by broader European initiatives, including the European Digital Media Observatory (EDMO) and the European Fact-Checking Standards Network (EFCSN). The EU is also fostering new projects, such as CALYPSO GR, aiming to promote collaboration among fact-checkers, subject matter experts, and the public to effectively identify and counter disinformation through technological initiatives and enhance media literacy (Dimitriadis, 2023).

Sources: For Germany: García-Gordillo, Rivas-de-Roca, & de-Lima-Santos (2025); Serrano, 2023. **For Armenia:** Mong, 2024; DW Akademie, 2022. **For France:** Dieudonné, 2017. **For Poland:** European Media and Information Fund (2025). **For Greece:** Dimitriadis (2023).

The interviews aimed not to highlight the specifics of fact-checking routines in each national context but to explore the challenges and benefits of fact-checking practices as a globally recognised trend. Since this study does not aim to compare findings across different national media environments, but rather to identify general perceptions of AI integration in fact-checking practices, no additional interviews were included in the research sample. For research purposes, the fact-checking initiatives included in the interviewee sample consist of media professionals dedicated to debunking disinformation across various platforms and formats (video, image, audio, and text) (Table 4).

Table 4. Research participants in semi-structured interviews

Interviewee' Code	Country	Role in national fact-checking initiative	Gender	Duration of the interview	Period of implementing the semi-structured interview
FC1	Germany	Editor and Fact-checker, Head of the fact-checking initiative	Male	44:08	November 2024
FC2	France	Photojournalist, Journalist, Fact-checker	Male	58:37	November 2024
FC3	Poland	Editor-in-Chief, Fact-checker and Vice President of the fact-checking initiative	Male	1:10:01	December 2024
FC4	Armenia	Fact checker, Investigative Journalist, Editor and Founder of the fact-checking initiative	Female	41:21	January 2025
FC5	Greece	Journalist/Editor-in-Chief in the fact-checking initiative	Male	1:56:51	October 2024
FC6	Greece	Fact-checker	Male	40:21	October 2024
FC7	Greece	Fact-checker, Founder of fact-checking initiative	Male	44:28	January 2025
FC8	Greece	Journalist, Fact-checking Product Manager focusing on AI and LLMs	Male	43:09	January 2025

RESEARCH RESULTS

Definition of Fact-Checking Practice and Motives Behind Engagement

According to research findings, fact-checking is mainly defined as an institutionalised sub-genre of journalism (50% strongly agree), exerting a corrective impact on mainstream media (44% strongly agree). However, it is considered questionable whether it can significantly influence the agenda-setting power of conventional media organisations (41% are neutral on this assumption, and another 41% agree or strongly agree that fact-checking does not have such an impact). Nevertheless, its role is seen as crucial in discrediting false news (91% strongly agree), upholding journalism's ideals (69% strongly agree), and enhancing people's critical mindset towards media content (63% strongly agree). In nearly all fact-checking initiatives studied, interviewees emphasise standard criteria for story selection: the virality of the content eligible for fact-checking, its widespread dissemination across digital platforms, its relevance to society or its impact on the audience, its urgency regarding the topics raised, its significance for the field of politics, and its potential harm to informed citizenship.

The motives behind fact-checking initiatives arise from the urgent need to help the public confront a persistent problem in modern society: the spread of false and misleading information. To do this, fact-checkers view their role as a non-traditional investigative journalism practice that builds the audience's resilience against the vulnerabilities of the contemporary public sphere, where disinformation is widespread. Some interviewees identify the infodemic that followed the COVID-19 pandemic and the 2020 U.S. presidential election as a key turning point, when communication and journalism actors decided to respond systematically to the surge of misinformation and disinformation facilitated by digital platforms. Even Elon Musk's acquisition of Twitter is regarded as another significant milestone in the rise of disinformation. Due to these events, citizens' intellectual and cognitive defenses are believed to be shaped not only by fact-checked content from independent organisations but also through media literacy training sessions, which all fact-checking initiatives support. Another reason cited is the need for fact-checkers to address the shortcomings of traditional journalism, which often lacks sufficient time, resources, or willingness to verify facts thoroughly. As one interviewee emphasises, in a context of a polarised public sphere and biased media coverage of political topics—often framed by politicians unaccountable for their words—we have emerged "to fight for people, for the quality of the public debate."

The complexity of the digital world— characterised by an abundance of dubious-quality information that requires verification through specialised digital tools —is believed to mean that not all journalists are qualified for comprehensive verification practices. Only those truly willing to invest time and resources into filtering the reliability of information circulating online are likely to pursue this. Such journalists systematically develop essential skills, including monitoring media content, digitally investigating sources, and managing or editing content. An interviewee emphasises the importance of fact-checking teams as a recognised aspect of the modern information society, stating that they help "deliver fact-based information on advanced disinformation campaigns and serve as a multiplier by sharing their knowledge and skills with colleagues, thereby enhancing their verification practices".

PERCEPTIONS OF AI-POWERED TECHNOLOGIES

The questionnaire survey indicates that using AI-based technologies for fact-checking is a standard professional practice among organisations in the research sample (72% of participants admit to adopting AI tools under their organisation's guidance and encouragement). These applications mainly assist fact-checkers in verifying image authenticity (illuminarti), transcribing audio and video files, and analysing manipulated text-based or audiovisual material. Other uses include detecting AI-generated content, translating content, proofreading, drafting reports (Gemini), or creating headlines and shortening previously published fact-checked material. In some cases, fact-checking organisations or media departments develop their own customised AI-powered tools, aimed at tasks such as face recognition in image databases, speech-to-text conversion, translation, and proofreading.

Among Generative AI tools, respondents highlight ChatGPT, Claude, Gemini, and Perplexity, which are regarded as effective in providing overall guidance through source discovery, analysis, summarising, and translating media content. Some participants emphasise that AI tools are used alongside human efforts, and media professionals should be aware of "hallucinations." Notably, no participant mentioned using AI tools for writing or debunking stories, indicating that these tools are seen as auxiliary aids that help, for example, fact-checkers streamline research, but do not replace their routines. A fact-checker referenced using ChatGPT as "a language editor for completed materials to ensure that no grammatical or spelling errors are present before publication."

The use of artificial intelligence tools in fact-checking routines is considered inevitable, as media professionals are eager to experiment with them and reap their benefits. According to the interviewees, their usefulness has mainly been shown in detecting distorted or fake news in the digital environment, and advances in IT technology are seen as ongoing and systematic. It is argued that the importance of automation cannot be fully understood without long-term and systematic application. An editor-in-chief, experienced with AI tools, stresses, "I think we will have to work on such tools for many decades before we can say they are effective".

Reservations for AI-Powered Tools Efficiency

Despite its innovative features, AI is not regarded as harmless technology and is associated with some responsibility for the informational inaccuracies prevalent in the modern media landscape, where participatory culture dominates content creation. This view is clearly expressed by an editor-in-chief, who notes that "in the age of AI, you are no longer sure that the source you're looking at is accurate or authentic". AI-based tools are often criticised for manipulating video, audio, or still images. Although the market for AI technologies is considered "competitive and powerful" and offers numerous options, these tools are sometimes criticised for lacking complete accuracy and proper development.

According to survey results, fact-checkers are hesitant about adopting automation at various stages of fact-checking. For example, in evidence curation, manual tasks are deemed more important than automated functions, and only the combination of automation and human input is regarded as highly significant (Table 5). Doubts about the effectiveness of automated methods are also evident in the claim verification process, with an acknowledgement that the machine-learning verdict is only moderately relevant to the quality of the justification. The development of machine learning may not be seen as the main barrier to expanding fact-checking routines (limited resources are more critical); however, it is mainly viewed as a significant obstacle, with few disagreeing. One respondent highlights the dangers of AI tools, focusing on AI image generation, which "are being used to spread false information, and with the advancement of reinforcement learning in machine learning models, the discrepancies in AI-generated content are becoming harder to detect".

Table 5. Assessment of significance regarding manual or automated means in the fact-checking process

On a scale of 0 to 10, how would you rate the following actions related to the fact-checking process? (0 = no important action at all & 10= highly important action)	Weighted Average	of zero or minimal importance (ratings 0-3)	of moderate importance (ratings 4-6)	of great or outstanding importance (ratings 7-10)
In fact-checking work, evidence must be curated through the manual efforts of fact-checkers	8.63/11	12	12	76
In fact-checking work, evidence must be curated through automated means (e.g. machine-generated data)	5.41/11	41	31	28
In fact-checking work, evidence must be curated through a combination of manual efforts by fact-checkers and automated means	7.97/11	12	19	69

continued on following page

Table 5. Continued

On a scale of 0 to 10, how would you rate the following actions related to the fact-checking process? (0 = no important action at all & 10= highly important action)	Weighted Average	of zero or minimal importance (ratings 0-3)	of moderate importance (ratings 4-6)	of great or outstanding importance (ratings 7-10)
In fact-checking work, the verdict coming from the machine-learning process is relevant for the quality of the justification process	**5.88/11**	**31**	**29**	**40**
In fact-checking work, justification production for claim verification is vital to be based on automated means in the form of system-generated textual explanations	**5.44/11**	**35**	**40**	**25**

Source: Author's own research.

Most survey respondents clearly recognise significant weaknesses in automated information search processes (78% agree or strongly agree) and doubt the so-called efficiency of AI technology in many areas (Figure 1). They are uncertain whether such technology can genuinely enhance the autonomy of fact-checkers, with 44% of respondents taking a neutral stance. In comparison, opinions are divided on whether AI-based fact-checking can effectively improve corrections when false or fake content is widely disseminated online (41% of participants agree or strongly agree, while 41% disagree or strongly disagree). Automating the fact-checking process is not seen as a cure-all for countering disinformation or misinformation.

Although automated tools - compared to manually based methods - are viewed by a notable portion of research participants (56%) as more effective at detecting mis/disinformed content, fact-checkers mostly find it difficult to imagine a media environment where professionals are replaced by robots (57% strongly agree or disagree, and 34% remain neutral).

Figure 2. Degree of agreement/disagreement regarding the relevance of automating the fact-checking process (%)

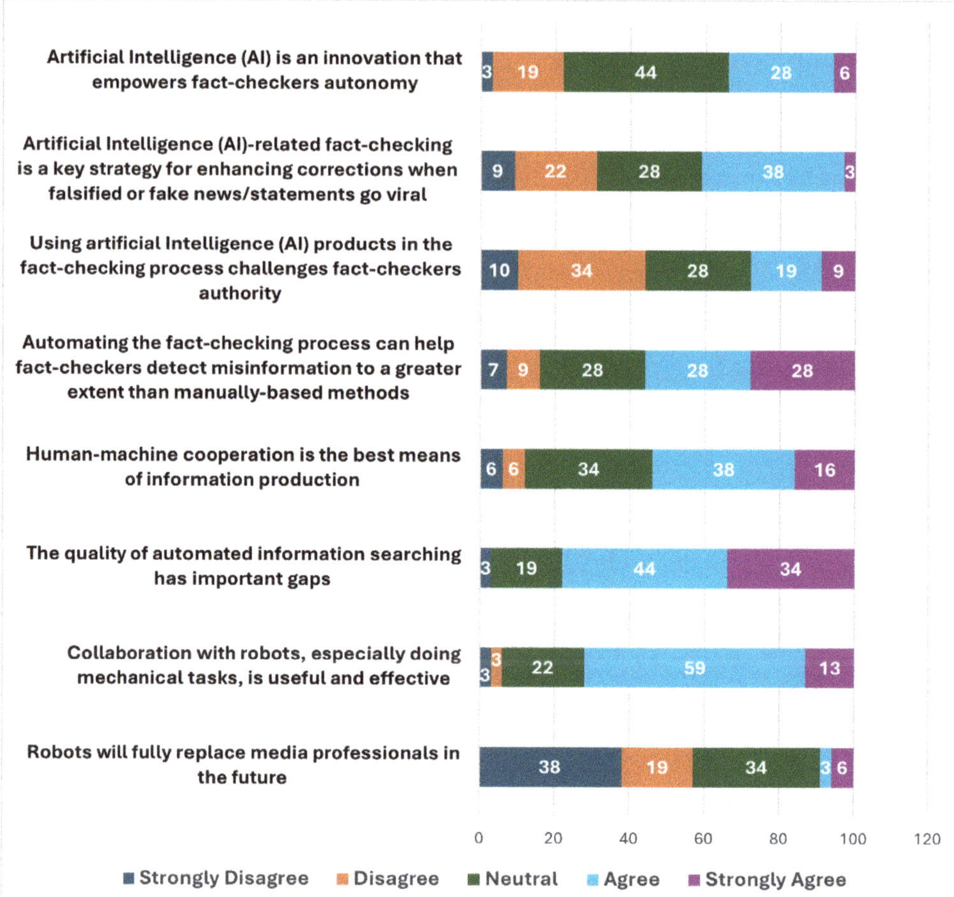

(Source: Author's own research)

Human-machine collaboration is primarily regarded as good practice in the information production process (54% agree or strongly agree). However, it is not a universally held view, as a distinct neutral stance on the same topic also exists (34%). In contrast, such collaboration is generally seen as valuable and efficient in mechanical tasks (72% agree and strongly agree).

Reticence towards AI-powered tools for fact-checking does not suggest a rejection of technological progress or a marginalisation of digitalisation's potential. This is demonstrated by the fact that most fact-checkers support combining traditional journalistic skills with computational abilities (75% agree or strongly agree, Figure 2) in a digital data environment, which is believed to be driving the journalism profession's shift towards data science skills (78% agree or strongly agree).

Fact-checkers envisage a future where automated processes dominate work routines due to investments in AI technologies; however, they remain sceptical about the credibility of a robot's finalised outcome. The dominance of artificial intelligence is questioned because it is not seen as capable of producing infallible suggestions in news verification. Some interviewees argue that the core of fact-checking lies in providing the audience with primary sources of information about a case, along with its context or background. Based on this reasoning, combining data checking through personal journalistic skills with AI-driven suggestions is regarded as best practice in a digital environment where detecting fake news can be time-consuming. Nonetheless, journalism has a duty to deliver accurate and timely documentation. An example of this ambivalence towards AI-based technologies and automation is the confession of an editor-in-chief who stated, "what we recommend in our training is always to combine checking content with the use of sophisticated advanced AI detectors and investigating the content itself with your visual skills and journalistic skills [...] you should be able as a journalist to double check".

Figure 3. Degree of agreement/disagreement regarding the familiarisation of journalism with technical/computational aspects of the digital field

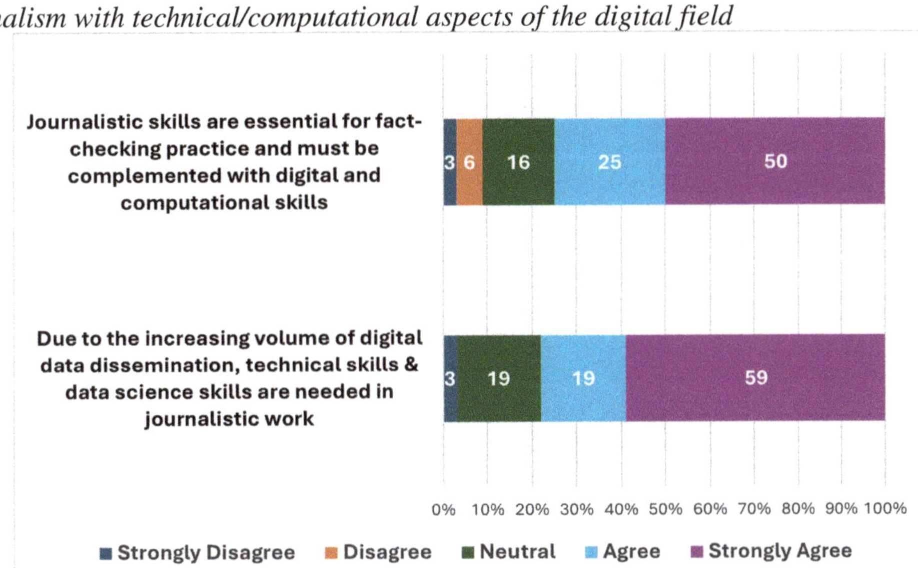

(Source: Author's own research)

Fact-checkers do not believe that AI-based tools undermine the authority of the journalism profession, as decisive steps in shaping the fact-checked content agenda still rest with media professionals rather than machines. Despite AI technology's vulnerabilities (Appendix, table 6), fact-checkers are mainly encouraged to improve their skills in open-source intelligence tools and utilise those available. These tools are seen as constantly evolving applications that, therefore, require users to remain vigilant in their learning. In some fact-checking initiatives, sharing experiences among fact-checkers regarding the use and development of automated tools is considered essential for ensuring the effective/practical impact of these tools on the final output.

DISCUSSION AND CONCLUSIONS

The challenge of spreading fact-based information amidst the rise of disinformation, mainly driven by the COVID-19 pandemic, seems to have led to the systematisation of fact-checking practices across different regions. Fact-checkers aim to fight disinformation by providing audiences with thoroughly verified content characterised by accurate sources and credible contextual details that help clarify narratives. The current study shows a change in the routines and professional profiles of fact-checkers within the research sample, resulting from the careful integration of AI into their practices, a finding also supported by previous research (Gutiérrez-Caneda & Vázquez-Herrero, 2024).

Most interviewees are enthusiastic but sceptical about using AI technologies in their fact-checking routines (Appendix, table 7). They recognise several benefits of AI integration in tasks related to investigation, summarisation, and translation of content, which save time for fact-checkers; however, they also believe that automation-based digital tools have limitations in their ability to produce accurate and comprehensive outputs.

In all fact-checking initiatives, AI-based solutions are primarily seen as tools assisting with fact-checking tasks. Although the newly established organisations in the research sample seldom develop AI-based tools themselves, they utilise these applications with caution and harbour reservations about their capabilities. Since AI is an expensive technology to grow, its use by fact-checking teams is supported by large tech companies (Google, Meta), which provide funding through various collaborative programmes.

Fact-checkers—at least for now—trust automated tools only partially and stress the importance of verifying their answers. They help analyse large datasets and suggest potential fake or manipulated content in texts or audiovisual media, but only under human supervision. Fact-checkers and researchers remain sceptical about fully automating the fact-checking process, as human intervention is still viewed as essential for several aspects of the task. This perspective is supported by earlier research on four fact-checking organisations in Nordic countries (Denmark, Finland, Norway, and Sweden), where fact-checkers, when assessing content accuracy and truthfulness, value human judgment and critical analysis. They engage with AI technologies cautiously and critically, recognising the tools' limitations in accuracy and reliability (Dierickx, Sirén-Heikel & Lindén, 2024, pp. 465-466). Despite this cautious attitude, AI-driven technologies can assist in various aspects of fact-checking, such as identifying claims from the expanding media landscape, locating previously verified statements, gathering supporting evidence from written or visual sources, summarising relevant material, and detecting claims spreading swiftly to prevent their virality (Hrckova et al., 2024, p. 3).

The results of this survey suggest that fact-checkers generally view the physical and social potential of new AI-based tools positively, a cognitive assessment (Norman, 1999) that encourages their creative use. At the same time, they form expectations for emerging automation-related technologies that cannot be fully realised until these tools reach a stage of widespread applicability. This highlights the importance of imagination in the development of affordances (Nagy & Neff, 2015). Despite a cautious attitude towards process automation, fact-checking is seen as a creative task that relies on digital tools and continuous updates of relevant knowledge. The challenges linked to automation are more pronounced in media ecosystems where training or educational opportunities in fact-checking practices are limited or absent. An example includes Greece and Balkan countries, where there is a perceived "not enough focus on digital journalism skills of media professionals." Conversely, the operational model of the Maldita.es platform in Spain, which offers a preparatory training course for new fact-checkers, is regarded as exemplary.

The challenge for fact-checkers is less about accessing AI-based technologies and more about developing a practical understanding of how to use these resources effectively without compromising the goals of fact-checking. Professionals involved in fact-checking should be encouraged to view automated processes as an ongoing resource. This can be achieved through the discipline of learning technology, which enhances fact-checkers' knowledge of various AI tools and helps them make informed decisions about their use. Both positive and negative aspects influence the possibilities associated with Information and Communication Technologies. The definition and classification of these possibilities (such as accessibility, speed of change, diversity, collaboration, reflection, multimodality, non-linearity, risk, fragility and uncertainty, immediacy, and surveillance) are proposed as useful frameworks to understand how these technologies can be effectively employed to support learning and teaching.

Research participants creatively interpret the affordances related to integrating automated processes into fact-checking tasks by perceiving various uses. However, they do not endorse the holistic applicability of AI technologies and are clearly sceptical about this possibility. Notably, in some cases, resorting to automated digital applications is seen as a helpful practice for detecting AI-generated content but not as a comprehensive approach for creating verified and well-documented material.

Despite reservations about the scope of use for these innovative tools, emerging affordances concerning AI applications offer valuable insights into how technology influences communication practices.

This study provides detailed yet nuanced answers to the question: "To what extent and in what ways are the affordances related to technologies enabling the automation of professional practices perceived and utilised by fact-checkers in several countries?" Despite the interesting findings discussed above, several limitations should be acknowledged. The findings focus on the perceived and imagined importance of integrating automation-based digital tools into the fact-checking profession rather than on the actual effectiveness of some tools already in use. Furthermore, the results do not examine the internal and external factors influencing the adoption of rapidly evolving automation technologies, nor do they highlight individual perspectives from several countries, which do not lead to a comprehensive or comparative understanding of the diverse experiences within fact-checking routines. Comparisons are possible based on a broader research sample, divided into clusters of countries according to the theory of media systems, considering the differences and discrepancies that distinguish them. Despite these limitations, the study's contribution lies in offering clear insights into how fact-checkers recognise or envisage the challenges associated with automation-based technologies within the changing field of fact-checking.

Future research should systematically compare how fact-checkers across different media systems and organisational contexts (media-based versus grassroots/independent entities) can maintain their independence and trustworthiness while upholding key principles such as accountability and transparency, even as they adopt automation in their work. Scalable automatic or computational fact-checking approaches have proved feasible by utilising network analytics and large-scale knowledge repositories, aiming to reduce the spread of unreliable information (Ciampaglia et al., 2015). Scholars studying strategies to manage information risks suggest that the rising trends of fake news and misinformation can be addressed if greater emphasis is placed on the positive role of emerging technologies (big data analytics, blockchain, artificial intelligence, social media, Internet of Things) in combating false information, especially during societal and business sector disruptions (Spanaki et al., 2023).

REFERENCES

Adair, B., Li, C., Yang, J., & Yu, C. (2017). Progress Toward “the Holy Grail”: The Continued Quest to Automate Fact-Checking. *Northwestern University*. https://cj2017.northwestern.edu/documents/progress-cj2017-paper-18.pdf

Adner, R. (2002). When are technologies disruptive? A demand-based view of the emergence of competition. *Strategic Management Journal, 23*, 667–688.

Akademie, D. W. (2022). “In Armenia, factchecking disinformation on politics and war”, May 12. Retrieved from https://akademie.dw.com/en/in-armenia-factchecking-disinformation-on-politics-and-war/a-61757306 (Accessed 25/8/2025).

Ali, W., & Hassoun, M. (2019). Artificial Intelligence and Automated Journalism: Contemporary Challenges and New Opportunities. *International Journal of Media* [IJMJMC]. *Journalism and Mass Communications*, *5*(1), 40–49. DOI: 10.20431/2454-9479.0501004

Amazeen, M. A. (2020). Journalistic interventions: The structural factors affecting the global emergence of fact-checking. *Journalism, 21*(1), 95–111. DOI: 10.1177/1464884917730217

Andersson, T. (2023). Nordic media systems. In Papathanassopoulos, S., & Miconi, A. (Eds.), *The media systems in Europe: Continuities and Discontinuities* (pp. 99–131). Springer., DOI: 10.1007/978-3-031-32216-7_5

Beattie, A., Edwards, A. P., & Edwards, C. (2020). A bot and a smile: Interpersonal impressions of chatbots and humans using emoji in computer-mediated communication. *Communication Studies*, *71*(3), 409–427. DOI: 10.1080/10510974.2020.1725082

Bongomin, O., Gilibrays Ocen, G., Oyondi Nganyi, E., Musinguzi, A., & Omara, T. (2020). Exponential Disruptive Technologies and the Required Skills of Industry 4.0. *Journal of Engineering*, *4280156*, 1–17. DOI: 10.1155/2020/4280156

Bongomin, O., Ocen, G. G., Nganyi, E. O., Musinguzi, A., & Omara, T. (2020). Exponential Disruptive Technologies and the Required Skills of Industry 4.0. *Journal of Engineering*, *4280156*, 1–17. DOI: 10.1155/2020/4280156

Bower, L. J., & Christensen, M. C. (1995). Disruptive technologies: Catching the wave. *Harvard Business Review*, *73*, 43–53. https://hbr.org/1995/01/disruptive-technologies-catching-the-wave

Braun, V., & Clarke, V. (2006). Using thematic analysis in psychology. *Qualitative Research in Psychology*, *3*(2), 77–101. DOI: 10.1191/1478088706qp063oa

Carnahan, D., & Bergan, D. E. (2021). Correcting the Misinformed: The Effectiveness of Fact-checking Messages in Changing False Beliefs. *Political Communication, 39*(2), 166–183. DOI: 10.1080/10584609.2021.1963358

Chemero, A., & Turvey, M. T. (2007). Complexity, hypersets, and the ecological perspective on perception-action. *Biological Theory, 2*, 23–36.

Cheong, P. H. (2020). Religion, Robots and Rectitude: Communicative Affordances for Spiritual Knowledge and Community. *Applied Artificial Intelligence, 34*(5), 412–431. DOI: 10.1080/08839514.2020.1723869

Chong, I., & Proctor, R. W. (2020). On the Evolution of a Radical Concept: Affordances According to Gibson and Their Subsequent Use and Development. *Perspectives on Psychological Science, 15*(1), 117–132. DOI: 10.1177/1745691619868207

Christensen, C. (1997). *The Innovator's Dilemma*. Harvard Business School Press.

Christensen, C. M. (1997). *The Innovator's Dilemma: The Revolutionary Book that Will Change the Way You Do Business*. Harvard Business School Press.

Christensen, C. M. (2014, January 1). *Disruptive Innovation*. Interaction Design Foundation - IxDF. https://www.interaction-design.org/literature/book/the-encyclopedia-of-human-computer-interaction-2nd-ed/disruptive-innovation

Ciampaglia, G., Shiralkar, P., Rocha, L. M., Bollen, J., Menczer, F., & Flammini, A. (2015). Computational Fact Checking from Knowledge Networks. *PLoS One, 10*(6), e0128193. DOI: 10.1371/journal.pone.0128193

Dias, N., & Sippitt, A. (2020). Researching Fact Checking: Present Limitations and Future Opportunities. *The Political Communication Quarterly, 91*(3), 605–613. DOI: 10.1111/1467-923X.12892

Dierickx, L., Opdahl, A. L., & Khan, S. A.. (2024). A data-centric approach for ethical and trustworthy AI in journalism. *Ethics and Information Technology, 26*, 64. DOI: 10.1007/s10676-024-09801-6

Dierickx, L., Sirén-Heikel, S., Lindén, C.-G., & Guerrero Rojas, D. C. (2024). Outsourcing, Augmenting, or Complicating: The Dynamics of AI in Fact-Checking Practices in the Nordics. *Emerging Media, 2*(3), 449–473. DOI: 10.1177/27523543241288846

Dieudonné, D. (2017). "Fact-checking the French election: lessons from CrossCheck, a collaborative effort to combat misinformation". Retrieved from https://blog.google/outreach-initiatives/google-news-initiative/fact-checking-french-election-lessons-crosscheck-collaborative-effort-combat-misinformation/ (Accessed 25/8/2025).

Dimitriadis, D. (2023). *Disinformation Landscape in Greece*, EU DisinfoLab, June. (Project funded by the Friedrich Naumann Foundation for Freedom). Retrieved from www.disinfo.eu/wp-content/uploads/2023/06/20230623_GreeceDisinfoFS .pdf (Accessed 25/8/2025).

Ehrlinger, L., Haunschmid, V., Palazzini, D., & Lettner, C. (2019). A DaQL to monitor data quality in machine learning applications. In *Lecture Notes in Computer Science* (pp. 227–237). Springer.

European Media and Information Fund. (2025). "Fact-Checking and Countering Disinformation: Poland's 2025 Presidential Election", March 25. Retrieved from https://gulbenkian.pt/emifund/projects/fact-checking-and-countering-disinformation -polands-2025-presidential-election/#:~:text=The%20project%20also%20 emphasizes%20disinformation,Updated%20on%2025%20march%202025 (Accessed 25/8/2025).

Floridi, L., Cowls, J., Beltrametti, M., Chatila, R., Chazerand, P., Dignum, V., Luetge, C., Madelin, R., Pagallo, U., Rossi, F., Schafer, B., Valcke, P., & Vayena, E. (2018). AI4People – An Ethical Framework for a Good AI Society: Opportunities, Risks, Principles, and Recommendations. *Minds and Machines*, *28*(1), 689–707.

Fridkin, K., Kenney, P. J., & Wintersieck, A. (2015). Liar, Liar, Pants on Fire: How Fact-Checking Influences Citizens' Reactions to Negative Advertising. *Political Communication*, *32*(1), 127–151. DOI: 10.1080/10584609.2014.914613

Gallofré Ocaña, M., & Opdahl, A. L. (2020). "Challenges and Opportunities for Journalistic Knowledge Platforms". *Proceedings of the CIKM 2020 Workshops*, October 19-20, Galway, Ireland.

García-Gordillo, M., Rivas-de-Roca, R., & de-Lima-Santos, M. (2025). The Fact-Checking Initiatives in the EU: A Diverse Ecosystem Against Disinformation. *Media and Communication*, *13*, 9421. Advance online publication. DOI: 10.17645/mac.9421

García-Marín, D., Elías, C., & Soengas-Pérez, X. (2022). "Big Data and Disinformation: Algorithm Mapping for Fact Checking and Artificial Intelligence". In J. Vázquez-Herrero, A. Silva-Rodríguez, M. C. Negreira-Rey, C. Toural-Bran, & X. López-García (Eds.), *Total Journalism*. Studies in Big Data, vol 97 (pp. 123-135). Springer, Cham. https://doi.org/DOI: 10.1007/978-3-030-88028-6_10

Garrett, R. K., & Weeks, B. E. (2013). "The promise and peril of real-time corrections to political misperceptions". In Conference Proceedings of *the 2013 Conference on Computer-supported Cooperative Work*. New York: Association for Computing Machinery (pp. 1047-1058). https://doi.org/DOI: 10.1145/2441776.244189

Gaver, W. W. (1991). Technology affordances. In Proceedings of the *SIGCHI Conference on Human Factors in Computing Systems: Reaching through Technology*. New York: Association for Computing Machinery (pp. 79–84).

Gholampour Rad, M. (2017). Disruptive innovation in media industry ecosystem and need for improving managerial cognitive capabilities in polymediation era. *Cogent Business & Management*, *4*(1). Advance online publication. DOI: 10.1080/23311975.2017.1352183

Gianni, R., Lehtinen, S., & Nieminen, M. (2022). Governance of responsible AI: From ethical guidelines to cooperative policies. *Frontiers of Computer Science*, *4*, 873437. DOI: 10.3389/fcomp

Gibson, J. J. (1979). *The ecological approach to visual perception*. Houghton Mifflin.

Girasa, R. (2020). *Artificial Intelligence as a Disruptive Technology: Economic Transformation and Government Regulation*. Palgrave Macmillan., DOI: 10.1007/978-3-030-35975-1

Graves, L. (2007). The affordances of blogging: A case study in culture and technological effects. *The Journal of Communication Inquiry*, *31*(4), 331–346.

Graves, L. (2016). *Deciding what's true: The rise of political fact-checking in American journalism*. Columbia University Press.

Graves, L., & Amazeen, M. A. (2019). Fact-checking as idea and practice in journalism. In Nussbaum, J. (Ed.), *Oxford Research Encyclopedia of Communication*. Oxford University Press., DOI: 10.1093/acrefore/9780190228613.013.808

Graves, L., & Anderson, C. (2020). Discipline and promote: Building infrastructure and managing algorithms in a "structured journalism" project by professional fact-checking groups. *New Media & Society*, *22*(2), 342–360. DOI: 10.1177/1461444819856916

Graves, L., & Cherubini, F. (2016). *The Rise of Fact-Checking Sites in Europe – Digital News Project 2016*. Reuters Institute for the Study of Journalism, with the support of Google Digital News Initiative. DOI: . Retrieved from https://reutersinstitute.politics.ox.ac.uk/sites/default/files/research/files/The%2520Rise%2520of%2520Fact-Checking%2520Sites%2520in%2520Europe.pdfDOI: 10.60625/risj-tdn4-p140

Gutiérrez-Caneda, B., & Vázquez-Herrero, J. (2024). Redrawing the Lines Against Disinformation: How AI Is Shaping the Present and Future of Fact-checking. *Tripodos*, (55), 55–74. DOI: 10.51698/tripodos.2024.55.04

Gutiérrez-Caneda, B., Vázquez-Herrero, J., & López-García, X. (2023). AI application in journalism: ChatGPT and the uses and risks of an emergent technology. *El Profesional de la Información*, *32*(5), e320514. DOI: 10.3145/epi.2023.sep.14

Hassim, M. N., & Mohamad Nasir, N. N. A. (2025). The Rise of AI in Journalism: Ethical and Practical Implications. *E- Journal of Media and Society, 8*(2), 46–58. https://doi.org/DOI: 10.24191/ejoms.v8i2.5841

Helberger, N., & Diakopoulos, N. (2022). The European AI Act and how it matters for research into AI in media and journalism. *Digital Journalism, online first*. https://doi.org/DOI: 10.1080/21670811.2022.2 082505

Hellmueller, L., Vos, T. P., & Poepsel, M. A. (2013). Shifting Journalistic Capital?: Transparency and Objectivity in the Twenty-First Century. *Journalism Studies*, *14*(3), 287–304. DOI: 10.1080/1461670X.2012.697686

Hopster, J. (2021). What are socially disruptive technologies? *Technology in Society*, *67*(101750), 1–8. DOI: 10.1016/j.techsoc.2021.101750

Horska, K. (2020). A New Test of Artificial Intelligence: Should the Media Industry Be Afraid? *Science and Education a New Dimension*, *VIII*(39), 26–29. DOI: 10.31174/send-hs2020-231viii39-06

Hrckova, A., Moro, R., Srba, I., Simko, J., & Bielikova, M. (2024). Autonomation, not Automation: Activities and Needs of Fact-checkers as a Basis for Designing Human-Centered AI Systems. *arXiv:2211.12143v2 [cs.CY]*. https://arxiv.org/html/2211.12143v2

Huang, Y., Lv, S., Tseng, K., Tseng, P., Xie, X., & Lin, R. F. (2023). Recent advances in artificial intelligence for video production system. *Enterprise Information Systems*, *17*(2). Advance online publication. DOI: 10.1080/17517575.2023.2246188

Humprecht, E. (2019). How Do They Debunk "Fake News"? A Cross-National Comparison of Transparency in Fact Checks. *Digital Journalism (Abingdon, England)*, *8*(3), 310–327. DOI: 10.1080/21670811.2019.1691031

Hutchby, I. (2014). Communicative affordances and participation frameworks in mediated interaction. *Journal of Pragmatics*, *72*, 86–89.

IFCN. (2025). The commitments of the Code of Principles: Eligibility to be a signatory. *International Fact-Checking Network – Poynter*. https://ifcncodeofprinciples.poynter.org/the-commitments (last access 10/1/2025).

Jekov, B., Petkova, P., Parusheva, Y., & Shoikova, E. (2022). Disruptive Technologies—Artificial Intelligence and Blockchain in Education. In Proceedings of the *11th Annual International Conference of Education, Research and Innovation (ICERI)*. IATED Academy (pp. 6784–6793).

Juneja, P., & Mitra, T. (2022). Human and Technological Infrastructures of Fact-checking. *Proceedings of the ACM on Human-Computer Interaction*, 6 (CSCW2), Article 418. https://doi.org/DOI: 10.1145/3555143

Kammer, A. (2020). Researching Affordances. In Hunsinger, J., Allen, M., & Klastrup, L. (Eds.), *Second International Handbook of Internet Research*. Springer., DOI: 10.1007/978-94-024-1555-1_7

Kaplan, J. (2025, January 7). "More Speech and Fewer Mistakes", about.meta.com. https://about.fb.com/news/2025/01/meta-more-speech-fewer-mistakes/

Karimi, J., & Walter, Z. (2015). The Role of Dynamic Capabilities in Responding to Digital Disruption: A Factor-Based Study of the Newspaper Industry. *Journal of Management Information Systems*, *32*(1), 39–81. DOI: 10.1080/07421222.2015.1029380

Laukyte, M. (2020). Disruptive Technologies and the Sport Ecosystem: A Few Ethical Questions. *Philosophies*, *5*(4), 24. DOI: 10.3390/philosophies5040024

Lazarski, E., Al-Khassaweneh, M., & Howard, C. (2021). Using NLP for Fact Checking: A Survey. *Designs*, *5*(3), 42. DOI: 10.3390/designs5030042

Lin, C. A. (2019). The Challenge of Information and Communication Divides in the Age of Disruptive Technology. *Journal of Broadcasting & Electronic Media*, *63*(4), 587–594. DOI: 10.1080/08838151.2019.1699677

Majchrzak, M., Faraj, S., Kane, G. C., & Azad, B. (2013). The Contradictory Influence of Social Media Affordances on Online Communal Knowledge Sharing. *Journal of Computer-Mediated Communication*, *19*(1), 1, 38–55. DOI: 10.1111/jcc4.12030

Majumdar, D., Banerji, P. K., & Chakrabarti, S. (2018). Disruptive technology and disruptive innovation: Ignore at your peril! *Technology Analysis and Strategic Management*, *30*(11), 1247–1255. DOI: 10.1080/09537325.2018.1523384

Melnyk, L., Dehtyarova, I., Kubatko, O., Karintseva, O., & Derykolenko, A. (2019). Disruptive technologies for the transition of digital economies towards sustainability. *Economic Annals-XXI*, *179*(9-10), 22–30. DOI: 10.21003/ea.V179-02

Mena, P. (2019). Principles and boundaries of fact-checking: Journalists' perceptions. *Journalism Practice*, *13*(6), 657–672. DOI: 10.1080/17512786.2018.1547655

Mong, A. (2024). “How Armenia’s Hetq media fights disinformation and divisions”, *DW Akademie*, April 14. Retrieved from https://akademie.dw.com/en/how-armenias-hetq-media-fights-disinformation-and-divisions/a-68771711 (Accessed 25/8/2025).

Monti, M. (2019). Automated Journalism and Freedom of Information: Ethical and Juridical Problems Related to AI in the Press Field. *Opinio Juris in Comparatione*, 1/2018. Retrieved from https://ssrn.com/abstract=3318460 (Accessed 25/8/2025).

Moreno-Gil, V., Ramon, X., & Rodríguez-Martínez, R. (2021). Fact-Checking Interventions as Counteroffensives to Disinformation Growth: Standards, Values, and Practices in Latin America and Spain. *Media and Communication*, *9*(1), 251–263. DOI: 10.17645/mac.v9i1.3443

Murphy, R. R. (1999). Case studies of applying Gibson’s ecological approach to mobile robots. *IEEE Transactions on Systems, Man, and Cybernetics*, *29*, 105–111.

Nagy, P., & Neff, G. (2015). Imagined Affordance: Reconstructing a Keyword for Communication Theory. *Social Media + Society*, *1*(2). Advance online publication. DOI: 10.1177/2056305115603385

Nagy, P., & Neff, G. (2023). Rethinking affordances for human-machine communication research. In Guzman, A. L., McEwen, R., & Jones, S. (Eds.), *Rethinking affordances for human-machine communication research* (Vol. 0, pp. 273–279). SAGE Publications Ltd., DOI: 10.4135/9781529782783

Norman, D. A. (1988). *The psychology of everyday things*. Basic Books.

Norman, D. A. (1999). Affordance, conventions, and design. *Interaction*, *6*(3), 38–43.

Norman, D. A. (2013). *The design of everyday things: Revised and Expanded Edition*. Basic Books.

Nyhan, B., & Reifler, J. (2012). *Misinformation and Fact-Checking: Research Findings from Social Science*, 28. Washington, D.C.: New America Foundation. https://www.issuelab.org/resources/15316/15316.pdf

Ogola, G. (2025). Between Fear and Hope: Generative AI, ChatGPT and Journalism. In Mutsvairo, B., & Orgeret, K. S. (Eds.), *The Palgrave Handbook of Global Digital Journalism*. Palgrave Macmillan., DOI: 10.1007/978-3-031-59379-6_13

Omenugha, N. O. (2019). Contextualizing Disruptive Technologies in Media Practice and Management. *Global Journal of Archaeology & Anthropology*, *8*(4). Advance online publication. DOI: 10.19080/GJAA.2019.08.555743

Păvăloaia, V.-D., & Necula, S.-C. (2023). Artificial Intelligence as a Disruptive Technology—A Systematic Literature Review. *Electronics (Basel), 12*(5), 1102. DOI: 10.3390/electronics12051102

Pavleska, T., Školkay, A., Zankova, B., Ribeiro, N., & Bechmann, A. (2020). Performance Analysis of Fact-Checking Organisations and Initiatives in Europe: A Critical Overview of Online Platforms Fighting Fake News. In Terzis, G., Kloza, D., Kużelewska, E., & Trottier, D. (Eds.), *Disinformation and Digital Media as a Challenge for Democracy* (pp. 217–246). Chapter, Intersentia., DOI: 10.1017/9781839700422.014

Picard, R. G. (2003). Cash Cows or Entrecôte: Publishing Companies and Disruptive Technologies. *TIC. Trends in Communication, 11*(2), 127–136. DOI: 10.1207/S15427439TC1102_04

Reynolds, S., & Nolan, J. (2025). Ethical Considerations in AI Journalism: Bias Detection and Mitigation. *ITSI Transactions on Electrical and Electronics Engineering, 12*(1), 18–23. Retrieved from https://journals.mriindia.com/index.php/itsiteee/article/view/144

Rodríguez-Hidalgo, C. (2020). Me and my robot smiled at one another: The process of socially enacted communicative affordance in Human-Machine Communication. *Human-Machine Communication, 1*, 55–69. DOI: 10.30658/hmc.1.4

Şahin, E., Çakmak, M., Doğar, M. R., Uğur, E., & Üçoluk, G. (2007). To Afford or Not to Afford: A New Formalization of Affordances Toward Affordance-Based Robot Control. *Adaptive Behavior, 15*(4), 447–472. DOI: 10.1177/1059712307084689

Salzmann, A., Guribye, F., & Gynnild, A. (2021). Adopting a mojo mindset: Training newspaper reporters in mobile journalism. *Journalism, 24* (2), 346-362. https://doi.org/ (Original work published 2023)DOI: 10.1177/1464884921996284

Sarr, S. F. S. (2024). Artificial Intelligence and Fact Checking in Africa: Between the Logic of Dependency and the Limits of Automation. [ESJ]. *European Scientific Journal, 20*(8), 91. DOI: 10.19044/esj.2024.v20n8p91

Schjøtt, A., & Bengtsson, M. (2024). De- and recoding algorithmic systems: The case of fact-checkers and fact-checked users. *Convergence, 0*(0). Advance online publication. DOI: 10.1177/13548565241289233

Schmidt, R. C. (2007). Scaffolds for social meaning. *Ecological Psychology, 19*(2), 137–151.

Schrock, A. R. (2015). Communicative Affordances of Mobile Media: Portability, Availability, Locatability, and Multimediality. *International Journal of Communication, 9*(18), 1229–1246.

Serrano, R. M. (2023). *Disinformation Landscape in Germany, EU DisinfoLab*, March (Project funded by the Friedrich Naumann Foundation for Freedom). Retrieved from www.disinfo.eu/wp-content/uploads/2023/03/20230224_GER_DisinfoFS.pdf

Shaw, A. (2017). Encoding and decoding affordances: Stuart Hall and interactive media technologies. *Media Culture & Society*, *39*(4), 592–602. DOI: 10.1177/0163443717692741

Singer, J. B. (2021). Border patrol: The rise and role of fact-checkers and their challenge to journalists' normative boundaries. *Journalism*, *22*(8), 1929–1946. DOI: 10.1177/1464884920933137

Školkay, A., & Filin, J. (2019). A Comparison of Fake News Detecting and Fact-Checking AI-Based Solutions. *Studia Medioznawcze*, *20*(4), 365–383. DOI: 10.33077/uw.24511617.ms.2019.4.187

Spanaki, K., Papadopoulos, T., Jayawickrama, U., Olan, F., & Liu, S. (2023). Editorial: fake news, misinformation, and supply chain disruptions: the role of emerging technologies. *Annals of Operations Research*, *327*, 601–604. DOI: 10.1007/s10479-023-05498-w

Subroto, V., Kusumajaya, R., & Tobing, W. (2024). The Business Model of the Mass Media Industry in the Era of Artificial Intelligence (AI) Development in Indonesia. *Journal of Management and Informatics*, *3*, 230–249. DOI: 10.51903/jmi.v3i2.31

van Dalen, A. (2024). Revisiting the Algorithms Behind the Headlines. How Journalists Respond to Professional Competition of Generative AI. *Journalism Practice*, •••, 1–18. DOI: 10.1080/17512786.2024.2389209

Vyas, D., Chisalita, C. M., & Van Der Veer, G. C. (2006). Affordance in interaction. In Proceedings of the *13th European Conference on Cognitive Ergonomics: Trust and Control in Complex Socio-technical Systems*. New York: Association for Computing Machinery (pp. 92-99).

Walter, N., Cohen, J., Holbert, R. L., & Morag, Y. (2019). Fact-Checking: A Meta-Analysis of What Works and for Whom. *Political Communication*, *37*(3), 350–375. DOI: 10.1080/10584609.2019.1668894

Wolfe, R., & Mitra, T. (2024). The Impact and Opportunities of Generative AI in Fact-Checking. *rXiv:2405.15985v1 [cs.HC]*. https://doi.org//arXiv.2405.15985DOI: 10.48550

World Economic Forum. (2024). *The Global Risks Report 2024*. Geneva: World Economic Forum. https://www3.weforum.org/docs/WEF_The_Global_Risks_Report_2024.pdf

Wu, Y., & Wang, H. (2022). Innovation Model of Journalism Education Based on Information Technology under the Background of New Media. *Mobile Information Systems*, *2022*, 1–13. DOI: 10.1155/2022/8367001

Yamanobea, N., Wana, W., Ramirez-Alpizar, I. G., Petitb, D., Tsuji, T., Akizukid, S., Hashimotoe, M., Nagataa, K., & Harada, K. (2017). A brief review of affordance in robotic manipulation research. *Advanced Robotics*, *31*(19–20), 1086–1101. DOI: 10.1080/01691864.2017.1394912

Young, D. G., Jamieson, K. H., Poulsen, S., & Goldring, A. (2018). Fact-Checking Effectiveness as a Function of Format and Tone: Evaluating FactCheck.org and FlackCheck.org. *Journalism & Mass Communication Quarterly*, *95*(1), 49–75. DOI: 10.1177/1077699017710453

Zayani, M. (2020). Digital Journalism, Social Media Platforms, and Audience Engagement: The Case of AJ+. *Digital Journalism (Abingdon, England)*, *9*(1), 24–41. DOI: 10.1080/21670811.2020.1816140

APPENDIX

Table 6. Key themes and subthemes raised by fact-checkers in interviews and in the survey's open-ended questions about fact-checking practices and the integration of automated digital tools.

Themes	Sub-themes	Quotes as Examples
Motivation for engagement with fact-checking practices	Distorted or fake news is perceived as a significant problem against the backdrop of a public sphere that expects it to be resolved.	"Fact-checking plays a major role in the educational process of providing and building strong misinformation resilience among society (members)"
	The high degree of partisanship and deficiency in how media organisations perform their role in society.	"Journalism is in a crisis state right now [...] Nobody really keeps the politicians accountable for their words [...] there are few organisations from the civil society sector that can fight for people, fight for the public debate, the quality of public debate"
	The protection that good journalism needs	"[...] I think the main reason was to continue the tradition of good journalism and research, to help a better journalism field"
Vulnerabilities of automated tools in fact-checking	The essence of fact-checking cannot be served comprehensively by AI-based technologies, but only by humans.	"We want to give the audience more than just the truth or the false. We want to deliver to them also contextual information about the case".
	Lack of comprehensive accuracy and reservations about the credibility of a robot's finalized outcome	"We use such tools, cross-check between different models of AI to ensure a consistent answer and then triple-check to be 100% sure that there is no problem (in terms of accuracy)".
	Partly blamed for the informational inaccuracy of the contemporary media environment	" In the age of AI, you're no longer sure that the source you're looking at is accurate or authentic. [...] No one tool is 100% accurate, and tool development can sometimes be lacking".

continued on following page

Table 6. Continued

Themes	Sub-themes	Quotes as Examples
Strong points of automated tools in fact-checking	A helpful accompanying means of combating disinformation	"We are testing and using verification (digital) tools, which often help us find solutions to these kinds of disinformation challenges.
	Helpful in investigating, summarising, and translating content	"We have developed an in-house AI-powered tool for translation and proofreading. This tool automatically translates and proofreads our verified fact-checks, streamlining the process for our team" & "We have used AI especially to monitor possible misinformation, transcribe audios and videos, and analyse manipulated videos, audios or photographs".
	Offering time-saving in the fact-checking process	"Our idea of implementing the AI-based tools and methods in our work is […] to find the areas that we think our work is tedious, our work is time-consuming, and there is a process that we can do something about it"
	Supplement to the editorial process	"ChatGPT support us in brainstorming sessions when we need to find some title idea or other idea for a headline. It's like a very small addition to the work we are doing"
	The ideal option is the combined use of AI-powered tools and human skills	"AI tools can help us do the fact-checking, but we won't let the tool do the fact-check, there will always be human intervention"
	They are helpful in detecting and providing scientific sources when debunking claims	"They (AI-empowered tools) are used complementary (with human skills), to help us identify and research claims, especially when research involves studying scientific papers. There, AI can help us spot key points in the scientific papers, explain complex data more simply […] and provide our readers with scientific sources to help validate or debunk a claim".

Source: Author's own research.

Table 7. Types of claims reflecting key perceptions of automated fact-checking practices.

1. Checking content based on combined methods, incorporating sophisticated AI-based tools and journalistic human skills, must be highly recommended in fact-checking initiatives
2. It is doubtful that core principles like accuracy and credibility can be served by automated digital tools applied to the fact-checking process
3. AI-based tools do not undermine the authority of the journalism profession, considering that key decisions (still) remain within the jurisdiction of the journalist/fact-checker
4. The future is envisioned as promising in terms of automated applications in fact-checking practices

continued on following page

Table 7. Continued

5. Fact-checkers feel encouraged to cultivate their skills in open-source intelligence tools
6. Eagerness to incorporate AI in more fact-checking practices in a way that renders the work routine more efficient and relevant
7. Automation and the use of AI-based technologies cannot substitute for the human perspective but supplement fact-checking work routine
8. Setting legal obligations for the platforms to comply with can enhance the position of fact-checking initiatives

Source: Author's own research.

Chapter 5
Segmenting Health Audiences Through Computational Analysis:
A Social Marketing Approach During the COVID-19 Pandemic in Mexico

Jose T. Olague
Universidad Autonoma de Baja California, Mexico

Cesario Armando Flores-Villanueva
Universidad Autónoma de Nuevo Leon, Mexico

Niki Athina Polymeri
https://orcid.org/0000-0002-3255-3254
Universidad Autonoma de Baja California, Mexico

ABSTRACT

This study presents a computational and behaviorally grounded approach to health communication, employing audience segmentation as a strategy to optimize the dissemination of public health messages during the COVID-19 pandemic. Drawing on a national-level database of the Mexican population (INEGI, 2020), we conducted an exploratory factor analysis and subsequent statistical modeling to identify three distinct attitudinal segments: indifferent and carefree, positive but evasive, and concerned and involved. These segments reflect heterogeneous emotional, cognitive, and behavioral patterns in response to the pandemic. The findings underscore the relevance of gender as a determinant of segment distribution, suggesting the need for

DOI: 10.4018/979-8-3693-8729-0.ch005

differentiated messaging strategies. This research contributes to the development of more precise, culturally resonant, and evidence-based social marketing campaigns that leverage data analytics to promote preventive health behaviors in crisis contexts.

INTRODUCTION

The lethality of SARS-CoV-2, the virus responsible for the COVID-19 pandemic, is no longer in question. By early 2021, the global impact of the virus was evident, with more than 112 million reported infections and over 2.48 million deaths worldwide (Johns Hopkins University, 2021). Among the various scientific dimensions under scrutiny, the behavioral and public health consequences of the pandemic on human populations have become increasingly central to scholarly inquiry (Arora & Grey, 2020). Communication management plays a pivotal role in this context, given its direct influence on health-related behaviors, especially in high-risk, crisis-driven scenarios such as global pandemics.

Long before COVID-19 was declared a pandemic in March 2020 (World Health Organization [WHO], 2020a), the international public health community had emphasized the importance of transparent and accurate information sharing as a fundamental element of pandemic preparedness. Empirical evidence indicates that timely and credible communication fosters public trust, which in turn is crucial for motivating adaptive behaviors and mitigating both viral spread and lethality (Hyvärinen & Vos, 2016; Rimal & Lapinski, 2009; Lazcano Ponce & Alpuche-Aranda, 2020; Siegrist & Zingg, 2014). In this regard, public information campaigns have been shown to significantly shape health-related behaviors (Ataguba & Ataguba, 2020; Brug et al., 2004; Kamate et al., 2010; Park, Cheong, Son, Kim, & Ha, 2010; Vaughan & Tinker, 2009; Wong & Sam, 2010).

Social marketing has contributed to public health efforts since the late 20th century by applying persuasive communication techniques to promote behaviors that prevent illness and enhance well-being (Evans, 2006). Central to this approach is a citizen-focused perspective that recognizes the individual as the final user of public health interventions (Grier & Bryant, 2005). Within this framework, audience segmentation emerges as a critical technique for the early design phase of communication strategies. By categorizing individuals based on shared sociodemographic, attitudinal, or behavioral characteristics, segmentation enhances message precision, relevance, and impact (Cirksena & Flora, 1995; Gómez, Loar, Kramer, & Garnett, 2019; Mathijssen, Janssen, Bon-Martens, & Goor, 2012; Neelakandan, Gopalakrishnan, Ramachandran, R., & Gopichandran, 2018).

Despite its demonstrated value, audience segmentation remains underexplored in Latin American contexts, particularly in Mexico. Addressing this gap is especially important in moments of public health emergency, where one-size-fits-all messaging strategies often fall short. In response to the COVID-19 outbreak, the Mexican federal government launched a broad communication campaign in March 2020, emphasizing hygiene practices and social distancing measures (Gobierno de México, 2020a; Gobierno de México, 2020b; Llano-Guibarra & Águila-Sánchez, 2020; Martínez-Estrella, 2020; Orellana-Centeno, 2020). In the early months of the pandemic (March–May 2020), media coverage in Mexico was dominated by daily federal press conferences, which served as the primary official communication channel. National television and radio networks amplified these messages, while independent outlets and social media platforms often framed the crisis with divergent narratives, influencing public perception of risk and trust in authorities. However, these efforts have not been immune to critique. Several scholars and civil society actors have questioned the effectiveness, coherence, and political tone of the federal response (COPARMEX, 2020; Elguera Echegaray, 2020; González Díaz, 2020; Semple, 2020), and no academic consensus has yet emerged regarding the communication strategy's overall efficacy.

In the Mexican context, the media ecosystem and framing dynamics present specific features that directly influence the effectiveness of public health communication. Studies of national media have shown that high concentration of media ownership, close ties between political and media actors, and the strong penetration of television, radio, and digital platforms such as Facebook and WhatsApp shape both the reach and reception of public health messages. The framing of information—often influenced by political and editorial agendas—can reinforce official narratives or generate polarized interpretations, thus affecting risk perception and trust in authorities. Moreover, the credibility assigned to sources and the differential exposure to traditional or digital media are linked to acceptance or rejection of public health measures. These factors highlight the need for segmentation strategies in public health to consider not only socioeconomic and demographic variables but also patterns of media consumption and political affinity, in order to tailor content and dissemination channels for maximum relevance and impact in each population segment (Rhpoz, 2019).

The present study contributes to this debate by presenting the findings of a nationwide research initiative conducted in Mexico. Using data drawn from a survey instrument incorporating behavioral and attitudinal variables related to the public's health responses during the pandemic, this study seeks to identify and describe population segments based on shared sociodemographic traits and distinct behavioral profiles. By applying computational techniques—specifically exploratory factor analysis and comparative segmentation modeling—we aim to demonstrate the relevance of data-driven approaches in crafting more nuanced and targeted communication strategies during times of crisis.

LITERATURE REVIEW

Factors Influencing Health Behavior During Pandemics

Health behavior encompasses the patterns, habits, and deliberate actions individuals adopt to maintain, restore, or improve physical and mental well-being (Kasl & Cobb, 1966). The emergence of global pandemics has intensified interest in the cognitive, emotional, and sociocultural factors that influence these behaviors under crisis conditions.

Systematic reviews conducted during the H1N1 outbreak (Bish & Michie, 2010; Tooher et al., 2013) identified three broad categories of health-related responses: (1) preventive behaviors (e.g., hand hygiene), (2) avoidance behaviors (e.g., avoiding crowds), and (3) management behaviors (e.g., seeking vaccination or medical advice). Across multiple outbreaks—including SARS in 2003 (Zwart et al., 2009), H1N1 (Cowling et al., 2010), and COVID-19 (Hagger et al., 2020)—consistent associations have been reported between these behaviors and demographic factors such as age, gender, educational level, and socioeconomic status, as well as cognitive and affective variables including knowledge, perceived susceptibility, perceived severity, and self-efficacy. Individuals with higher educational attainment, greater awareness of disease transmission, and prior health experience tend to exhibit stronger adherence to recommended practices (Lin et al., 2011; Park et al., 2010; Serino et al., 2011).

Recent research also highlights the role of socioeconomic status in shaping health attitudes and behaviors. Ishikawa, Nishiuchi, Hayashi, and Viswanath (2012) demonstrated that educational attainment and income levels significantly influence individuals' search for, exposure to, and trust in health information, leading to differentiated patterns of preventive behavior adoption. Similar findings have been observed in Mexico's ENSANUT surveys, which report disparities in health-related knowledge and practices across socioeconomic strata. These insights underscore the necessity of integrating socioeconomic variables into health communication strategies to enhance both reach and effectiveness.

Knowledge and risk perception are closely interrelated: higher awareness of transmission pathways and heightened perceptions of susceptibility or severity often lead to increased compliance with hygiene and distancing measures (Bish & Michie, 2010; Cowling et al., 2010; Lin et al., 2011; Park et al., 2010; Serino et al., 2011). Trust in authorities, cues to action, and social norms reinforce these effects (Bish & Michie, 2010; Sibley et al., 2020; Liu & Mesch, 2020), while contextual and cultural factors—such as regional identity (Zwart et al., 2009) and patriotic values (Sibley et al., 2020)—can further shape public responses.

Emotional states exert differentiated influences on behavior. Fear and anxiety may motivate physical distancing but do not consistently enhance hygiene practices

(Cowling et al., 2010; Lau et al., 2010; Harper et al., 2020). Elevated concern and worry have been linked to greater engagement in both preventive and management-oriented actions (Serino et al., 2011). This interplay between affective and cognitive dimensions highlights the need to address mental as well as physical health in pandemic communication and policy (Fofana et al., 2020; Idoiaga et al., 2020; Ornell et al., 2020; Petzold et al., 2020; Restubog et al., 2020).

Media consumption patterns and individual attitudes can be powerful predictors, sometimes outweighing traditional sociodemographic variables (Pedersen & Favero, 2020). Confidence in media sources, public trust in institutions, and exposure to misinformation directly influence behavioral responses (Sibley et al., 2020; Liu & Mesch, 2020). These findings emphasize that effective interventions should integrate demographic, psychosocial, emotional, and contextual variables into health behavior models, ensuring that strategies reflect the complexity of human responses during public health crises.

SOCIAL MARKETING AND HEALTH COMMUNICATION

According to Kotler and Lee (2008), social marketing is an approach designed to influence individual behaviors to achieve positive change. Such influence may involve persuading people to adopt healthy behaviors, reject harmful practices, modify existing habits, or abandon detrimental routines. Health communication, in turn, focuses on disseminating information and persuasive messages to encourage the adoption of practices that maintain or improve individual and community health (Schiavo, 2014). Both share the common goal of promoting healthier behaviors at the population level and have become increasingly intertwined since the 1980s (Parvanta, 2011; Parvanta & Freimuth, 2000). In essence, social marketing provides a strategic, audience-centered, and behaviorally focused framework for health communication, leveraging marketing principles to enhance the effectiveness of public health campaigns and ensuring that interventions are both strategic and culturally sensitive.

From a theoretical perspective, the integration of social marketing with health communication rests on two guiding principles from marketing and behavioral science. First, communication strategies should be grounded in an appropriate theory of behavior change for the target audience (Parvanta & Parvanta, 2011; Slater & Flora, 1991). Theories such as Social Cognitive Theory (Bandura, 1985) — which highlights observational learning, self-efficacy, and environmental influences — and the Reasoned Action Approach (Fishbein & Ajzen, 2015) — which focuses on attitudes, subjective norms, and perceived control — help identify key determinants of behavior (e.g., beliefs, social pressures, skills) that can be addressed in campaign design. Using such frameworks allows practitioners to anticipate barriers, leverage facilitators, and craft messages that are more persuasive and relevant.

Second, health communication should apply audience segmentation and positioning techniques drawn from commercial marketing (Parvanta & Parvanta, 2011; Slater & Flora, 1991). Segmentation involves dividing a broad population into smaller subgroups that share relevant characteristics so interventions can be tailored to their needs and preferences. Positioning entails framing the message or health offering in ways that resonate with each segment's motives and cultural context. A "one-size-fits-all" approach often fails to engage diverse audiences; instead, careful segmentation and culturally attuned positioning increase a campaign's impact by making messages personally relevant. The close link between segmentation and theory-of-change is a cornerstone of effective strategy, requiring audience research before message dissemination to determine both the most applicable theory and the optimal segmentation criteria.

Health communication is a cyclical, iterative process rather than a one-time effort. Schiavo (2014) describes three stages: (1) Planning, (2) Implementation and Monitoring, and (3) Evaluation with feedback and refinement. In the planning phase, practitioners define objectives, select an appropriate theory of change, and profile their target audiences, including segmentation based on demographic, behavioral, and psychosocial factors (Thomas, 2006; Slater & Flora, 1991; Boslaugh, Kreuter, Nicholson, & Naleid, 2005). This typically follows a situational analysis to ensure strategies fit the cultural and social context. Implementation then delivers the planned strategy through chosen media channels and activities, while monitoring reception across segments. Finally, evaluation assesses effectiveness, with feedback informing refinements to audience profiles and message targeting. This cyclical model ensures that segmentation remains dynamic, adapting to evolving audience needs and maximizing communication impact over time.

Figure 1. The health communication cycle

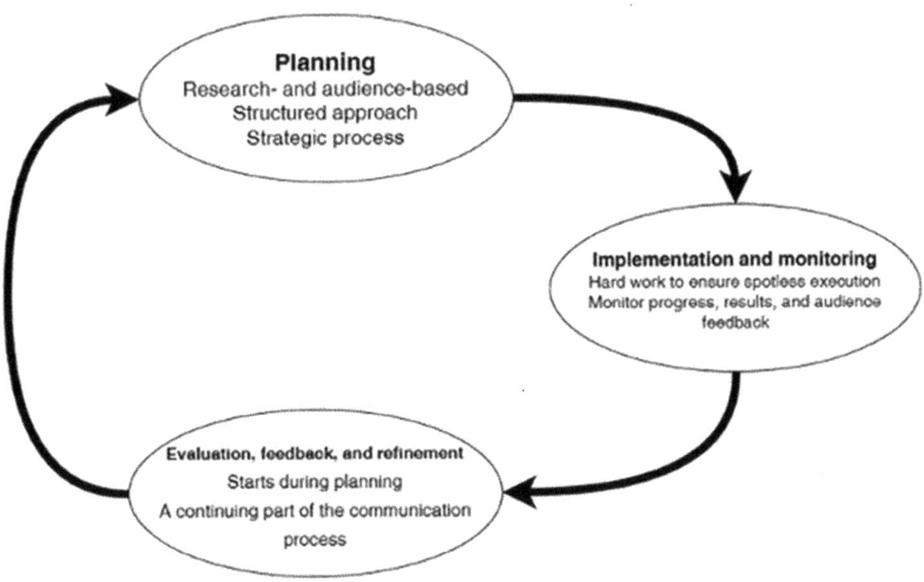

Adapted from Schiavo (2014)

AUDIENCE SEGMENTATION IN HEALTH COMMUNICATION

At the start of the planning process, audience segmentation is recognized as a key technique for effective health communication (Thomas, 2006; Slater & Flora, 1991; Boslaugh et al., 2005). It involves dividing the broader public into relatively homogeneous subgroups that share attributes, enabling communicators to design interventions tailored to their characteristics rather than relying on general messages. Common segmentation criteria include geographic location, demographic attributes, physical or mental conditions, psychosocial factors such as attitudes and values, and observable health-related behaviors (Thackeray & McCormack Brown, 2005; Vladutiu, Nansel, Weaver, Jacobsen, & Kreuter, 2006; Yankelovich & Meer, 2006). These may range from age, gender, and education level to risk perception, cultural background, technology usage, lifestyle, or trust in medical authorities. Behavioral segmentation can distinguish, for example, between individuals by vaccination status or adherence to mask-wearing and distancing practices. Using multiple criteria allows communicators to identify segments that would be overlooked if relying solely on broad demographics.

Research consistently shows that psychological and emotional factors strongly influence health behaviors, reinforcing the need to go beyond surface demographics. Granzin, Olsen, and Painter (1998) demonstrated that emotions, attitudes, and beliefs significantly affect whether individuals adopt behaviors to maintain or recover health. Accordingly, segmentation incorporating attitudinal and behavioral variables can more accurately predict and influence audience responses. This approach has been applied to tailor messages about emerging infectious diseases based on public knowledge and threat perception (Freimuth, Linnan, & Potter, 2000), improve nutrition and exercise campaigns by targeting motivational profiles (Forthofer & Bryant, 2000), and adjust risk communication during public health emergencies (Glik, 2007). Reviews of mass media health campaigns confirm that segmentation and message tailoring are key contributors to effectiveness (Noar, 2006), though producing multiple message versions presents practical challenges (Stellefson, Hanik, Chaney, & Chaney, 2008). Overall, segmentation enhances persuasion by aligning messages with the specific concerns, values, and contexts of each subgroup (Noar, Grant Harrington, & Shemanski Aldrich, 2009).

Evidence from various health domains underscores its value. Seminal studies by Slater and Flora (1991) and Williams and Flora (1995) showed that segmenting audiences by health lifestyle profiles or behavior patterns improved intervention design, including culturally tailored messaging for Hispanic communities. Later work integrated behavioral theory, as in Maibach et al.'s (1996) American Health-Styles project, which identified distinct groups with different behavior patterns and psychosocial profiles, demonstrating that demographic segmentation alone was insufficient. Similarly, Granzin, Olsen, and Painter (1998) and Boslaugh et al. (2005) confirmed that lifestyle and psychosocial variables—such as self-efficacy, social support, and risk perception—are critical for identifying receptive segments.

The importance of segmentation has been particularly evident during global health crises. Analyses of the 2009 H1N1 pandemic revealed that undifferentiated, one-size-fits-all messages often failed to engage key audience segments, especially when they did not address varying emotional reactions such as fear, denial, or apathy (Crosier, McVey, & French, 2014). Lessons from H1N1 have carried into the COVID-19 response, where segmentation proved vital for delivering messages suited to diverse communities (Ihm & Lee, 2021). Ihm and Lee (2021) identified distinct groups in South Korea based on multiple variables, highlighting the role of information and communication technology (ICT) use and skills in vulnerability to misinformation and responsiveness to health measures. Digital literacy and media consumption patterns thus represent emerging segmentation criteria, enabling strategies that combine traditional methods with computational communication tools.

By analyzing social media behavior, mobile phone usage, or participation in online communities, health communicators can deliver micro-targeted messages through the most relevant channels (Korda & Itani, 2013).

Despite its benefits, audience segmentation remains underdeveloped in some contexts. In Latin America, there is a notable scarcity of research systematically applying segmentation to health campaigns. Few studies have examined how Mexican or Latin American populations can be segmented based on health-related attitudes and behaviors during emergencies such as COVID-19 (Ihm & Lee, 2021; Crosier et al., 2014).

Filling this gap is not merely an academic exercise but a practical necessity. Identifying subgroups with distinct psychological, behavioral, and socioeconomic profiles can inform targeted interventions that address unique challenges and misperceptions. By advancing culturally attuned, data-driven segmentation—enhanced by digital analytics—health communication initiatives can achieve greater reach, resonance, and behavioral impact across diverse segments of society. This gap in segmentation research within the Mexican context provided the foundation for the present study. By applying computational techniques to a nationally representative dataset, this research seeks to identify distinct audience profiles and explore how behavioral, cognitive, and emotional variables—together with sociodemographic factors—can inform targeted public health communication strategies during pandemic scenarios.

METHODOLOGY

Sampling and Participants

A national probabilistic survey was conducted in September 2020 to examine behavioral and psychosocial responses to COVID-19 in the Mexican population. The sampling frame covered all regions of the country and employed a stratified, multistage, probabilistic design based on Nielsen regional zones, ensuring geographic and demographic representativeness. In the first stage, geographical stratification was performed at the regional level; in the second stage, clusters of municipalities and localities were selected; in the third stage, households within selected clusters were chosen via simple random sampling; and in the final stage, an adult respondent was randomly selected within each household following the Kish grid method.

This design ensured proportional representation across urban and rural areas, socioeconomic strata, and age groups, which is critical in audience segmentation studies to prevent bias toward overrepresented populations. The final sample consisted of 1,211 adults, with a 95% confidence level and a margin of error of ±2.8%. The nonresponse rate was 18%, which is within the acceptable range for large-scale social surveys during pandemic conditions. The demographic breakdown was as follows: 49.1% male (n = 594) and 50.9% female (n = 617); age groups—18–29 years (29%), 30–39 (22%), 40–49 (18%), 50–59 (22%), and 60+ (8%). Socioeconomic classification followed the standard Mexican ABC+/C/C–/D/D+ typology: 19% ABC+, 34% C/C–, and 47% D/D+. Educational attainment ranged from primary or no formal education (<1%) to postgraduate degrees (6.6%), with the majority holding university (52.4%) or upper secondary qualifications (32%).

Instrument and Variables

The survey instrument was designed from a multidisciplinary review of public health, health communication, and social psychology literature, ensuring both theoretical and empirical validity. Items were presented on 5-point Likert-type scales, with anchors adapted to each construct (e.g., from "strongly disagree" to "strongly agree," or "never" to "always").

The main constructs measured included:

1. *COVID-19 Knowledge and Misinformation* – Adapted from WHO guidelines and prior misinformation studies (WHO, 2020b; Jamieson & Albarracín, 2020; Bridgman et al., 2020). Participants rated the accuracy of both correct and incorrect statements, allowing simultaneous measurement of knowledge and misinformation susceptibility.
2. *Health Behaviors (Preventive and Avoidant)* – Based on the COVIDiSTRESS Global Survey (2020) and Agüero et al (2011), covering behaviors such as mask use, hand hygiene, minimizing outings, and avoiding crowded places.
3. *Affective States* – Measured with an adapted Positive and Negative Affect Schedule (PANAS; Watson et al., 2001), generating separate indices for positive emotions (e.g., calm, hopeful) and negative emotions (e.g., fear, anxiety).
4. *Risk Perception* – Adapted from prior risk communication research (Morton & Duck, 2001; Muñiz & Corduneanu, 2020), capturing perceived vulnerability and severity.
5. *Perceived Stress* – Measured using the validated Spanish version of the Perceived Stress Scale (PSS-10; Cohen et al., 1983).

Reliability was assessed using Cronbach's alpha, with coefficients ranging from $\alpha = .70$ to $\alpha = .85$, meeting or exceeding standard thresholds for internal consistency in behavioral research.

ANALYTICAL STRATEGY

Data were cleaned, weighted, and analyzed using SPSS v26. Weighting adjusted for sampling probabilities and demographic distributions based on national census data, ensuring representativeness. The analytical process consisted of:

1. Descriptive statistics – To explore the distribution of variables and assess normality.
2. Exploratory Factor Analysis (EFA) – Applied to all multi-item constructs using principal axis factoring with varimax rotation to identify latent dimensions underlying the measured items. The EFA served two purposes: (a) to validate the construct structure prior to segmentation and (b) to reduce dimensionality for clustering. Factor retention followed Kaiser's criterion (eigenvalues > 1) and scree plot inspection, with sampling adequacy confirmed by KMO (> .80) and Bartlett's test of sphericity ($p < .001$).
3. Cluster Analysis – Conducted in two steps:
    - o Hierarchical clustering (Ward's method, squared Euclidean distance) to identify the optimal number of clusters based on the dendrogram and the largest relative increase in the agglomeration coefficient.
    - o K-means clustering on standardized factor scores (z-scores) to produce final, stable clusters. This two-step procedure is standard in segmentation research (Hair et al., 1999; Kucukemiroglu, 1999) as it combines the exploratory nature of hierarchical methods with the stability and efficiency of non-hierarchical methods.

Cluster validation involved cross-tabulating cluster membership with demographic variables, testing for statistically significant differences (χ^2 tests), and examining effect sizes. This ensured that identified segments were both statistically robust and substantively meaningful for communication strategy design.

This combination of factor analysis and cluster modeling is widely recognized in audience segmentation for public health (Slater & Flora, 1991; Parvanta & Parvanta, 2011), as it allows communicators to identify profiles that integrate cognitive, emotional, and behavioral dimensions—key determinants for tailoring messages, channels, and framing strategies.

RESULTS

Regarding the descriptive results of the study variables, the corresponding values are presented in Table 1. All measured constructs demonstrated an acceptable level of reliability, with Cronbach's alpha coefficients exceeding the commonly accepted threshold of 0.70, indicating strong internal consistency and construct reliability (Hair, Jr., Black, Babin, & Anderson, 1999).

Table 1. Constructs, items, measurements and reliability

	M_{scale}	SD_{scale}	α
Health behavior (prevention)	4.25	0.99	0.826
COVID-19 knowledge	4.16	1.01	0.725
Health behavior (avoidance)	4.13	1.10	0.845
COVID-19 misinformation	3.89	1.29	0.844
Perceived risk	3.63	1.08	0.785
Emotions (positive)	3.23	1.22	0.740
Perceived stress (negative)	3.13	1.15	0.849
Perceived stress (positive)	2.61	1.02	0.776
Emotions (negative)	2.46	1.24	0.891
M = Mean *DE = Standard Deviation* *α = Cronbach's Alpha*			

The segmentation process of the sample was carried out using factor analysis as the primary technique, ensuring compliance with data quality assumptions and standards (Kucukemiroglu, 1999; Flores Villanueva, Garza Villegas, & Olague, 2017; Rondán Cataluña, Navarro García, Díez de Castro, Rodríguez Rad, & Guisado González, 2010; Hair, Jr., Black, Babin, & Anderson, 1999). To identify the grouping structure appropriate for the objective of this study, only those factors with eigenvalues greater than one were retained.

To assign each case in the sample to one of the identified groups or factors, the regression method was applied. Each case was classified into the factor for which it had the highest factorial loading. To assess whether the groups differed significantly across the study's key variables, the non-parametric Kruskal–Wallis test was employed.

After six iterations, the rotated component matrix confirmed the presence of three distinct factors. These three factors were constructed based on the nine study variables and yielded a total explained variance of 68.33%. The detailed results of the factor loadings and groupings can be consulted in Table 2.

Table 2. Factor analysis results

	FACTOR	FACTOR	FACTOR
	1	2	3
Preventive behavior	0.833		
Avoidant behavior	0.822		
COVID-19 knowledge	0.703		
Perceived risk	0.690		
Emotions (negative)		0.867	
Stress (negative)		0.813	
COVID-19 misinformation		-0.608	
Stress (positive)			0.814
Emotions (positive)			-0.798
Extraction method: Principal component analysis.			
Rotation method: Varimax rotation with Kaiser normalization			
Note: Rotation converged in 6 iterations.			

To confirm that the identified groups were significantly different across the analyzed variables, a mean comparison test was conducted. The significant variables and corresponding results are presented in Table 3.

Table 3. Means of subgroup variables

	SEGMENT 1	SEGMENT 2	SEGMENT 3	*Total*	
Cases	416	412	383	1211	
Percentage	34%	34%	32%	100%	
Groups	1	2	3		
				M	*SD*
Sociodemographic Variables					
Sex (S)*	*1.49*	1.48	**1.56**	1.51	0.50
Age	38.88	39.77	40.15	39.60	13.83
Education level	5.49	5.58	5.58	5.55	0.81
Socioeconomic level	4.06	3.97	3.92	3.98	1.51
Study Variables					
Knowledge (KN)*	*3.99*	4.21	**4.27**	4.16	0.69
Avoidant behavior (AB)*	*3.95*	**4.24**	4.21	4.13	0.71

continued on following page

Table 3. Continued

	SEGMENT 1	SEGMENT 2	SEGMENT 3	*Total*	
Preventive behavior (PB)*	*4.06*	4.34	**4.36**	4.25	0.68
Misinformation (MI)*	3.86	*3.79*	**4.00**	3.88	0.86
Negative emotions (NE-)	2.49	2.46	2.43	2.46	0.86
Positive emotions (PE+)*	3.17	**3.44**	*3.04*	3.22	0.98
Perceived risk (PR)*	*3.50*	3.68	**3.71**	3.63	0.84
Negative stress (S-)*	3.10	*3.04*	**3.25**	3.13	0.86
Positive stress (S+)*	2.65	*2.39*	**2.79**	2.60	0.78
*Significant difference ($p < 0.05$). Kruskal–Wallis variance test					
Minimum values in *italics* and maximum values in **bold**					

Based on the values obtained through the analysis, the Mexican population can be segmented into the following three profiles:

Segment 1: Indifferent and Unconcerned Individuals

This group represents slightly more than one-third of the population (34%). It is primarily composed of male participants. Members of this segment demonstrate the lowest levels of cognitive involvement regarding the pandemic and perceive the least risk associated with COVID-19. Their stress levels are average, and they exhibit the lowest adherence to both preventive and avoidant health behaviors.

Segment 2: Positive Avoiders

Also comprising approximately 34% of the population, this group is again predominantly male. Individuals in this segment report low levels of misinformation and display above-average knowledge about COVID-19. They tend to experience low stress levels and show higher levels of positive emotional engagement. They are the most consistent in practicing social distancing and perform preventive behaviors above the population mean.

Segment 3: Concerned and Highly Involved Individuals

This segment makes up slightly less than one-third of the population (32%) and is predominantly female. Participants in this group are marked by high cognitive involvement, reporting the highest levels of both knowledge and misinformation exposure. They also engage in preventive and avoidant health behaviors at above-average levels. This group displays the highest levels of stress and concern regarding the pandemic.

DISCUSSION AND CONCLUSIONS

This study provides empirical support for the application of computational segmentation methodologies to public health communication during crises. Through the integration of multivariate statistical techniques and theory-driven constructs, we identified three psychologically and behaviorally distinct segments within the Mexican population during the COVID-19 pandemic: indifferent and unconcerned individuals, positive avoiders, and concerned and highly involved individuals. These findings confirm the heterogeneity of pandemic responses and highlight the critical role of attitudinal, emotional, and cognitive variables in shaping health behavior—going beyond traditional sociodemographic predictors.

The segmentation structure observed is consistent with previous theoretical and empirical frameworks emphasizing that behavioral responses to health crises are neither uniform nor random but structured along psychological and informational lines (Bish & Michie, 2010; Tooher et al., 2013; Harper et al., 2020). Segment 1, characterized by low involvement, diminished risk perception, and minimal adherence to preventive or avoidant behaviors, represents a communicative challenge. Despite relatively low levels of misinformation, this group exhibits a notable disengagement from both cognitive and behavioral responses to the pandemic. This aligns with prior findings that risk underestimation and emotional detachment hinder the adoption of health-promoting behaviors (Cowling et al., 2010; Morton & Duck, 2001). Accordingly, this group demands high-priority targeting through strategic, emotionally engaging interventions that amplify perceived relevance and urgency without resorting to fear appeals, which have been shown to backfire in disengaged populations (Siegrist & Zingg, 2014).

Segment 3, in contrast, reflects a highly involved public that demonstrates both elevated knowledge and a proactive behavioral profile, despite also registering high stress and misinformation levels. This paradox of simultaneous high involvement and misinformation exposure underlines the informational complexity introduced by digital media ecosystems (Bridgman et al., 2020; Ihm & Lee, 2021). The segment's strong emotional and cognitive investment suggests receptivity to corrective messaging strategies, particularly those rooted in affirmation-based and clarity-enhancing frameworks. Leveraging this group's active engagement, public communication could use this segment as a vector of peer diffusion, as supported by Social Cognitive Theory's emphasis on observational learning and modeling (Bandura, 1985).

Segment 2 occupies an intermediary but crucial position. Its members exhibit above-average knowledge and behavioral adherence, especially in avoidant actions such as social distancing, and report higher positive affect. This profile aligns with theoretical propositions that suggest emotional valence and self-efficacy serve as buffers to stress and predictors of health compliance (Sibley et al., 2020; Hagger et al., 2020). From a communication strategy standpoint, this group could benefit from reinforcement-oriented messaging, sustaining their health behaviors while gradually shifting emphasis toward prevention-oriented actions. The segment's emotional profile also opens opportunities for prosocial or hope-based messaging strategies—approaches that have shown promise in sustaining long-term behavioral adherence (Petzold et al., 2020).

Building on these findings, each segment presents specific opportunities and challenges for the design of communication strategies. For Segment 1 – Indifferent and Unconcerned Individuals, the priority is to overcome disengagement by making the pandemic personally relevant without resorting to fear-based approaches, which tend to backfire in low-involvement audiences. Messages for this group should use salience framing, linking protective behaviors to immediate personal and family benefits, and be delivered through high-reach entertainment media such as television, radio, and widely used social platforms like Facebook and WhatsApp. Local community leaders, popular sports figures, and entertainment personalities can serve as effective messengers, leveraging familiarity and trust to increase receptivity.

For Segment 2 – Positive Avoiders, the communication focus should be on reinforcing and sustaining their above-average adherence to social distancing and preventive practices. Gain-framed and affirmation-based messages that emphasize the value and social contribution of their current behaviors can help maintain engagement, while gradually introducing additional preventive actions. An encouraging, prosocial tone is recommended, highlighting shared responsibility and collective benefits. This group is likely to respond well to digital channels such as Instagram and Facebook, as well as community networks and public service announcements. Credible figures, including health professionals, educators, and respected volunteers, can amplify these messages.

Segment 3 – Concerned and Highly Involved Individuals requires a different approach. This group's high knowledge levels and proactive health behaviors are accompanied by elevated stress and exposure to misinformation, making them receptive to corrective, clarity-enhancing messages. Framing should focus on accuracy and transparency, validating their concerns while providing actionable solutions. Trusted, detail-oriented channels—such as digital news outlets, webinars, specialized health portals, and official government platforms—are well suited for reaching this segment. Given their high involvement and strong commitment to protective behaviors, they can also be mobilized as peer educators and diffusion agents within their social networks, amplifying the reach of accurate information and modeling recommended practices.

A key contribution of this study lies in confirming gender as a significant predictor of segment membership. The data support earlier findings regarding the gendered nature of pandemic responses, where women tend to demonstrate higher cognitive involvement and preventive health behavior, albeit accompanied by higher stress and misinformation exposure (Lin et al., 2011; Park et al., 2010). This reinforces the need for gender-sensitive segmentation in health communication campaigns, particularly in sociocultural contexts where gender roles affect health information processing and compliance (Boslaugh et al., 2005).

Importantly, this research contributes to bridging a significant gap in the literature: the scarcity of segmentation-based health communication studies in Latin American contexts. While segmentation is widely recommended as a strategic pillar of health campaigns (Slater & Flora, 1991; Parvanta & Parvanta, 2011), its application in Mexico has been limited. Our study stands as one of the few national-level, empirically grounded efforts to systematically categorize public health audiences during a pandemic based on psychographic and behavioral data.

Methodologically, the study's application of exploratory factor analysis and subsequent cluster modeling aligns with international standards for audience segmentation in public health (Hair et al., 1999; Kucukemiroglu, 1999). The total explained variance of 68.33% and the internal consistency coefficients above .70 confirm the robustness of the measurement model and the quality of the identified segments.

The findings also bring to the forefront the role of the digital divide as a determinant of both information access and health behavior during crises. Disparities in connectivity, device availability, and digital skills can significantly shape the extent to which individuals are exposed to accurate health information or, conversely, to misinformation. In Mexico, where internet penetration and digital literacy vary widely across regions and socioeconomic strata, these gaps may exacerbate health inequities by limiting the reach and effectiveness of digitally mediated public health messages. For segments with limited online access, traditional media channels and community-based outreach remain essential to ensure equitable message delivery. Conversely, highly connected groups—while having greater access to timely information—may face higher exposure to misinformation, underscoring the need for tailored digital literacy interventions alongside health communication efforts.

However, even with precise audience segmentation and theoretically grounded strategies, the practical implementation of tailored messages faces significant constraints. Limited institutional resources can restrict both the diversity and frequency of message dissemination, particularly in contexts where health communication competes with other urgent policy priorities. Moreover, high levels of media saturation during prolonged crises such as COVID-19 can lead to audience fatigue, diminishing attention and recall regardless of message quality. Resistance from certain segments—whether due to entrenched beliefs, political polarization, or distrust in institutions—can further undermine the uptake of recommended behaviors. These challenges highlight the necessity of integrating segmentation insights with realistic assessments of operational capacity, adapting frequency, channel selection, and narrative style to balance message impact with sustainability over time.

Nonetheless, limitations must be acknowledged. The cross-sectional design restricts causal inference, and the online survey format introduces a potential bias toward individuals with greater access to digital technologies. These limitations are inherent to rapid data collection in crisis contexts but must be accounted for in future research replication efforts. Despite these constraints, the study provides a scalable and replicable analytical framework for real-time audience segmentation in public health emergencies.

The findings underscore the value of computational segmentation as a tool for targeted, evidence-based health communication. They suggest that differentiated messaging—tailored to emotional engagement, risk perception, and behavioral readiness—can significantly improve the efficacy of public health interventions during pandemics. As countries confront evolving health crises, and as misinformation continues to challenge communication efforts, strategies that are simultaneously data-driven and human-centered become not only desirable but necessary. By grounding communication in segmentation and behavioral insight, social marketing initiatives can move from mass messaging to meaningful impact.

FUNDING INFORMATION

This chapter was developed as part of the research project entitled "Analysis of media coverage of the COVID-19 pandemic in Mexico and its impact on the development of attitudes and behaviors among the public" (Project No. 312437), funded by the National Council of Science and Technology (CONACYT) of Mexico under the 2020-1 Call for Scientific Research, Technological Development, and Innovation in Health in Response to the COVID-19 Contingency.

REFERENCES

Agüero Santangelo, F., Nebot Adell, M., Pérez Giménez, A., López Medina, M. J., & García Continente, X. (2011). Actitudes y comportamientos preventivos durante la pandemia de gripe (H1N1) 2009 en España. *Revista Española de Salud Pública, 85*(1), 73–80. https://www.mscbs.gob.es/biblioPublic/publicaciones/recursos_propios/resp/revista_cdrom/vol85/vol85_1/RS851C_portada.pdf PMID: 21750845

Arora, T., & Grey, I. (2020). Health behaviour changes during COVID-19 and the potential consequences: A mini-review. *Journal of Health Psychology, 25*(9). Advance online publication. DOI: 10.1177/1359105320937053 PMID: 32551944

Ataguba, O., & Ataguba, J. E. (2020). Social determinants of health: The role of effective communication in the COVID-19 pandemic in developing countries. *Global Health Action, 13*(1), 1788263. DOI: 10.1080/16549716.2020.1788263 PMID: 32657669

Bandura, A. (1985). *Social Foundations of Thought and Action: A Social Cognitive Theory*. Prentice Hall.

Bish, A., & Michie, S. (2010). Demographic and attitudinal determinants of protective behaviours during a pandemic: A review. *British Journal of Health Psychology, 15*, 797–824. DOI: 10.1348/135910710X485826 PMID: 20109274

Boslaugh, S. E., Kreuter, M. W., Nicholson, R. A., & Naleid, K. (2005). Comparing demographic, health status and psychosocial strategies of audience segmentation to promote physical activity. *Health Education Research, 20*(4), 430–438. DOI: 10.1093/her/cyg138 PMID: 15572439

Bridgman, A., Merkley, E., Loewen, P. J., Owen, T., Ruths, D., Teichmann, L., & Zihilin, O. (2020). The causes and consequences of COVID-19 misperceptions: Understanding the role of news and social media. *Harvard Kennedy School Misinformation Review, 1*. Advance online publication. DOI: 10.37016/mr-2020-028

Brug, J., Aro, A. R., Oenema, A., Zwart, O., Richardus, J. H., & Bishop, G. D. (2004). SARS Risk Perception, Knowledge, Precautions, and Information Sources, the Netherlands. *Emerging Infectious Diseases, 10*(8), 1486–1489. 10.3201%2Feid1008.040283 PMID: 15496256

Cirksena, M. K., & Flora, J. A. (1995). Audience segmentation in worksite health promotion: A procedure using social marketing concepts. *Health Education Research, 10*(2), 211–224. DOI: 10.1093/her/10.2.211

Cohen, S., Kamarck, T., & Mermelstein, R. (1983). A Global Measure of Perceived Stress. *Journal of Health and Social Behavior, 24*(4), 385–396. DOI: 10.2307/2136404 PMID: 6668417

COPARMEX. (2020, junio 22). *Confederación Patronal de la República Mexicana*. Retrieved febrero 23, 2021, from Gobierno de México mala estrategia y errática comunicación frente al Covid-19: https://coparmex.org.mx/54308-2/

COVIDiSTRESS Global Survey Network. (2020, mayo 10). *COVIDiSTRESS Global Survey*. https://doi.org/DOI: 10.17605/OSF.IO/Z39US

Cowling, B. J., Ng, D. M., Ip, D. K., Liao, Q., Lam, W. W., Wu, J. T., & Fielding, R. (2010). Community Psychological and Behavioral Responses through the First Wave of the 2009 Influenza A(H1N1) Pandemic in Hong Kong. *The Journal of Infectious Diseases, 202*(6), 867–876. DOI: 10.1086/655811 PMID: 20677945

Crosier, A., McVey, D., & French, J. (2014). 'By failing to prepare you are preparing to fail': Lessons from the 2009 H1N1 'swine flu' pandemic. *European Journal of Public Health, 25*(1), 135–139. DOI: 10.1093/eurpub/cku131 PMID: 25125575

Elguera Echegaray, E. (2020). *C&E*. Retrieved febrero 23, 2021, from Comunicación y estrategia política en tiempos de COVID-19: https://www.ceonline.com.mx/post/comunicacion-y-estrategia-politica-en-tiempos-de-covid-19

Evans, W. D. (2006). How social marketing works in health care. *BMJ (Clinical Research Ed.), 332*(7551), 1207–1210. 10.1136%2Fbmj.332.7551.1207-a PMID: 16710002

Fishbein, M., & Ajzen, I. (2015). *Predicting and Changing Behavior: The Reasoned Action Approach*. Routledge.

Flores Villanueva, C. A., Garza Villegas, J. B., & Olague, J. T. (2017). Identifying strategic groups: An assessment in Mexican franchises. *Revista Ibero-Americana de Estratégia, 16*(2), 20–34. DOI: 10.5585/riae.v16i2.2352

Fofana, N. K., Latif, F., & Sarfraz, S., Bilal, Bashir, M. F., & Komal, B. (2020). Fear and agony of the pandemic leading to stress and mental illness: An emerging crisis in the novel coronavirus (COVID-19) outbreak. *Psychiatry Research, 291*, 113230. DOI: 10.1016/j.psychres.2020.113230 PMID: 32593067

Forthofer, M. S., & Bryant, C. A. (2000). Using Audience-Segmentation Techniques to Tailor Health Behavior Change Strategies. *American Journal of Health Behavior*, 36 - 43. https://doi.org/DOI: 10.5993/AJHB.24.1.6

Freimuth, V., Linnan, H. W., & Potter, P. (2000). Communicating the Threat of Emerging Infections to the Public. *Perspectives*, 337 - 347. https://doi.org/DOI: 10.3201/eid0604.000403

Glik, D. C. (2007). Risk Communication for Public Health Emergencies. *Annual Review of Public Health*, *28*(1), 33–54. DOI: 10.1146/annurev.publhealth.28.021406.144123 PMID: 17222081

Gobierno de México. (2020a). *Coronavirus*. Retrieved octubre 1, 2020, from Todo sobre el coronavirus: https://coronavirus.gob.mx/

Gobierno de México. (2020b, marzo 23). *Gobierno de México*. Retrieved mayo 1, 2020, from Jornada Nacional de Sana Distancia: https://www.gob.mx/cms/uploads/attachment/file/541687/Jornada_Nacional_de_Sana_Distancia.pdf

Gómez, A., Loar, R., Kramer, A. E., & Garnett, G. P. (2019). Reaching and targeting more effectively: The application of market segmentation to improve HIV prevention programmes. *Journal of the International AIDS Society*, *22*(S4), e25318. DOI: 10.1002/jia2.25318 PMID: 31328397

González Díaz, M. (2020, marzo 17). *BBC News*. Retrieved febrero 23, 2021, from Mundo: https://www.bbc.com/mundo/noticias-america-latina-51921323

Granzin, K. L., Olsen, J. E., & Painter, P. J. (1998). Marketing to consumer segments using health-promoting lifestyles. *Journal of Retailing and Consumer Services*, *5*(3), 131–141. DOI: 10.1016/S0969-6989(96)00059-8

Grier, S., & Bryant, C. A. (2005). Social marketing in public health. *Annual Review of Public Health*, *26*, 319–339. DOI: 10.1146/annurev.publhealth.26.021304.144610 PMID: 15760292

Hagger, M. S., Smith, S. R., Keech, J. J., Moyers, S. A., & Hamilton, K. (2020). Predicting Social Distancing Intention and Behavior During the COVID-19 Pandemic: An Integrated Social Cognition Model. *Annals of Behavioral Medicine*, *54*, 713–727. DOI: 10.1093/abm/kaaa073 PMID: 32914831

Hair, J. F.Jr, Black, W. C., Babin, B. J., & Anderson, R. E. (1999). *Análisis Multivariante* (Prentice, E., & Cano, D., Trans.). Prentice Hall.

Harper, C. A., Satchell, L. P., Fido, D., & Latzman, R. D. (2020). Functional Fear Predicts Public Health Compliance. *International Journal of Mental Health and Addiction*. Advance online publication. DOI: 10.1007/s11469-020-00281-5 PMID: 32346359

Hyvärinen, J., & Vos, M. (2016). Communication Concerning Disasters and Pandemics: Co-producing community resilience and crisis response. In A. Schwartz, M. Seeger, & C. Auer (Eds.), *The Handbook of International Crisis Communication Research* (p. 552). Hoboken, EE.UU.: Wiley - Blackwell.

Idoiaga, N., Berasategi, N., Eiguren, A., & Picaza, M. (2020). Exploring Children's Social and Emotional Representations of the COVID-19 Pandemic. *Frontiers in Psychology*, *11*, 1952. DOI: 10.3389/fpsyg.2020.01952 PMID: 32922334

Ihm, J., & Lee, C.-J. (2021). Toward More Effective Public Health Interventions during the COVID-19 Pandemic: Suggesting Audience Segmentation Based on Social and Media Resources. *Health Communication*, *36*(1), 98–108. DOI: 10.1080/10410236.2020.1847450 PMID: 33198528

Ishikawa, Y., Nishiuchi, H., Hayashi, H., & Viswanath, K. (2012). Socioeconomic status and health communication inequalities in Japan: A nationwide cross-sectional survey. *PLoS One*, *7*(7), e40664. DOI: 10.1371/journal.pone.0040664 PMID: 22808229

Jamieson, K., & Albarracin, D. (2020). The Relation between Media Consumption and Misinformation at the Outset of the SARS-CoV-2 Pandemic in the US. *Harvard Kennedy School Misinformation Review*, *1*. Advance online publication. DOI: 10.37016/mr-2020-012

Johns Hopkins University. (2021). *Center for Systems Science and Engineering (CSSE)*. Retrieved octubre 14, 2020, from COVID-19 Dashboard: https://gisanddata.maps.arcgis.com/apps/opsdashboard/index.html#/bda7594740fd40299423467b48e9ecf6

Kamate, S. K., Agrawal, A., Chaudhary, H., Singh, K., Mishra, P., & Asawa, K. (2010). Public knowledge, attitude and behavioural changes in an Indian population during the Influenza A (H1N1) outbreak. *Journal of Infection in Developing Countries*, *4*(1), 7–14. DOI: 10.3855/jidc.501 PMID: 20130372

Kasl, S. V., & Cobb, S. (1966). Health Behavior, Illness Behavior, and Sick-Role Behavior. *Archives of Environmental Health: An International Journal*, *12*(2), 246–266. DOI: 10.1080/00039896.1966.10664421 PMID: 5322534

Korda, H., & Itani, Z. (2013). Harnessing social media for health promotion and behavior *change. Health Promotion Practice*, *14*(1), 15–23. DOI: 10.1177/1524839911405850 PMID: 21558472

Kotler, P., & Lee, N. R. (2008). *Social marketing: Influencing behaviors for good.* Sage Publications.

Kucukemiroglu, O. (1999). Market segmentation by using consumer lifestyle dimensions and ethnocentrism. An empirical study. *European Journal of Marketing*, *33*(5/6), 470–487. DOI: 10.1108/03090569910262053

Lau, J. T., Griffiths, S., Choi, K. C., & Tsui, H. Y. (2010). Avoidance behaviors and negative psychological responses in the general population in the initial stage of the H1N1 pandemic in Hong Kong. *BMC Infectious Diseases*, *10*, 139. DOI: 10.1186/1471-2334-10-139 PMID: 20509887

Lazcano Ponce, E., & Alpuche-Aranda, C. (2020). Alfabetización en salud pública ante la emergencia de la pandemia por Covid-19. *Salud Pública de México*, *62*(3), 331–340. DOI: 10.21149/11408 PMID: 32304371

Lin, Y., Juang, L., Nie, S., Liu, Z., Yu, H., Yan, W., & Xu, Y. (2011). Knowledge, Attitudes and Practices (KAP) related to the Pandemic (H1N1) 2009 among Chinese General Population: A Telephone Survey. *BMC Infectious Diseases*, *11*, 128. DOI: 10.1186/1471-2334-11-128 PMID: 21575222

Liu, X.-J., & Mesch, G. S. (2020). The Adoption of Preventive Behaviors during the COVID-19 Pandemic in China and Israel. *International Journal of Environmental Research and Public Health*, *17*, 7170. DOI: 10.3390/ijerph17197170 PMID: 33007985

Llano-Guibarra, N. I., & Águila-Sánchez, J. C. (2020). Conferencias de prensa y COVID-19: explorando la respuesta gubernamental mexicana desde la comunicación en salud. *Revista Española de Comunicación en Salud*(1), 128-141. https://doi.org/doi.org/10.20318/recs.2020.5451

Maibach, E. W., Maxfield, A., Ladin, K., & Slater, M. (1996). Translating Health Psychology into Effective Health Communication: The American Healthstyles Audience Segmentation Project. *Journal of Health Psychology*, *1*(3), 261–277. DOI: 10.1177/135910539600100302 PMID: 22011991

Martínez-Estrella, C. G. (2020). Uso de personajes y metáforas en la gestión de la crisis sanitaria del COVID-19. Revisión de la comunicación de sanidad pública en España y México. *Revista Española de Comunicación en Salud*(1), 319-327. https://doi.org/doi.org/10.20318/recs.2020.5458

Mathijssen, J., Janssen, M., Bon-Martens, M., & Goor, I. (2012). Adolescents and alcohol: An explorative audience segmentation analysis. *BMC Public Health*, *12*, 742. DOI: 10.1186/1471-2458-12-742 PMID: 22950946

Morton, T., & Duck, J. (2001). Communication and health beliefs. Mass and interpersonal influences on perceptions of risk to self and others. *Communication Research*, *28*(5), 602–626. DOI: 10.1177/009365001028005002

Muñiz, C., & Corduneanu, L. E. (2020). Audience segmentation and climate change: The Chilean case. *Environmental Communication*, *14*(7), 941–957. DOI: 10.1080/17524032.2020.1725084

Muñiz, C., & Corduneanu, V. I. (2020). Percepción de riesgo y consumo mediático durante el inicio de la pandemia de COVID-19 en México. *Más poder local*(41), 44-47. Retrieved from http://maspoderlocal.es/files/articulos/mexico-percepcion-riesgo-pandemia-mpl41.pdf

Neelakandan, B. V., Gopalakrishnan, S., Ramachandran, H. R. A., & Gopichandran, V. (2018). Health Beliefs Regarding Uptake of Hypertension Screening in an Urban Setting: A Population Segmentation Strategy. *International Journal of Medicine and Public Health, 8*(1), 11-17. https://https://doi.org/DOI: <ALIGNMENT.qj></ALIGNMENT>10.5530/ijmedph.2018.1.3

Noar, S. M. (2006). A 10-Year Retrospective of Research in Health Mass Media Campaigns: Where Do We Go From Here? *Journal of Health Communication: International Perspectives*, *11*(1), 21–42. DOI: 10.1080/10810730500461059 PMID: 16546917

Noar, S. M., Grant Harrington, N., & Shemanski Aldrich, R. (2009). The Role of Message Tailoring in the Development of Persuasive Health Communication Messages. *Annals of the International Communication Association*, *33*(1), 73–133. DOI: 10.1080/23808985.2009.11679085

Orellana-Centeno, J. (2020). El coronavirus 19 (COVID-19) en México. *Revista Chilena de Salud Pública*, *24*(1), 72–73. DOI: 10.5354/0719-5281.2020.57588

Ornell, F., Schuh, J. B., Sordi, A. O., & Kessler, F. H. (2020). "Pandemic fear" and COVID-19: Mental health burden and strategies. *The British Journal of Psychiatry*, *42*(3), 232–235. DOI: 10.1590/1516-4446-2020-0008 PMID: 32267343

Park, J. H., Cheong, H.-K., Son, D.-Y., Kim, S.-U., & Ha, C.-M. (2010). Perceptions and behaviors related to hand hygiene for the prevention of H1N1 influenza transmission among Korean university students during the peak pandemic period. *BMC Infectious Diseases*, *10*, 222. DOI: 10.1186/1471-2334-10-222 PMID: 20663229

Parvanta, C. (2011). Introduction to Public Health Communication and Informatics. In Parvanta, C. F., Nelson, D. E., Parvanta, S. A., & Harner, R. N. (Eds.), *Essentials of Public Health Communication* (pp. 3–18). Jones & Bartlett Learning.

Parvanta, C. F., & Freimuth, V. (2000). Health Communication at the Centers for Disease Control and Prevention. *American Journal of Health Behavior, 24*(1), 18–25. DOI: 10.5993/AJHB.24.1.4

Parvanta, S., & Parvanta, C. F. (2011). Persuasive Health Communications: The Role of Theory. In Parvanta, C. F., Nelson, D. E., Parvanta, S. A., & Harner, R. N. (Eds.), *Essentials of Public Health Communication* (pp. 147–158). Jones & Bartlett Learning.

Pedersen, M. J., & Favero, N. (2020). Social Distancing during the COVID-19 Pandemic: Who Are the Present and Future Noncompliers? *Public Administration Review, 5*, 805–814. DOI: 10.1111/puar.13240 PMID: 32836442

Petzold, M. B., Bendau, A., Plag, J., Pyrkosch, L., Maricic, L. M., Betzler, F., & Strölle, A. (2020). Risk, resilience, psychological distress, and anxiety at the beginning of the COVID-19 pandemic in Germany. *Brain and Behavior, 10*, e01745. DOI: 10.1002/brb3.1745 PMID: 32633464

Restubog, S. L., Ocampo, A. C., & Wang, L. (2020). Taking control amidst the chaos: Emotion regulation during the COVID-19 pandemic. *Journal of Vocational Behavior, 119*, 103440. DOI: 10.1016/j.jvb.2020.103440 PMID: 32390659

Rhpoz, L. (2019). Framing communication in Mexico: Media, politics, and public perception. *Communicatio Socialis, 13*(1), 45–65. DOI: 10.15581/003.13.1.45-65

Rimal, R. N., & Lapinski, M. K. (2009). Why health communication is important in public health. *Bulletin of the World Health Organization, 87*, 247. DOI: 10.2471/BLT.08.056713 PMID: 19551226

Rondán Cataluña, F. J., Navarro García, A., Díez de Castro, E. C., Rodríguez Rad, C., & Guisado González, M. (2010). Estudio del performance de los grupos estratégicos en el sistema de franquicia español. *Investigaciones Europeas de Dirección y Economía de la Empresa, 16*(2), 43–62. DOI: 10.1016/S1135-2523(12)60111-7

Schiavo, R. (2014). *Health Communication. From theory to practice* (2nd ed.). Jossey-Bass.

Semple, K. (2020, julio 14). El gobierno de México llama a la calma y adopta medidas mesuradas frente al virus. *The New York Times*. Retrieved febrero 23, 2021, from https://www.nytimes.com/es/2020/03/16/espanol/deportes/mexico-coronavirus-amlo-futbol.html

Serino, L., Meleleo, C., Maurici, M., Bagnato, B., Sorbara, D., Zaratti, L., & Franco, E. (2011). Knowledge and worry as basis for different behaviors among university students: The case of Pandemic Flu H1N1v. *Journal of Preventive Medicine and Hygiene, 52*, 144–147. DOI: 10.15167/2421-4248/jpmh2011.52.3.269 PMID: 22010546

Sibley, C. S., Greaves, L. M., Satherley, N., Wilson, M. S., Overall, N. C., Lee, C. H., & Barlow, F. K. (2020). Effects of the COVID-19 Pandemic and Nationwide Lockdown on Trust, Attitudes Toward Government, and Well-Being. *The American Psychologist, 75*(5), 618–630. DOI: 10.1037/amp0000662 PMID: 32496074

Siegrist, M., & Zingg, A. (2014). The role of public trust during pandemics: Implications for crisis communication. *European Psychologist, 19*(1), 23–32. DOI: 10.1027/1016-9040/a000169

Slater, M. D., & Flora, J. A. (1991). Health Lifestyles: Audience Segmentation Analysis for Public Health Interventions. *Health Education Quarterly, 18*(2), 221–233. DOI: 10.1177/109019819101800207 PMID: 2055779

Stellefson, M. L., Hanik, B. W., Chaney, B. H., & Chaney, D. J. (2008). Challenges for Tailored Messaging in Health Education. *Health Education. American Journal of Health Education, 39*(5), 303–313. DOI: 10.1080/19325037.2008.10599054

Thackeray, R., & McCormack Brown, K. (2005). Social Marketing's Unique Contributions to Health Promotion Practice. *Health Promotion Practice, 6*(4), 365–368. DOI: 10.1177/1524839905278848 PMID: 16247912

Thomas, R. K. (2006). *Health Communication*. New York: Springer Science+Business Media Inc.

Tooher, R., Collins, J. E., Street, J. M., Braunack-Mayer, A., & Marshall, H. (2013). Community knowledge, behaviours and attitudes about the 2009 H1N1 Influenza pandemic: A systematic review. *Influenza and Other Respiratory Viruses, 7*(6), 1316–1327. DOI: 10.1111/irv.12103 PMID: 23560537

Vaughan, E., & Tinker, T. (2009). Effective Health Risk Communication About Pandemic Influenza for Vulnerable Populations. *American Journal of Public Health, 99*(2), 324–331. DOI: 10.2105/AJPH.2009.162537 PMID: 19797744

Vladutiu, C. J., Nansel, T. R., Weaver, N. L., Jacobsen, H. A., & Kreuter, M. W. (2006). Differential strength of association of child injury prevention attitudes and beliefs on practices: A case for audience segmentation. *Injury Prevention : Journal of the International Society for Child and Adolescent Injury Prevention, 12*, 35–40. DOI: 10.1136/ip.2004.007153 PMID: 16461418

Watson, D., Clark, L. A., & Tellegen, A. (2001). Development and Validation of Brief Measures of Positive and Negative Affect: The PANAS Scales. *Journal of Personality and Social Psychology*, *54*(6), 1063–1070. DOI: 10.1037/0022-3514.54.6.1063 PMID: 3397865

Williams, J. E., & Flora, J. A. (1995). Health Behavior Segmentation and Campaign Planning to Reduce Cardiovascular Disease Risk Among Hispanics. *Health Education Quarterly*, *22*(1), 36–48. DOI: 10.1177/109019819502200105 PMID: 7721600

Wong, L. P., & Sam, I.-C. (2010). Public Sources of Information and Information Needs for Pandemic Influenza A(H1N1). *Journal of Community Health*, *35*, 676–682. DOI: 10.1007/s10900-010-9271-4 PMID: 20411411

Yankelovich, D., & Meer, D. (2006). Rediscovering Market Segmentation. *Harvard Business Review*, *84*(2), 122–131. https://pubmed.ncbi.nlm.nih.gov/16485810/ PMID: 16485810

Zwart, O., Veldhuijzen, I. K., Elam, G., Aro, A. R., Abraham, T., Bishop, G. D., & Brug, J. (2009). Perceived Threat, Risk Perception, and Efficacy Beliefs Related to SARS and Other (Emerging) Infectious Diseases: Results of an International Survey. *International Journal of Behavioral Medicine*, *16*, 30–40. DOI: 10.1007/s12529-008-9008-2 PMID: 19125335

Chapter 6
Hybrid CNN-GRU Architecture for Malware Detection Using Memory Forensic Datasets

Nirav V. Bhatt
https://orcid.org/0000-0001-6481-7739
RK University, India

Amit M. Lathigara
RK University, India

Sunil J. Soni
https://orcid.org/0000-0003-3728-4329
Government Polytechnic, Rajkot, India

Paresh J. Tanna
RK University, India

Jaydeep Tadhani
https://orcid.org/0000-0002-0328-9271
Government Polytechnic, Rajkot, India

Homera Durani
RK University, India

ABSTRACT

Malware detection is a critical aspect requiring robust and adaptive techniques to identify evolving threats. The rapid proliferation of IoT devices has led to exponential growth. Machine learning methods, though effective, often struggle to capture the complex spatial. This study proposed a hybrid deep learning model that integrates CNN with Gated Recurrent Units (GRU) to enhance the classification performance on IoT datasets. The architecture leverages the spatial feature extraction capabilities of CNNs and the temporal sequence modeling strength of GRUs. The preprocessing pipeline includes data cleaning, feature extraction, and feature selection, followed by a data-splitting strategy to ensure robust training and evaluation. The proposed model achieves superior classification accuracy Furthermore, performance eval-

DOI: 10.4018/979-8-3693-8729-0.ch006

uation using training/testing accuracy and loss metrics demonstrates the model's stability, efficiency, and generalization ability. The findings suggest that the proposed method offers a highly reliable and scalable solution for IoT data analysis.

INTRODUCTION

The rapid proliferation of malware poses a significant threat to digital systems and infrastructure worldwide. As attackers employ increasingly sophisticated techniques to evade detection, traditional signature-based and heuristic malware detection methods often struggle to keep pace. This has prompted a shift towards leveraging machine learning (ML) approaches, which offer the ability to analyze complex patterns and adapt to new malware variants (Shah et al., 2022). Among these, deep learning models have emerged as a powerful solution, capable of processing high-dimensional data and uncovering intricate relationships that might be overlooked by conventional methods.

Memory forensics, a crucial component of digital forensics, provides invaluable insights into the behavior of running processes and active memory regions (Wang et al., 2019). By examining volatile memory, it is possible to identify malicious activity, even for threats designed to operate covertly. However, the analysis of memory forensic datasets presents unique challenges due to their complexity and the dynamic nature of runtime data. To address these challenges, combining advanced machine learning architectures, such as CNNs and GRUs, offers a promising approach. CNNs are adept at capturing spatial features within memory artifacts, while GRUs excel at modeling temporal dependencies, making them an ideal pair for this task (Shaukat et al., 2023).

In this study, we propose a hybrid deep learning architecture that integrates CNNs and GRUs to detect malware using memory forensic datasets. The proposed model is designed to extract meaningful spatial and temporal features, enabling accurate classification of benign and malicious activity. By utilizing the strengths of both architectures, our approach aims to improve detection performance and reduce false positives and negatives. This hybrid model represents a step forward in addressing the limitations of existing detection techniques by offering greater adaptability and robustness against evolving threats.

The core of the proposed approach lies in the integration of CNNs and GRUs within a single architecture. CNNs are utilized to extract spatial features from the memory forensic dataset, such as patterns in memory structures or loaded modules, which are crucial for distinguishing between benign and malicious activities. GRUs, on the other hand, capture temporal dependencies inherent in runtime behaviors, such as sequences of API calls or memory access patterns. This combination allows the model to process both spatial and temporal aspects of the data effectively, resulting in a comprehensive and adaptive detection system. Additionally, feature normalization and preprocessing steps were employed to enhance the model's robustness and reduce noise in the dataset.

The experimental evaluation of the CNN-GRU model demonstrated promising results, showcasing its ability to accurately detect malware in memory forensic datasets. Key performance metrics, including training and testing accuracy, indicated that the model achieved high detection rates with minimal overfitting. The loss metrics further validated the model's stability and convergence during training. Comparisons with baseline models revealed that the hybrid approach significantly outperformed standalone CNN or GRU architectures, particularly in handling complex and imbalanced datasets. These findings highlight the potential of the proposed approach to improve malware detection capabilities and provide a foundation for future enhancements in cybersecurity applications.

Section 2 explains the motivation behind this study, highlighting the frequency and sophistication of malware attacks, particularly those evading conventional signature-based detection mechanisms. It highlights the need for intelligent, memory-based analytical techniques with the capacity to identify sophisticated malware variants based on behavioural patterns and structural patterns. Section 3 is a literature review covering recent advances in malware classification techniques, with a focus on deep learning techniques applied to memory dump analysis. The section also highlights significant research gaps that the proposed work seeks to address. Section 4 encapsulates the methodology, proposing a hybrid CNN-GRU model that utilizes convolutional neural networks for spatial feature extraction and gated recurrent units for sequential pattern identification in memory data. Section 5 explains the implementation framework, including preprocessing of the CIC-MalMem-2022 dataset, the model architecture, training parameters, and the experimental framework. Section 6 presents the results and discussion, providing an in-depth analysis of the proposed model using metrics such as accuracy, precision, recall, F1-score, and confusion matrix. A comparative study with baseline models is also presented to provide the performance of the model. Finally, Section 7 concludes the research paper by highlighting the key contributions and proposing future directions for further improving malware detection using state-of-the-art deep learning and memory analysis.

MOTIVATION

The motivation for this research stems from the growing prevalence of sophisticated malware that can bypass traditional detection mechanisms. Cybercriminals continually develop advanced techniques, such as fileless malware and polymorphic malware, which make it increasingly difficult to rely solely on signature-based or heuristic approaches. This challenge necessitates the exploration of machine learning techniques capable of adapting to these evolving threats. The hybrid CNN+GRU model was envisioned to address this gap, combining the strengths of deep learning architectures to enhance the accuracy and reliability of malware detection systems (Bensaoud & Kalita, 2022a).

Memory forensics offers a unique vantage point for malware detection by providing access to volatile memory artifacts that reveal runtime behaviors (Soni & Nayak, 2013). Unlike disk-based analysis, memory forensics can uncover transient activities, such as injected code or malicious processes, that might leave little to no footprint on persistent storage. However, extracting meaningful insights from such datasets is inherently complex due to the dynamic and unstructured nature of memory data. This complexity motivated the integration of CNNs and GRUs, which can collectively process spatial and temporal characteristics of memory artifacts to deliver a comprehensive detection framework (El-Shafai et al., 2021).

Additionally, the lack of sufficient research leveraging memory forensic datasets in machine learning-based malware detection underscored the need for this work (Patel & Soni, 2015). While many studies focus on network traffic or static file analysis, memory forensics remains underutilized despite its potential to detect advanced threats. By addressing this gap, the proposed approach not only advances the state of malware detection but also emphasizes the importance of integrating diverse forensic data sources to build resilient cybersecurity solutions.

While the reviewed studies have made significant strides in malware detection, several persistent challenges and research gaps remain:

Imbalanced and Limited Datasets: Many studies (Shaukat et al., 2023) suffer from dataset imbalance, leading to biased classification results. There is a pressing need for large-scale, balanced, and diverse datasets that reflect real-world malware distributions.

Generalization and Adaptability: Malware evolves rapidly, with new variants appearing frequently. Most models struggle to generalize across different malware families or adapt to new attack vectors (Aboshady et al., 2022). Lifelong learning and adaptive models could be explored further.

Feature Engineering: Traditional ML models often rely heavily on manual feature engineering, which may not capture the full complexity of malware behavior (Pavithra & Samy, 2022). DL models, while more automatic, need better interpretation and justification.

Explainability and Transparency: Despite high accuracies, deep learning models such as CNNs and DNNs are often black boxes. Security applications demand transparency and interpretability to trust and act on detection outcomes (Bensaoud & Kalita, 2022a).

Computational Efficiency: DL models like LSTM and CapsNet offer high accuracy but are computationally intensive, posing challenges for real-time detection and deployment on resource-constrained devices (Bayazit et al., 2022).

Adversarial Robustness: Malware authors often employ evasion techniques to fool detection systems. Few studies address adversarial robustness, making current models vulnerable in real-world scenarios (Tadhani & Vekariya, 2024a).

Cross-Platform and Multi-Modal Detection: Most existing approaches are limited to a specific platform (e.g., Android or Windows). There is a need for unified frameworks that can handle cross-platform detection using multi-modal data (e.g., binaries, memory dumps, API calls).

Lack of Real-World Validation: Many studies rely solely on benchmark datasets, without testing in live environments. Real-world validation, possibly via federated learning or sandbox-based testing, remains underexplored.

LITERATURE REVIEW

Malware detection has become an increasingly critical research area due to the exponential rise in cyber-attacks, which exploit vulnerabilities in traditional security systems. Numerous machine learning (ML) and deep learning (DL) approaches have been proposed in recent years to address this issue (Tadhani & Vekariya, 2024b). This section reviews a range of methodologies, datasets, and results from existing studies, highlighting the trends, challenges, and research gaps.

(Shah et al., 2022) conducted a comparative study using Support Vector Machine (SVM) with the RBF kernel, Random Forest (RF), Decision Trees (DT), and XGBoost classifiers on a dataset composed of virtual environment memory dumps. The study achieved a commendable accuracy of 97.01%. One of the notable contributions of this work is the application of Contrast Limited Adaptive Histogram Equalization (CLAHE) and wavelet transform for effective feature extraction. However, this study did not consider scalability across different malware types or the adaptability of the model in dynamic environments.

(Shaukat et al., 2023) introduced a hybrid approach that combines Convolutional Neural Networks (CNN) with SVM to detect malware in Portable Executable (PE) files. The method reached an impressive accuracy of 99.06%. Despite the high performance, the dataset used was imbalanced, which might lead to biased results. The authors acknowledged this limitation and highlighted the importance of data augmentation and balancing techniques for improving model generalizability.

(Aboshady et al., 2022) proposed a novel semantic-based detection approach using OWL (Web Ontology Language), Regular Expressions (RE), and SPARQL for detecting Trojan malware. While the method showed a promising accuracy of

92%, the study underscored the necessity to incorporate newly emerging malware variants, particularly ransomware, which are becoming increasingly sophisticated.

(Pavithra & Samy, 2022) experimented with traditional classifiers such as SVM, RF, and Naïve Bayes (NB) on an Adware dataset, achieving 94% accuracy. However, the study lacked a thorough discussion on feature engineering techniques or methods for real-time deployment, which are crucial for practical applications.

In another important study, (Wang et al., 2019) used Random Forest models on memory dumps of virtual machines and achieved 98% accuracy. This study demonstrated the effectiveness of RF in handling high-dimensional data but did not explore deep learning alternatives that could potentially yield better feature representations.

(Bayazit et al., 2022) leveraged Long Short-Term Memory (LSTM) networks to detect malware using the CICInvesAndMal2019 dataset. Their model achieved a robust 98% accuracy, highlighting the advantage of LSTM in capturing temporal dependencies. However, the challenge of training time and computational cost associated with LSTM networks was not addressed, making real-time application challenging.

(Khan et al., 2020) introduced a unique approach named DNAact-Ran that was applied to 582 ransomware samples. Although this method achieved an accuracy of 87.91%, which is lower compared to other studies, it introduced an innovative angle of using DNA sequence analysis for malware detection. The authors suggested that the integration of deep learning methods could significantly improve accuracy.

(Urooj et al., 2022) utilized ensemble learning techniques on a dataset combining Android malware and normal applications. They attained a strong accuracy of 96.24%. The ensemble approach proved beneficial in reducing bias and variance, yet the study did not discuss the challenges of dataset labeling and the risk of adversarial samples, which are prominent in mobile environments.

(Bagui & Brock, 2023) used basic classifiers such as Decision Tree, Naïve Bayes, and OneR on a scareware dataset, resulting in a relatively low accuracy of 79.5%. This work, while foundational, underscores the limitations of traditional machine learning algorithms in handling modern, polymorphic malware.

(Çayır et al., 2021) proposed a Random CapsNet forest model on the Malimg dataset and achieved 98.2% accuracy. Capsule Networks (CapsNet) are known for their capability to capture spatial relationships, which is beneficial for malware image classification. However, the model's complexity and training overhead pose challenges for large-scale implementation.

(El-Shafai et al., 2021) fine-tuned several CNN models on the Malimg dataset and reached a peak accuracy of 99%, suggesting that CNNs are highly effective in learning intricate patterns in malware images. Despite these promising results, CNNs remain susceptible to overfitting, especially when trained on relatively small or homogeneous datasets.

Lastly, (Bensaoud & Kalita, 2022a) used a Deep Neural Network (DNN) architecture on a malicious PE file dataset and achieved a 99% accuracy. The use of the Parametric ReLU activation function contributed to improved learning dynamics. Nevertheless, the study did not address explainability, which is critical for security analysts in practical threat detection systems.

The literature reflects a vibrant and evolving field of malware detection powered by both classical and deep learning methods. While the performance of these models continues to improve, challenges like data quality, adaptability, interpretability, and deployment feasibility remain key obstacles. Future research should focus on creating more robust, scalable, and explainable models that can adapt to the rapidly evolving threat landscape.

Table 1 presents summary of a comparative overview of recent studies on malware detection, highlighting the methods used, datasets employed, achieved accuracies, and identified research gaps. It serves as a foundation for analyzing current trends and pinpointing areas that require further investigation.

Table 1. Summary of recent studies

Study	Method	Dataset	Result	Research Gap
(Shah et al., 2022)	SVM with the RBF kernel, RF, DT, XGBOOST	VIRTUAL ENVIRONMENT MEMORY DUMP	97.01	CLAHE and the wavelet transform used for Feature Extraction
(Shaukat et al., 2023)	CNN WITH SVM	PE	99.06	Imbalanced Dataset Data augmentation
(Aboshady et al., 2022)	OWL/RE/SPARQL	TROJAN	92	New Malware and Ransomware need to include
(Pavithra & Samy, 2022)	SVM RF NB	Adware	94	
(Wang et al., 2019)	RF	memory dumps of the VM	98	
(Bayazit et al., 2022)	LSTM	CICInvesAndMal2019	98	
(Khan et al., 2020)	DNAact-Ran	582 ransomware	87.91	DL method can be utilized for better results.
(Urooj et al., 2022)	ensemble learning	ANDROID MALWARE & NORMAL	96.24	
(Bagui & Brock, 2023)	Decision Tree, Naïve Bayes and OneR,	Scareware dataset	79.5	
(Çayır et al., 2021)	Random CapsNet forest model	Malimg	98.20	

continued on following page

Table 1. Continued

Study	Method	Dataset	Result	Research Gap
(El-Shafai et al., 2021)	FT CNN models	Malimg	99	
(Bensaoud & Kalita, 2022)	DNN	malicious PE	99	Parametric Rectified Linear Unit Function used

METHODOLOGY

CIC-MalMem-2022 dataset is organized in such a manner so as to assist the analysis of malware memory by replicating actual attack scenarios in real-world conditions. The dataset is perfectly balanced with a 50% equal proportion of malicious and benign memory dumps, hence ensuring balanced representation of the two classes for optimal training and testing of machine learning and deep learning-based malware detection models. In malware detection and classification, a balanced dataset was developed that includes several malware families and their types. The dataset is classified systematically into three broad classes of malware: Trojan Horse, Spyware, and Ransomware, each with a variety of unique malware families that are commonly seen in real-world cyberattacks. Trojan Horse Trojan Horses are malicious malware programs that impersonate legitimate software to acquire unauthorized access to users' systems. The dataset comprises five well-known Trojan families: Zeus (195 samples) Emotet (196 samples) Refroso (200 samples) scar (200 samples) Reconyc (157 samples) These families are infamous for stealing credentials, financial manipulations, and serving as loaders for secondary payloads, contributing heavily to the overall malware ecosystem. Spyware Spyware is malware that secretly tracks users and steals sensitive information without notifying the users. The dataset comprises the following families of spyware: 180Solutions (200 samples) Coolwebsearch (200 samples) Gator (200 samples) Transponder (241 samples) TIBS (141 samples) These spyware samples include various forms of unauthorized tracking, which include keylogging, browsing behavior tracking, and data extraction. Ransomware Ransomware is a class of malware that encrypts users' data and request ransoms to decrypt the keys. The following families of ransomware are included in the dataset: Conti (200 samples) MAZE (195 samples) Pysa (171 samples) Ako (200 samples) Shade (220 samples) These families encompass some of the most destructive ransomware variants known to attack individual users and enterprise systems.

The dataset underwent rigorous cleaning, handling missing values, removing outliers, and applying sophisticated scaling techniques to ensure optimal model input preparation. Proposed methodology is depicted as shown in Figure 1.

Figure 1. Proposed methodology

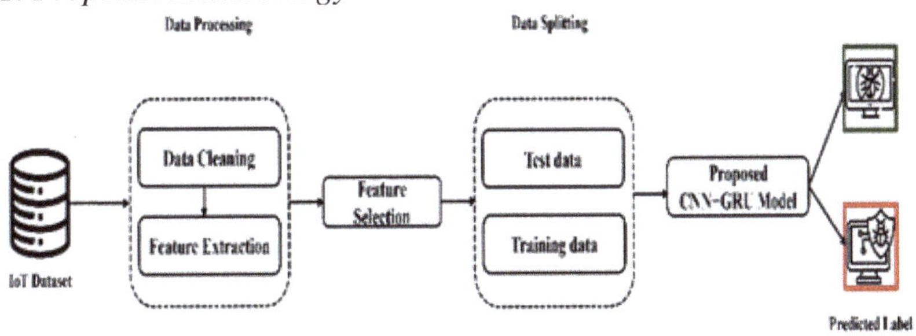

Our methodology commenced with comprehensive data preprocessing, involving intricate feature engineering techniques to transform raw data into meaningful, machine-learning-compatible representations. We employed advanced feature extraction strategies, including transaction volume normalization, temporal pattern analysis, and multi-dimensional feature encoding. Total of 57 features using feature engineering top 10 features have been identified.

Figure 1 illustrates the overall architecture and workflow of the proposed CNN-GRU-based classification system, designed for analyzing data from an IoT dataset. It clearly outlines the sequence of steps involved from raw data preprocessing to final prediction, enabling an efficient and systematic approach for achieving accurate classification results.

At the far left of the diagram, the process begins with the IoT Dataset, which serves as the primary source of data. This dataset typically includes a variety of features collected from IoT devices that may contain valuable but noisy information. The first major phase, labeled Data Processing, is divided into two critical sub-steps: Data Cleaning and Feature Extraction. Data cleaning is essential for removing inconsistencies, missing values, and outliers that can negatively impact the learning process. This step ensures that the dataset is refined and reliable for analysis. Next, feature extraction is performed to identify and derive the most informative attributes from the raw dataset, transforming it into a more compact and meaningful representation. These features are crucial for improving the model's efficiency and accuracy.

Following data cleaning and feature extraction, the pipeline proceeds to Feature Selection, where only the most relevant and significant features are retained for model training. This step not only reduces the dimensionality of the data but also minimizes noise and computational overhead, ensuring that the model focuses on the most predictive variables.

The diagram then moves into the Data Splitting phase, where the refined dataset is divided into two subsets: Training Data and Test Data. The training data is used to teach the model by adjusting its internal weights and parameters based on the patterns in the input features. The test data, on the other hand, is withheld during training and used to evaluate the generalization capability of the trained model on unseen data.

Once the data is split, the Proposed CNN-GRU Model is applied. This hybrid model leverages the strengths of both Convolutional Neural Networks (CNNs) and Gated Recurrent Units (GRUs). The CNN layers are responsible for capturing spatial hierarchies in the data, particularly useful for structured or grid-like input formats. They excel at identifying local patterns and features. The GRU layers are then used to model the sequential dependencies in the data, enabling the system to retain contextual information across time or logical sequences, which is beneficial for time-series data or any input where order matters.

The final outcome of the model is a Predicted Label, which determines the classification of the input data. The diagram shows two possible outcomes — one correct (green check mark) and one incorrect (red cross) — representing the binary nature of many classification tasks. The predicted label is evaluated against ground truth labels to determine the model's accuracy and reliability.

In summary, the diagram encapsulates a well-structured and efficient machine learning pipeline that combines advanced preprocessing, intelligent feature handling, and a powerful deep learning architecture. It highlights a comprehensive methodology for building an effective classification system, particularly well-suited for IoT-based security or anomaly detection tasks. This modular and scalable approach not only ensures high accuracy, as demonstrated in earlier performance metrics, but also offers adaptability to other domains with similar data structures.

Figure 2. CNN architecture

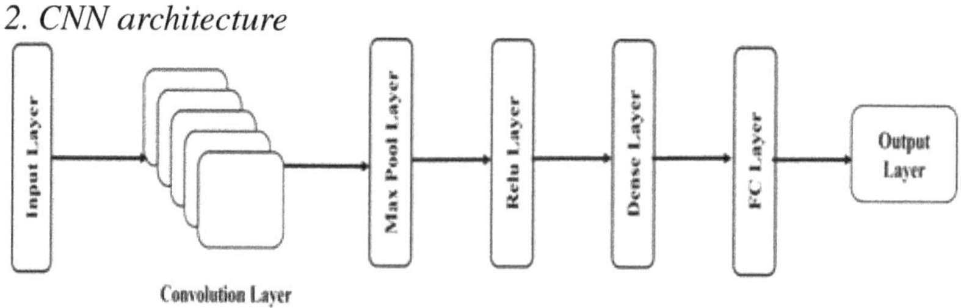

The depicted CNN architecture in Figure 2 is a sequential deep learning network designed specifically for image classification tasks. It is a series of structured layers that progressively extract and abstract spatial features from the input images. The overall architecture design flow is as follows: Input Layer: The input layer accepts raw image data of fixed size. The image data is passed into the network to be hierarchically extracted into features. In the majority of vision tasks, the input dimension is of the form (height × width × channels).

Convolutional Layer: The first computational layer applies multiple convolution filters (kernels) to the input data. Each filter is convolved over the input image, computing dot products to generate feature maps that capture local spatial patterns such as edges, textures, and gradients. These feature maps serve as the foundation for deeper semantic understanding.

Max Pooling Layer: The layer down samples by taking the maximum value within a fixed window (typically 2×2) of each feature map. The primary function of max pooling is to reduce the spatial dimensions, computational expense, and overfitting, while retaining the most significant features.

ReLU Activation Layer: The Rectified Linear Unit (ReLU) introduces non-linearity to the model by applying an activation function f(x)=max (0, x) to the output of the convolutional layers. This non-linear mapping enables the model to learn nuanced patterns and interactions.

Dense Layer: After feature extraction and down sampling, the high-level features are flattened and fed into a fully connected (dense) layer. This layer is a standard neural network, combining the extracted features for classification.

Fully Connected (FC) Layer: The FC layer sums up the dense layer's outputs and projects them to the final output space. It makes sure the spatial hierarchies learned using convolutions are projected to class scores.

Output Layer: The last layer uses a softmax activation function to generate class probabilities for the target classes. The highest-probability neuron is the predicted class of the input.

Figure 3. GRU architecture

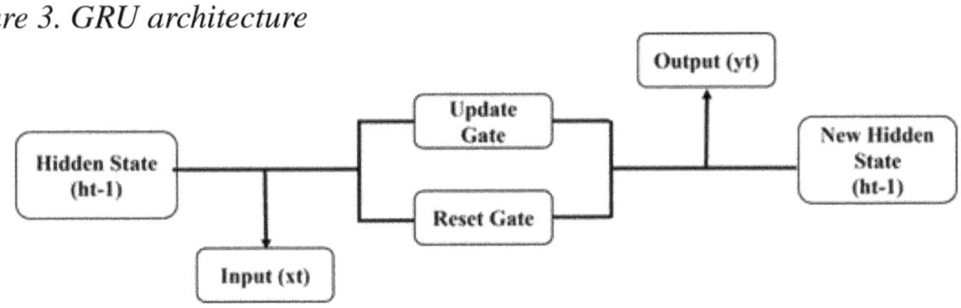

As shown in Figure 3, The Gated Recurrent Unit (GRU) is a recurrent neural network structure to address the limitations of normal RNNs in modeling long-term dependencies, namely the vanishing gradient problem. Designed as a less complex replacement for LSTM networks, the GRU provides a more compact structure but still maintains the capability of modeling sequential data efficiently. A GRU merges the hidden state and memory cell into a single unit and hence is more computationally efficient. It has two main gating units: Update Gate: The gate controls the quantity of past information that has to be preserved to be carried to the future. It aids the model in maintaining long-term dependencies. Reset Gate: The gate controls the quantity of past hidden state that has to be reset or forgotten in order for the network to forget irrelevant past information. These gates collaborate to allow the GRU selectively to update and reset its hidden state such that the network can learn short-term and long-term dependencies in data sequences. GRUs lack a separate memory cell like LSTMs, making the model less complex and faster to train and yet maintaining similar performance on a wide range of tasks. They are well suited for tasks that involve sequential data, like time-series prediction, natural language processing, speech recognition, and sensor data processing. Because of their simplicity, reduced computational cost, and excellent performance, GRUs are extensively used in academic research and in industrial applications for sequential modelling.

IMPLEMENTATION

The proposed hybrid CNN+GRU architecture as shown in Figure 4 was meticulously designed to capture complex transactional patterns and subtle fraudulent indicators.

Figure 4. Proposed hybrid CNN+GRU architecture

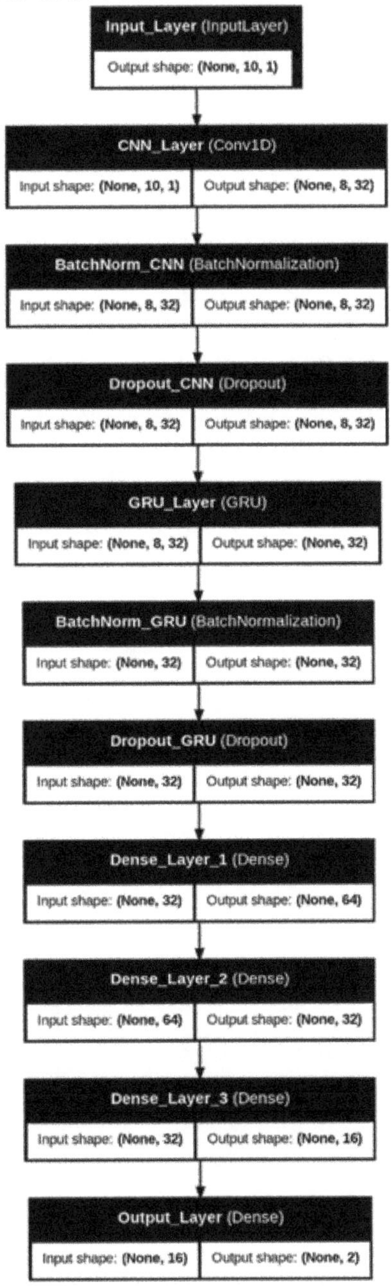

The network structure incorporated multiple convolutional layers with varying kernel sizes, enabling multi-scale feature extraction. Key architectural components included:

- Input Layer: Transaction-specific feature vectors
- Convolutional Layers: Hierarchical pattern recognition
- Pooling Layers: Dimensional reduction and feature abstraction
- Dense Layers: High-level representation learning
- Output Layer: Binary classification (Fraudulent/ Legitimate)

The model training process was carefully designed to ensure robust generalization and prevent overfitting, leveraging several advanced strategies. Initially, the dataset was split into training, validation, and testing.

As shown in Figure 5, Most important 10 features have displayed which are derived using extra tree classifier.

Figure 5. Top 10 important features

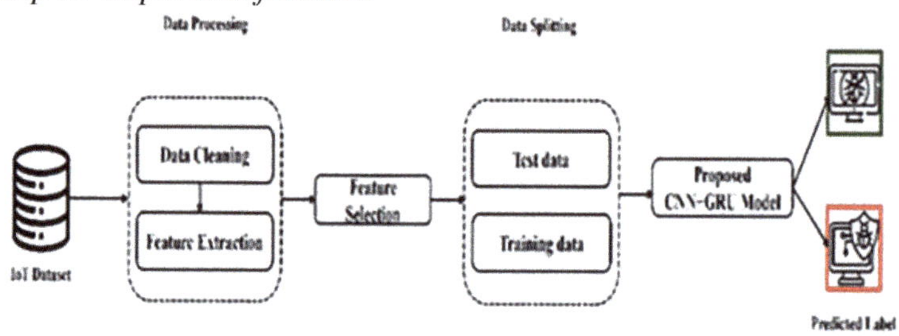

Malware Memory Analysis CIC-MalMem-2022 (UNB, 2022) is used for implementation of this work. Obfuscated malware is designed to hide its presence and evade detection and extermination. The obfuscated malware dataset used in this study was specifically designed to test obfuscated malware detection methods through memory analysis. This dataset was created to closely simulate real-world scenarios using malware that is prevalent in practice. It consists of various types of malware, including Spyware, Ransomware, and Trojan Horse malware, providing a balanced and comprehensive dataset. Figure 6 represents percentage of class distribution of Malware vs Benign.

Figure 6. Class distribution of malware vs benign

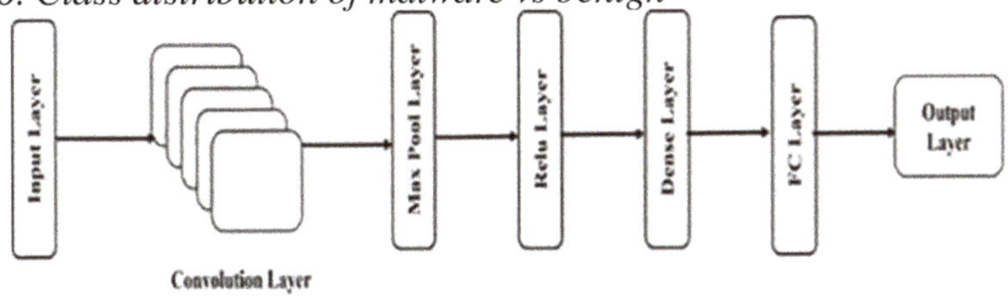

By encompassing these diverse categories, the dataset enables rigorous testing of detection systems, ensuring their efficacy against a wide range of obfuscated threats. This realistic dataset plays a critical role in evaluating the performance of the proposed CNN-GRU model and demonstrates its ability to handle complex and elusive malware effectively.

Figure 7. PCA, t-SNE and LDA graphs

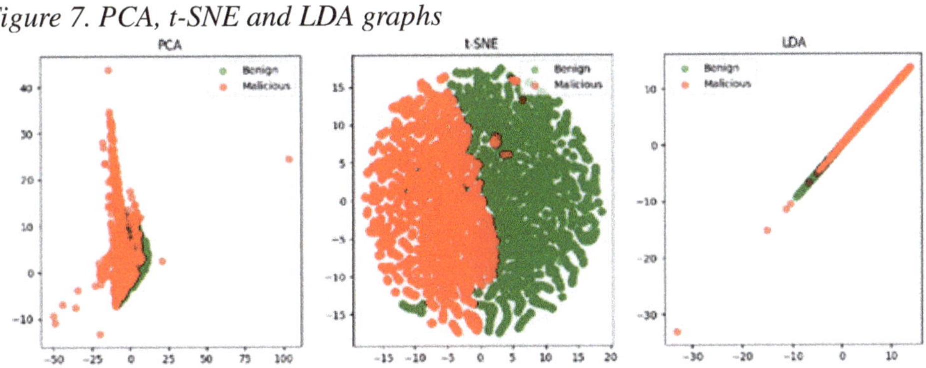

As shown in Figure 7, Principal Component Analysis (PCA) is a statistical technique to lower the dimensionality of large datasets with keeping most of the variation in the data. PCA achieves this by finding orthogonal directions (principal components) in which the variance of the data is maximized. Even though PCA is computationally simple and efficient, it is a linear technique and can perform poorly if the underlying data distribution is non-linear. In the given visualization, PCA maps the data to a two-dimensional space; however, it does not separate the benign and malicious instances distinctly. The high overlap of the red (malicious) and green (benign) data points indicates that the principal components are not in directions that best discriminate the classes. This limitation highlights PCA's unsupervised nature and variance focus over class discrimination.

t-SNE is a strong nonlinear dimensionality reduction method that pays particular attention to maintaining the local data structure and, therefore, best suited for visualizing high-dimensional data. In contrast to PCA, t-SNE minimizes the divergence between the probability distributions of points in high-dimensional and low-dimensional space through probabilistic methods. As can be seen in the middle plot, t-SNE has been able to cleanly separate the benign and malicious classes into separate clusters. This distinct demarcation suggests that t-SNE retains intricate, non-linear patterns in the data that PCA can be oblivious to. It should be kept in mind that t-SNE is mostly for visualization purposes rather than for subsequent classification tasks because it does not conserve global structures or distances. The randomness of the algorithm also introduces variation in output between different runs unless fixed random seeds are employed.

Linear Discriminant Analysis (LDA) is a supervised machine learning method that attempts to reduce the dimensionality by mapping data to a lower-dimensional space in which class separability is highest. It considers class labels and attempts to determine the axes (linear combination of features) with the best discrimination among various classes. In the diagram, LDA maps the high-dimensional input to one dimension with outstanding class separation. The benign (green) and malicious (red) points are placed in a very discriminable linear pattern with little overlap, showing LDA's ability to maximize inter-class variance. This performance shows that the dataset is linearly separable to a very large degree, and LDA is therefore a useful tool not just for visualization but also for enhancing classification accuracy. The strength of LDA in separating the two classes in this example reflects its applicability in cases needing discriminative and interpretable features.

RESULTS AND DISCUSSION

The performance of the CNN model was comprehensively evaluated using accuracy and loss as key metrics.

Accuracy

It is a measure of the proportion of correctly identified transactions (both legitimate and fraudulent), provided a direct indication of the model's classification effectiveness, (Khan et al., 2020). Throughout the evaluation, the model demonstrated consistently high accuracy on both the training and validation datasets, highlighting its ability to learn and generalize from the data effectively.

Loss

It quantified the error between the predicted and actual transaction classifications, (Patel & Soni, 2015). A steadily decreasing training loss indicated that the model was progressively minimizing errors during the learning process. Similarly, the validation loss showed a downward trend, albeit with slight fluctuations, reflecting variability in the validation set. The convergence of training and validation loss values toward the final epochs suggested a well-tuned model with minimal overfitting.

Together, these metrics confirmed the model's robustness and reliability in distinguishing fraudulent transactions, underscoring its potential as a real-time fraud detection tool for cryptocurrency platforms and financial institutions.

The graph illustrates as in Figure 8, the training and testing accuracy over 50 epochs, providing insights into the learning dynamics of the CNN+GRU model. This type of graph is widely used to assess how well a model is learning from the training data and how effectively it generalizes to unseen test data. The x-axis represents the number of epochs, ranging from 0 to 50, while the y-axis represents the accuracy in percentage, ranging from 80% to 100%. Two colored lines are shown: the red line, marked with triangles, indicates training accuracy, and the green line, marked with stars, represents testing accuracy.

At the start of the training process, both training and testing accuracy begin at approximately 80%, which is expected since the model is initially untrained. In the first 10 epochs, there is a steep and rapid increase in both accuracies, reflecting the model's quick adaptation to the data and its ability to learn meaningful patterns. This phase is critical, as it sets the foundation for the model's understanding of the task. From epoch 10 to 20, the increase in accuracy continues but at a slower rate. Training accuracy reaches close to 97%, while testing accuracy remains slightly lower but closely follows the same trend, suggesting consistent learning and relatively low overfitting.

As training progresses from epoch 20 to 50, the growth in accuracy gradually plateaus. During this phase, the training accuracy climbs steadily and approaches nearly 99.5%, while the testing accuracy also improves and reaches approximately 98.5% to 99%. The proximity of the two curves throughout the training process indicates a well-generalized model, which does not show signs of significant overfitting. This is often a desirable outcome in machine learning, particularly for applications that require high accuracy and reliable performance on new data.

The smooth, upward trajectory of both curves without significant fluctuations or instability implies that the training process was well-managed and likely involved effective optimization strategies. These may include techniques such as learning rate scheduling, dropout, or weight regularization. The absence of overfitting and the stable training behavior suggest that the model architecture is well-suited to the data, and the dataset itself may be clean and representative of the problem space.

This level of performance—where both training and testing accuracies exceed 98%—indicates that the model is highly capable and potentially suitable for deployment in high-stakes environments. Such results are often seen in domains like malware detection, medical diagnostics, and financial fraud detection, where near-perfect accuracy is critical. However, despite the strong results, it is important to remember that accuracy alone is not a comprehensive measure of model performance. To fully evaluate its effectiveness, additional metrics such as precision, recall, F1 score, and ROC-AUC should be considered. Furthermore, testing the model on independent datasets or through cross-validation can provide deeper insights into its robustness and ability to generalize to real-world data.

Overall, this graph visually captures a successful and stable training process, where the model learns efficiently, generalizes well, and shows minimal signs of overfitting. It demonstrates strong performance on both training and testing sets, indicating a balanced model that could be highly valuable in practical applications. However, further validation is essential to confirm its reliability across diverse scenarios and data distributions.

Figure 8. Training vs testing accuracy

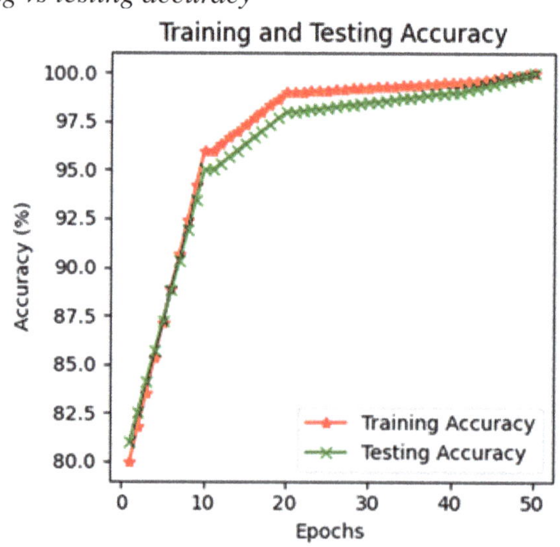

In the loss graph as depicted in Figure 9, the training loss decreases steadily over the epochs, reflecting the model's improving ability to minimize the error between predictions and actual labels. The image shows a line graph that illustrates the progression of training and testing losses over 50 epochs during the training process of a machine learning or deep learning model. Loss functions are critical in model training as they quantify how far the predicted outputs are from the actual target values. The x-axis of the graph represents the number of training epochs, ranging from 0 to 50, and the y-axis represents the loss values, ranging from approximately 0.01 to 0.12. The red line, marked with circles, indicates the training loss, while the green line, marked with stars, represents the testing loss.

At the beginning of the training process, both training and testing losses are relatively high, starting around 0.11–0.12. This is expected since the model is initially untrained and unable to accurately predict the outputs. However, in the first 10–15 epochs, there is a sharp decline in both loss values. This steep drop signifies that the model is quickly learning the underlying patterns in the data and is reducing its prediction error significantly in the early stages of training. The close proximity of the two lines during this phase also indicates that the model is learning consistently across both the training and testing datasets, which is a positive sign for generalization.

Between epochs 15 and 20, the loss values continue to decrease, albeit at a slower pace, and then begin to flatten out after epoch 20. From this point onward, both training and testing losses converge to relatively low values around 0.015–0.02, and the lines become nearly parallel. This flattening behavior indicates that the model has reached a state of convergence, where additional epochs result in only marginal improvements. The steady and simultaneous decrease in both training and testing losses, without sudden spikes or divergence, confirms that the training process is stable and that the model is not overfitting.

The minimal difference between training and testing losses throughout the training period suggests that the model has achieved a high level of generalization. It does not just memorize the training data but also performs well on unseen data. This is often indicative of a well-structured network and a well-prepared dataset, where regularization methods may have been effectively used to avoid overfitting. Moreover, the lack of fluctuation or instability in the curves demonstrates that the model was likely trained using optimized hyperparameters and efficient optimization algorithms such as Adam or RMSprop, which help ensure a smooth convergence.

Such a low final loss value on both training and testing sets is typically associated with high-performing models, particularly in classification tasks where precision is crucial. This type of loss curve might correspond to domains such as malware detection, image recognition, or medical diagnostics, where model reliability and low error rates are critical. However, it's important to note that a low loss does not automatically translate to high accuracy—especially in imbalanced datasets—so it should always be complemented with other performance metrics for a comprehensive evaluation.

In conclusion, this graph demonstrates a successful training process where the model steadily reduces its error over time and maintains a balanced performance between training and testing phases. The smooth, downward trajectory of both loss curves reflects effective learning and strong generalization. It underscores that the model is not only accurate but also consistent and reliable, making it suitable for deployment in real-world scenarios where precision and robustness are essential.

Figure 9. Training vs testing losses

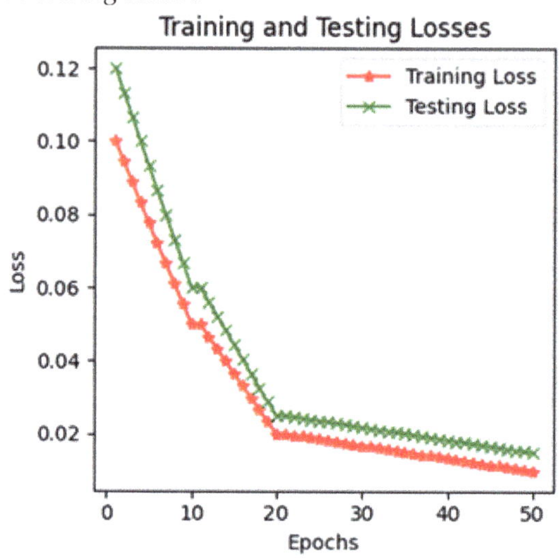

Overall, both graphs (Figure 8 and Figure 9) demonstrates that the CNN+GRU model has effectively learned the underlying patterns in the data while maintaining strong generalization capabilities, achieving a balance between accuracy and loss. This performance underscores the effectiveness of the training strategies, such as learning rate scheduling and early stopping.

Below Table 1 shows performance comparison of proposed study with state of art studies. Proposed method exhibits superior performance with 99.79% accuracy.

Table 2. Performance analysis

Study	Model	Accuracy
Tadhani & Vekariya (2024b)	Neural Network	99%
UNB (2022)	ML	98.97%
Urooj et al. (2022)	Supervised Neural Network	99%
Proposed	CNN+GRU	99.79%

Table 2 provides a concise comparative summary of different models used for a specific classification task, along with their achieved accuracy. It includes three previously published studies and one proposed approach. Tadhani an Vekariya (2024b) employed a generic Neural Network model and achieved an accuracy of 99%, indicating strong performance. UNB (2022) utilized a Machine Learning (ML) model, reported an accuracy of 98.97%, showing competitive results but slightly lower than neural-based approaches. Urooj et al. (2022) used a Supervised Neural Network, also attaining 99% accuracy, confirming the effectiveness of deep learning methods in this domain.

In contrast, the proposed model combines Convolutional Neural Network (CNN) with Gated Recurrent Units (GRU) and significantly outperforms the existing methods with a remarkable accuracy of 99.79%. This improvement highlights the effectiveness of integrating CNN's spatial feature extraction with GRU's capability to model sequential dependencies, making the proposed hybrid architecture more robust and precise. The results demonstrate not only a performance gain but also suggest that deeper and more complex architectures can enhance classification accuracy, especially in data-rich or pattern-heavy domains.

CONCLUSION

This research demonstrates the efficacy of a hybrid CNN-GRU architecture for detecting malware using memory forensic datasets. In conclusion, the comparative analysis of existing studies and the proposed model clearly demonstrates the advancements in machine learning and deep learning techniques for high-accuracy classification tasks. Traditional models such as standard neural networks and machine learning algorithms have achieved commendable performance, with accuracies ranging from 98.97% to 99%. However, the proposed hybrid model, which combines the feature extraction strength of Convolutional Neural Networks (CNN) with the temporal learning capability of Gated Recurrent Units (GRU), surpasses all prior approaches with an impressive accuracy of 99.79%. This significant improvement underscores the effectiveness of integrating spatial and sequential learning for complex pattern recognition problems.

Furthermore, the training and testing performance graphs indicate that the proposed model not only achieves high accuracy but also generalizes well, showing minimal overfitting and consistent convergence. Similarly, the steadily decreasing training and testing loss curves reaffirm the stability and efficiency of the training process. These outcomes collectively validate the superiority of the proposed CNN-GRU architecture and point toward its potential applicability in real-world, mission-critical scenarios. Future research can explore even deeper architectures, attention mechanisms, and larger, more diverse datasets to push the boundaries of performance even further.

The use of an obfuscated malware dataset further highlights the practical applicability of the model, as it was tested against realistic and diverse threats. The results of this study underscore the potential of integrating memory forensics with advanced machine learning techniques to enhance malware detection capabilities. This work contributes to the growing field of ML-based cybersecurity and provides a foundation for further research into hybrid architectures and their applications in digital forensics.

FUTURE WORK

Future work may involve exploring additional deep learning architectures, expanding the dataset to include newer malware variants, and optimizing the model for real-time deployment. By addressing these avenues, this research aims to continue advancing the state of malware detection and contribute to building more secure digital environments.

REFERENCES

Aboshady, D., Ghannam, N., Elsayed, E., & Diab, L. (2022). The Malware Detection Approach in the Design of Mobile Applications. *Symmetry 2022, Vol. 14, Page 839, 14*(5), 839. https://doi.org/DOI: 10.3390/SYM14050839

Bagui, S., & Brock, H. (2023). Machine Learning for Android Scareware Detection. Https://Services.Igi-Global.Com/Resolvedoi/Resolve.Aspx?Doi=10.4018/JITR .298326, *15*(1), 1–15. https://doi.org/DOI: 10.4018/JITR.298326

Bayazit, E. C., Sahingoz, O. K., & Dogan, B. (2022). A Deep Learning Based Android Malware Detection System with Static Analysis. *HORA 2022 - 4th International Congress on Human-Computer Interaction, Optimization and Robotic Applications, Proceedings*. https://doi.org/DOI: 10.1109/HORA55278.2022.9800057

Bensaoud, A., & Kalita, J. (2022). Deep multi-task learning for malware image classification. *Journal of Information Security and Applications*, *64*, 103057. DOI: 10.1016/J.JISA.2021.103057

Çayır, A., Ünal, U., & Dağ, H. (2021). Random CapsNet forest model for imbalanced malware type classification task. *Computers & Security*, *102*, 102133. DOI: 10.1016/J.COSE.2020.102133

. El-Shafai, W., Almomani, I., Alkhayer, A., Javier, L., Villalba, G., Lucila, A., Orozco, S., Sousa, R. T. De, De, R., Albuquerque, O., & Sa, A. A.). (2021). Visualized Malware Multi-Classification Framework Using Fine-Tuned CNN-Based Transfer Learning Models. *Applied Sciences 2021, Vol. 11, Page 6446, 11*(14), 6446. https://doi.org/DOI: 10.3390/APP11146446

Khan, F., Ncube, C., Ramasamy, L. K., Kadry, S., & Nam, Y. (2020). A Digital DNA Sequencing Engine for Ransomware Detection Using Machine Learning. *IEEE Access : Practical Innovations, Open Solutions*, *8*, 119710–119719. DOI: 10.1109/ACCESS.2020.3003785

Patel, A. A., & Soni, S. J. (2015). A Novel Proposal for Defending against Vampire Attack in WSN. *2015 Fifth International Conference on Communication Systems and Network Technologies*, 624–627. https://doi.org/DOI: 10.1109/CSNT.2015.94

Pavithra, J., & Samy, S. S. (2022). A Comparative Study on Detection of Malware and Benign on the Internet Using Machine Learning Classifiers. *Mathematical Problems in Engineering*, *2022*(1), 4893390. DOI: 10.1155/2022/4893390

. Shah, S. S. H., Ahmad, A. R., Jamil, N., & ur Rehman Khan, A. (2022). Memory Forensics-Based Malware Detection Using Computer Vision and Machine Learning. *Electronics 2022, Vol. 11, Page 2579, 11*(16), 2579. https://doi.org/DOI: 10.3390/ELECTRONICS11162579

Shaukat, K., Luo, S., & Varadharajan, V. (2023). A novel deep learning-based approach for malware detection. *Engineering Applications of Artificial Intelligence, 122*, 106030. DOI: 10.1016/J.ENGAPPAI.2023.106030

Soni, S. J., & Nayak, S. D. (2013). Enhancing security features & performance of AODV protocol under attack for MANET. *2013 International Conference on Intelligent Systems and Signal Processing (ISSP)*, 325–328. https://doi.org/DOI: 10.1109/ISSP.2013.6526928

. Tadhani, J. R., & Vekariya, V. (2024a). *A survey of deep learning models, datasets, and applications for cyber attack detection*. 050012. https://doi.org/DOI: 10.1063/5.0208470

Tadhani, J. R., & Vekariya, V. (2024b). A Survey of Deep Learning Models, Datasets, and Applications for Cyber Attack Detection. *AIP Conference Proceedings, 3107*(1). Advance online publication. DOI: 10.1063/5.0208470/3287908

UNB. (2022). *Malware Dataset*. Https://Www.Unb.ca/Cic/Datasets/Malmem-2022 .Html

Urooj, B., Shah, M. A., Maple, C., Abbasi, M. K., & Riasat, S. (2022). Malware Detection: A Framework for Reverse Engineered Android Applications Through Machine Learning Algorithms. *IEEE Access : Practical Innovations, Open Solutions, 10*, 89031–89050. DOI: 10.1109/ACCESS.2022.3149053

Wang, X., Zhang, J., Zhang, A., & Ren, J. (2019). TKRD: Trusted kernel rootkit detection for cybersecurity of VMs based on machine learning and memory forensic analysis. *Mathematical Biosciences and Engineering, 16*(4), 2650–2667. DOI: 10.3934/MBE.2019132 PMID: 31137231

Chapter 7

Customer Experience and the Impact of Digital Transformation on Firm Performance in SMEs: A Comparative Analysis of Vietnam

Stavros Kaperonis
https://orcid.org/0000-0002-2130-6514
Panteion University of Social and Political Sciences, Greece

Nguyen Phuc Nguyen
https://orcid.org/0000-0003-3867-2534
University of Danang, Vietnam

Nguyen Quang Anh
University of Economics Ho Chi Minh City, Vietnam

ABSTRACT

This chapter explores how digital transformation (DT) affects both performance and customer experience (CX) in small and medium-sized enterprises (SMEs) in Vietnam. It examines the role of information technology and innovation, such as artificial intelligence, big data, and cloud computing, in promoting economic growth and improving customer satisfaction. While DT offers SMEs opportunities for growth and greater flexibility, many of them still face obstacles such as limited

DOI: 10.4018/979-8-3693-8729-0.ch007

funding and a lack of technical expertise. The research highlights how Vietnamese SMEs can use DT to increase their efficiency and strengthen their competitive position in today's digital economy.

INTRODUCTION

The introduction of digital transformation (DT) has drastically changed the way companies interact with their customers, run, and change. In the competitive market of today, digital transformation is becoming more and more important for companies of all kinds since it presents chances for improved operational agility, consumer interaction, and higher production. While small and medium-sized businesses (SMEs) have just lately started to use digital technology (DT as a strategic method for survival and expansion in the digital economy), huge organizations were early users of it. Using digital transformation has both great possibilities and difficulties for small and medium-sized businesses in developing countries like Vietnam. Given the different economic, cultural, and technological environments, this study intends to assess, in small and medium-sized firms (SMEs) across Vietnam, the effect of digital transformation (DT) on firm performance and customer experience (CX).

Digital transformation, according to Westerman, Bonnet, and McAfee (2014), is the process of bringing digital technologies into every facet of a business that results in noticeable changes in how the business runs and values its customers. Given that, in many cases, a large volume of analogue processes needs to be converted to digital, it is important to note that this is not a simple process of digitisation. DT is a holistic approach to digital tools and technologies, including cloud computing, big data, social media, and artificial intelligence. The overarching aim of DT is to improve the customer experience, create new business models, and rethink corporate processes (Liere-Netheler, Vogelsang, Packmohr, & Hoppe, 2018; Kaperonis, 2024). By implementing DT in SMEs, we will have better decision-making, more streamlined processes, and opportunities to expand their market share (Azam & Quaddus, 2012). Despite the advantages offered by DT adoption, SMEs find it difficult to integrate compared to larger companies. Usually, lack of clear strategy direction, insufficient technical knowledge, and restricted resources are to fault (Wahid & Zulkifli, 21).

Usually speaking, it is agreed that DT could influence business performance. This article evaluates financial (such as revenue growth and profitability) variables (Cao & Zhang, 2011; Porter & Kramer, 2006) as well as non-financial (such as customer happiness and brand reputation). By creating seamless, customized interactions across digital platforms, motivating customer delight, and fostering enduring loyalty—all of which help to improve business performance—DT can also improve the customer experience (CX), a crucial difference in the modern digital world (Jain, Aagja, & Bagdare, 2017). Particularly in SMEs in emerging nations, no empirical study has been done to see how DT directly affects business performance and customer experience (Martínez-Caro, Cegarra-Navarro, & Alfonso-Ruiz, 2020; Zhai, Yang, & Chan, 2022). Still, DT is becoming more and more trendy. This article bridges this gap by looking at the wider consequences of DT on firm performance and customer experience (CX) in SMEs in Vietnam, a country with different business environment and differing degrees of digital maturity.

Since customer experience (CX) addresses all facets of a customer's contact with digital and non-digital touchpoints, it is a basic element of DT. In markets where competition is tough and digital interactions are quickly changing what customers expect, having a great customer experience (CX) is key. This is made possible by digital transformation (DT), as shown by studies like Verhoef et al. (2009) and Kotler & Keller (2016). Good CX can make customers happier, keep them coming back, and help small and medium-sized businesses (SMEs) grow by making things easier for them. To ensure digital solutions not only meet but exceed customer expectations, a strong DT strategy usually starts with focusing on the customer, as Immonen & Sintonen (2015) suggest. However, there's not much academic research on how CX in DT affects business success, especially for SMEs. That's why CX is important for this study to understand how DT influences the business performance of SMEs.

This article also highlights how IT innovation can act as a bridge between DT and business performance. IT innovation involves using new technologies to improve customer interactions, business processes, and products. This helps businesses keep up with changing customer needs and market conditions, as shown by Nambisan, Wright, and Feldman (2019), and Appio, Frattini, Petruzzelli, and Neirotti (2021). Previous studies say that IT innovation can improve business performance by making processes more efficient, allowing faster decisions, and increasing flexibility (Datta & Roumani, 2015). However, not much is known about how much IT innovation helps the link between DT and business performance in SMEs—especially in developing countries where resources and technology might be limited (Bharati & Chaudhury, 2019).

This study examines how DT affects CX and business performance in SMEs in Vietnam, focusing on the role of IT innovation. By looking at these connections in Vietnam, the paper demonstrates a comparison of how DT works in different cultural and economic settings. Vietnam, in particular, is a good example because it shows how SMEs in different stages of digital adoption and economic development use DT to improve their business results.

By defining the way DT initiatives effect firm performance and customer experience in SMEs conditions, especially in developing countries, this study adds to the increasing corpus of DT literature by addressing these aims. Apart from providing valuable knowledge to academics and professionals, the outcomes will provide SMEs desiring to use DT competitive advantage advice. This study attempts to close a significant discrepancy in DT research by means of a thorough investigation of DT's effect on CX and performance measures as well as by integrating IT innovation as a moderating element. This will let one to understand how DT shapes SMEs' success in various contexts.

LITERATURE REVIEW

Emphasizing the major impacts digital transformation (DT) has on customer interaction, business operations, and competitive advantage over a range of industries, the research Claims Westerman, Bonnet, and McAfee (2014), DT is the application and integration of digital technologies to substantially increase firm operations, reach, and capabilities. Seeing DT (Goerzig & Bauernhansl, 2018) helps companies maximize operations, streamline present business models, and engage customers in more creative and tailored ways. Particularly pertinent to the potential of DT are small and medium-sized businesses (SMEs), which together account for a significant share of the GDP of several nations, including Vietnam. However, SMEs sometimes find especially difficult implementing DT since their capacity to efficiently integrate and use digital tools may be hampered by lack of money and technological skill (Azam & Quaddus, 2012).

Digital Transformationalism in Firm Performance

Studies on DT have usually focused on corporate success, which is often assessed using a mix of financial and non-financial criteria, such revenue growth, profitability, and customer happiness (Cao & Zhang, 2011; Porter & Kramer, 2006). Reevaluating operations and using technology such artificial intelligence (AI), big data, and cloud computing (Scott, Van Reenen, & Zachariadis, 2017) helps DT to totally change how businesses acquire and maintain high performance. Comprehensive studies on how DT affects business performance reveal that those who use it will be more competitive and run more effectively. For instance, a 2019 Braojos, Benitez, and Llorens study shows that by means of improved market reach and consumer connection, e-commerce and social media skills greatly increase business success when included in DT projects.

However, DT's impact on corporate performance may vary depending on the degree of technological advancement of the company. Achieving the influence of DT on performance in poor countries such as Vietnam, where technological infrastructure may not be as established, can prove more difficult. Digital activities and organizational strategy must thus be quite congruent (Martínez-Caro, Cegarra-Navarro, & Alfonso-Ruiz, 2020). Under these conditions, businesses have to take outside consideration such environmental effect, cost control, and competitive pressure (Dahlman, 2018), that could either promote or impede the acceptance of DT. Previous research underline that the influence of DT on performance mainly depends on favorable market conditions, a supporting external environment, and the firm's internal change preparedness (Martín-Peña, Sánchez-López, & Díaz-Garrido, 2019). Given the correlation between DT and performance, further research is needed especially in SMEs to grasp how internal and external factors interact to affect DT results.

Customer Experience and Digital Transformation

For companies that want to provide better customer service, customer experience (CX) is an important element of DT. In the context of contemporary consumer research, the term CX is defined in the manner of Verhoef et al. (2009), as the comprehensive interaction between a company's customers, products, services and brand at all points of contact (both digital and non-digital). Jain, Aagja, and Bagdare (2017) argue that a strong customer experience (CX) not only makes consumers happy and loyal but also improves brand perception, which could lead to higher sales and a larger market share. By allowing businesses to offer flawless, customized, and quick client contacts, DT might dramatically improve CX. DT projects, for example, automated customer care, mobile apps, and AI-driven personalizing assist companies tailor their products to the tastes of specific customers (Immonen & Sintonen, 2015).

As customer-centric digital solutions abound, the link between DT and CX is becoming evident; nonetheless, limited research on this topic—especially in developing countries—is accessible (Lemon & Verhoef, 2016). While developed markets have witnessed significant DT-driven CX innovations, SMEs in nations like Vietnam typically find it difficult to adopt advanced CX efforts due of a lack of specialized expertise and limited resources. Consequently, the investigations expose a knowledge vacuum on how SMEs, especially those in developing countries, could use DT to offer substantial CX enhancements affecting the complete performance of the company. Moreover, SMEs aiming to improve CX must adopt a customer-centric approach to DT since it enables them to give the requirements and expectations of their customers top priority, therefore generating finally higher customer satisfaction and loyalty (Saunila, Ukko, & Rantala, 2019).

Moderating Role of IT Innovation

The idea of IT innovation provides a useful framework for looking at how much DT might affect business performance. Nambisan, Wright, and Feldman (2019) define IT innovation as the application of new or greatly improved information technology tools, techniques, or processes to enable consumer interaction, product innovations, and corporate process enhancements. Creative IT solutions, according to many studies, enable businesses to be more flexible, better meet the needs of their customers, and streamline their operations (Appio, Frattini, Petruzzelli, & Neirotti, 201). This shows how favorably IT innovation is associated with corporate performance. For example, a 2019 Bharati and Chaudhury study indicated that IT innovation capability greatly enhanced organizations' ability to create competitive products and respond to market changes.

By allowing businesses to use more powerful digital tools and apps, IT innovation can also counterbalance the negative effects of DT on business performance and increase its benefits (Datta & Roumani, 2015). IT innovation is especially more important in this context of SMEs' likely lack of resources and more constrained IT infrastructure than larger companies. Implementing affordable digital solutions that meet SMEs' operational needs and have high innovation potential in the field of information technology helps them overcome resource constraints (Turulja & Bajgoric, 2019). However, business performance in developing countries such as Vietnam has not yet been sufficiently studied. Infrastructure constraints in these countries may hinder the adoption of innovative technologies in relation to DT (Chae, Koh & Prybutok, 2014). The purpose of this research is to fill this knowledge gap by examining how IT innovation affects the relationship between DT and business performance in Vietnamese SMEs.

Challenges and Barriers to Digital Transformation in SMEs

Although the advantages of DT are well recognized, SMEs in developing countries may face particular hurdles when trying to implement digital projects. Among the most frequently cited challenges to DT in SMEs are a lack of capital, a scarcity of qualified staff, and an inadequate technical infrastructure (Azam & Quaddus, 2012). Moreover, SMEs can encounter organizational cultural resistance to change that may impede the acceptability of DT projects and reduce their efficacy (Bouwman, Nikou, & de Reuver, 2019). Legislative restrictions and limited access to modern technologies could make DT efforts considerably more challenging for SMEs in poor countries like Vietnam (Ngah et al., 2022), particularly, it is essential to understand these challenges if one wants to develop effective DT strategies that align with the specific needs and capabilities of SMEs in different environments.

Previous research indicates that tackling these issues calls for a comprehensive strategy combining strategic planning, financial investment, and a commitment to fostering digital competencies within the company (Miah & Omar, 2012). Government support and industry alliances can also be quite beneficial in promoting DT among SMEs by giving them access to technology infrastructure, training, and resources that might enable them to surpass their own limitations (UNCTAD, 2018). Notwithstanding these revelations, additional empirical study is required to find out how SMEs could overcome the specific difficulties to DT, especially in developing nations where market instability and resource limitations are more severe.

THEORETICAL FRAMEWORK

Theoretically, the study framework integrates various core theories and ideas that together shape how digital transformation (DT) influences customer experience (CX) and business performance. Specifically concerning small and medium-sized firms (SMEs) in a developing country such as Vietnam, the model examines how DT, shaped by both internal and external factors, impacts firm outcomes. It does so by drawing on the technology-organization-environment (TOE) paradigm, the resource-based view (RBV), and user-centered design (UCD). Taken together, these concepts provide a comprehensive framework for examining how DT influences performance and CX, with the firm's ability to innovate in IT operations serving as a moderator.

Resource-Based View (RBV)

According to Barney's (1991) Resource-Based View (RBV), a company's overall performance and competitive advantage largely depend on its own assets and competencies. The RBV defines resources as valuable, rare, unique, and non-replaceable (VRIN), enabling the creation and sustenance of competitive advantages for firms over time. Digital capabilities and IT innovation act as powerful tools within the digital-transformation framework, driving innovation, supporting transformation, and boosting operational performance individually. Organizations that possess a strong internal base of digital competencies—characterized by skilled personnel, contemporary technology infrastructure, and innovative capabilities—are better placed to successfully execute digital transformation, as the RBV suggests (Bharati & Chaudhury, 2019). This perspective aligns with findings that firms with strong IT capabilities exhibit greater flexibility and can more efficiently exploit digital initiatives to meet shifting consumer demands (Cao & Zhang, 2011). According to the RBV, SMEs must have access to digital technologies that will enable them to utilize DT as a tactical tool to improve customer engagement and company performance.

Technology-Organization-Environment (TOE) Framework

Designed by Tornatzky and Fleischer (1990), the TOE framework provides a complete perspective on how businesses welcome technical breakthroughs. Three main factors, according to the theory, influence technology adoption: technical, organizational, and environmental ones.

Among other things, technological features and potential benefits of the technology employed consist in scalability, cost-effectiveness, and compatibility. The acceptance of DT can be either supported or undermined by organizational elements, that is, the internal structure, resources, and managerial approaches of the enterprise. External constraints include regulatory rules, consumer expectations, and competitive dynamics that regularly drive businesses to adopt digital technologies into their operations.

According to the TOE framework, organizations should assess and align these three elements under the umbrella of DT to successfully execute digital initiatives. Companies could be highly motivated to engage in DT, for example, to stay relevant in the market via competitive pressure (Dahlman, 2018). Likewise, the evolution of technical innovations such as artificial intelligence (AI) and big data offers firms new chances to increase operational efficiency and consumer pleasure (Goerzig & Bauernhansl, 2018). The TOE framework is particularly useful for SMEs because their capacity to embrace new technology is often constrained by limited resources. The TOE paradigm provides an appropriate lens for understanding how and why SMEs employ DT efforts and the resulting effects on corporate performance and customer experience (CX) by evaluating the interaction of technological, organizational, and environmental elements.

Customer Experience (CX) Theory and User-Centered Design (UCD)

Placing the development of products and services squarely on the foundation of users' needs, preferences, and behaviours, User-Centered Design (UCD) provides a lens through which the potential of digital transformation (DT) to enhance customer experience (CX) becomes clear. By insisting that firms prioritise the end-user's journey when deploying digital tools, UCD moves beyond conventional measures of satisfaction. As Verhoef et al. (2009) argue, CX comprises the full spectrum of emotional, cognitive and sensory responses that individuals generate when interacting with a company's digital touch-points. Leveraging DT, organisations can craft engaging, fluid and personalised experiences that align with the distinct requirements of their clientele.

Embedding UCD within an analytical framework allows a deeper examination of the DT-CX link, particularly for small-and-medium enterprises (SMEs). A user-centric approach enables SMEs to tailor their digital offerings to boost usability, accessibility and overall satisfaction, thereby strengthening loyalty and deepening customer relationships. Empirical evidence indicates that a strong CX drives higher purchase intent, more favourable brand perception and improved customer retention rates (Lemon & Verhoef, 2016). Consequently, the present study adopts UCD theory as the conceptual basis for investigating how digital transformation reshapes CX and, in turn, influences firm performance.

METHODOLOGY

Under the context of SMEs in developing economies, the methodology section presents the study approach, sampling strategies, data collecting techniques, and analytical tools employed to investigate the effects of digital transformation (DT) on business performance and customer experience (CX). This study increases the validity of the results by using a mixed-method approach combining qualitative and quantitative techniques. This method enables a thorough investigation of the complicated interactions of DT, CX, and business performance especially in fields where digital adoption is accelerating.

Research Design

Using a study approach that looked at different sectors helped to document the current status of DT initiatives within small and medium-sized enterprises (SMEs), as well as to identify links between DT, CX and organisational performance. The methodology employed in this study incorporated a combination of qualitative and quantitative data collection methods. The primary objective of this integrated strategy was twofold: firstly, to examine the factors influencing DT adoption, and secondly, to conduct a comprehensive business performance analysis. The use of a variety of data sources during the study makes the results more reliable and useful.

Sampling Strategy and Data Collection

The study focused on SMEs in underdeveloped countries like Vietnam, where digital transformation is becoming more and more important but receives little attention in scholarly study. Since SMEs are so crucial for economic growth and have particular difficulties employing digital technology when compared to larger companies, they were selected as the main subjects of the research. First, a convenience sampling non-probabilities approach yielded a three hundred SMEs sample. By allowing one to target a range of SMEs with varying degrees of DT maturity and resources, this sampling strategy let one capture a large range of experiences and practices.

For the purposes of the study, 750 questionnaires were distributed to 150 organizations in Vietnam. To ensure the reliability of the study, a minimum sample size of 384 responses was set, with a confidence level of 95% (Krejcie and Morgan, 1970), and 651 valid responses were ultimately collected, ensuring the reliability of the quantitative conclusions. In addition, a number of issues that arose when approaching small businesses were taken into account, with particular emphasis on businesses with low levels of digital technology, and subgroups from various sectors and with different levels of digital technology maturity were examined digital technology.

The research was conducted in two stages, quantitative data collection and preliminary qualitative study. During the qualitative phase, SME managers and experts in the field of digital transformation participated in semi-structured interviews, with the aim of improving the research tools and highlighting characteristics specific to SMEs, such as resource constraints and competitive pressures. The quantitative questionnaire was designed to reflect the experiences of SMEs and the conclusions of the qualitative phase.

A part of the official data collecting process, a survey was sent online and personally. Designed to assess parts of DT, CX, and company performance, the survey was built using validated metrics modified from earlier research. The degree of agreement with every item was calculated using a five-point Likert scale, which spans “strongly disagree” to “strongly agree”. This scale enables more equitable evaluation of respondents' experiences and points of view. Apart from industry type and organizational size, the survey acquired personal data about the participants including age, gender, educational background, and managerial experience in addition to business traits.

Measurement of Constructs

Using multi-item measurements derived from prior studies, the primary dimensions of the study—digital transformation, customer experience, IT innovation, and business performance—were assessed. Using assessment tools gathered from research by Hess et al. (2016) and Martín-Peña et al. (2019), DT was operationalized as the degree to which businesses employ digital technology into their operations and relationships with consumers. Considering sensory, cognitive, and emotional components, customer experience (CX) (Verhoef et al., 2009) was gauged by consumer impressions of their contacts with the company. Using both financial and non-financial indicators including market share, growth, and customer happiness, the company's performance became operationalized (Venkatraman & Ramanujam, 1986). One mitigating element explaining the company's ability to absorb and adapt new technology for corporate improvements was IT creativity.

Data Analysis

Programs such as SPSS and Smart PLS helped to evaluate data therefore enabling both descriptive and inferential statistical analysis. First used to profile the sample and spot data trends were descriptive statistics. Using scores from Cronbach's alpha and composite reliability, one could evaluate the validity and dependability of the structures; a suitable internal consistency criteria of 0.7 was devised. Structural equation modeling (SEM) let the proposed linkages between DT, CX, business performance, and IT innovation be explored concurrently, so evaluating several dependent relationships at once. Since SEM offers insight of both direct and indirect effects and can manage complex models incorporating mediating and moderating variables, it was chosen.

With reference to IT innovation, firm performance, customer experience (CX), and digital transformation (DT), the created research model for this study looks at the interactions among small and medium-sized enterprises (SMEs). The Technology-Organization-Environment (TOE) paradigm (Tornatzky & Fleischer, 1990) posits that DT is the main independent variable directly affecting customer experience as well as firm performance. Including CX as a mediating variable helps the approach to reflect the special contribution user-centered digital strategies perform in enhancing customer satisfaction, engagement, and loyalty—factors that lead to better company performance. CX's mediating role highlights the prospect that DT by itself might not ensure performance improvements—that is, these advantages are realized when DT improves the complete customer experience.

One mitigating element included into the link between DT and corporate performance is IT innovation. Higher degrees of IT innovation suggest that businesses fit better to implement digital initiatives to improve consumer connections, boost flexibility, and simplify operations. The model investigates whether a high degree of IT innovation enhances the impact of DT on business performance in view of the idea that creative IT practices enable more successful digital transformation and increased responsiveness to market needs.

Four antecedent elements—technological development, cost reduction, competitive pressure, and environmental influence—that affect DT adoption also find place in the research model. Although cost minimization shows the financial drive for digital solutions, technology advancement is meant to help DT to be accepted. Furthermore, noted to be favorably influencing the adoption of DT are environmental aspects including market and regulatory conditions as well as competitive pressure, which motivates SMEs to innovate and simplify operations to remain competitive.

This approach is to give total awareness of how performance of SMEs is influenced by digital transformation, IT innovation, and user interface. By means of both direct and indirect consequences, one can grasp the strategic alignment of digital and creative capacities for corporate performance in dynamic business contexts.

Summary of Methodological Choices

To guarantee complete and trustworthy insights on the effects of digital transformation (DT), customer experience (CX), and IT innovation on firm performance within small- and medium-sized enterprises (SMEs), the research methodology applied in this study includes a methodical approach to sampling and measurement in order to The study employs sampling techniques as well as construct measuring approaches appropriate for the particular research model and hypotheses to accomplish this. The research process was conducted in two steps: preliminary and formal. Preliminary qualitative research was conducted using group discussion techniques to adjust the scale and form the survey questions.

Formal quantitative research was conducted using a non-probability convenience sampling technique, and paper questionnaires were sent directly to enterprises. The target respondents were the staff and managers of SMEs in Vietnam. In this study, the authors used a 5-point Likert scale to measure the participants' responses. In the survey questionnaire, "1" implies "strongly disagree," and "5" implies "strongly agree." SPSS and Smart PLS software were used to analyze the data.

RESULTS

The findings of the study assist to better understand how digital transformation (DT), customer experience (CX), and IT innovation affect the performance of small and medium-sized firms (SMEs) in Vietnam. The data analysis conducted using SPSS and Partial Least Squares Structural Equation Modeling (PLS-SEM) using Smart PLS software employed both descriptive and inferential statistical approaches to assess the relationships inside the advised research model and verify the hypotheses.

Descriptive Statistics

The data collected from the questionnaires completed by the participants in this study was input into SPSS to produce statistical results. Descriptive statistical techniques, such as frequencies and percentages, were employed to illustrate the demographic characteristics of the participants.

The participants were described according to five variables: sex, age, working department, working position, and academic level. The sample results illustrate the distribution of participants according to their demographic characteristics.

First, in terms of sex, males accounted for 76.8%, approximately three times more than females (23.2%).

Next, the results show the participants' ages: 30.2% of the participants were less than 25 years old, 46.7% were between 25 and 40 years old, and 23.1% were over 40 years old.

Regarding working departments, onsite construction comprised the largest share, accounting for 44.1% of the participants surveyed. The office sector accounted for 35.5% and the technical sector represented 20.4% of the total sample.

The sample shows that most study participants (81.8%) were employees or staff, and 18.2% were leaders, vices/heads of departments, or equivalents.

The sample showed that 9.5% of the participants held a higher degree (postgraduate, such as a master's or Ph.D. degree), 60.2% held a bachelor's degree (graduate), and 30.3% were educated at college level. Therefore, most participants in this study were well -educated, enabling the assessment of the impact of DT on firm performance.

ANALYSIS RESULTS OF MEASUREMENT MODEL

Convergence Validity

First, we evaluated the reliability of the scale through two typical indicators: Cronbach's alpha index (CA) and composite reliability (CR). Next, the convergence value of the latent variables was assessed based on the external loading coefficient and the average variance extracted (AVE) (Hair Jr et al., 2017).

From Table 1, it can be seen that the Cronbach's alpha coefficient (CA) of the variables have good reliability, with CA $\in$ 0.7÷ 0.95 (Nunnally & Bernstein, 1994). Specifically, cost minimization (CM), competitive pressure (CP), customer experience (CX), digital transformation (DT), environmental influence (EI), firm performance (FP), and IT innovation (IT) had Cronbach's alpha of 0.880, 0.834, 0.888, 0.788, 0.896, 0.896, and 0.911, respectively.

Table 1. Analysis results CA, CR, and AVE

	Cronbach's Alpha (CA)	Composite Reliability (CR)	Average Variance Extracted (AVE)
CM	0.880	0.917	0.735
CP	0.834	0.901	0.752
CX	0.888	0.918	0.691
DT	0.788	0.863	0.613
EI	0.896	0.923	0.705
FP	0.896	0.924	0.708
IT	0.911	0.944	0.849

On the contrary, the CR of the research variables was greater than 0.7, confirming the suitability and reliability of the scale. The coefficient of the AVE of the research variables was also greater than 50%. Thus, all the research variables had convergent values.

Table 2 shows that all observed variables converge on their own factors with external load coefficients from 0.702 to 0.927, reaching the acceptable level (> 0.7), and are statistically significant with $p < 0.05$.

Table 2. Analysis results outer loadings (OL)

	Original sample	Sample mean	Standard deviation	T value	P value
CM1 <- CM	0.840	0.839	0.014	58.454	0.000
CM2 <- CM	0.847	0.847	0.012	71.242	0.000
CM3 <- CM	0.864	0.863	0.013	68.984	0.000
CM4 <- CM	0.878	0.878	0.011	77.239	0.000
CP1 <- CP	0.798	0.798	0.021	38.905	0.000
CP2 <- CP	0.908	0.907	0.008	108.838	0.000
CP3 <- CP	0.892	0.892	0.010	90.336	0.000
CX1 <- CX	0.825	0.825	0.018	44.824	0.000
CX2 <- CX	0.861	0.860	0.015	58.883	0.000
CX3 <- CX	0.820	0.820	0.025	33.394	0.000
CX4 <- CX	0.842	0.842	0.012	71.039	0.000
CX5 <- CX	0.808	0.808	0.018	43.766	0.000
DT1 <- DT	0.797	0.797	0.018	45.348	0.000
DT2 <- DT	0.830	0.830	0.014	57.499	0.000
DT3 <- DT	0.796	0.796	0.022	36.968	0.000
DT4 <- DT	0.702	0.701	0.029	24.455	0.000

continued on following page

Table 2. Continued

	Original sample	Sample mean	Standard deviation	T value	P value
EI1 <- EI	0.819	0.818	0.017	46.918	0.000
EI2 <- EI	0.850	0.849	0.014	62.440	0.000
EI3 <- EI	0.828	0.827	0.018	46.112	0.000
EI4 <- EI	0.856	0.856	0.011	75.870	0.000
EI5 <- EI	0.845	0.846	0.013	66.592	0.000
FP1 <- FP	0.870	0.869	0.011	81.591	0.000
FP2 <- FP	0.870	0.870	0.011	82.213	0.000
FP3 <- FP	0.821	0.820	0.017	49.317	0.000
FP4 <- FP	0.838	0.837	0.014	57.933	0.000
FP5 <- FP	0.806	0.805	0.018	45.221	0.000
IT1 <- IT	0.920	0.919	0.006	145.692	0.000
IT2 <- IT	0.927	0.926	0.006	146.342	0.000
IT3 <- IT	0.918	0.918	0.007	137.600	0.000
TA1 <- TA	0.825	0.824	0.017	47.400	0.000
TA2 <- TA	0.882	0.882	0.012	73.188	0.000
TA3 <- TA	0.875	0.875	0.013	66.427	0.000
TA4 <- TA	0.837	0.837	0.016	50.815	0.000

Discriminant Validity

The evaluation of discriminant value is usually based on three criteria: the cross-loading coefficient of the observed variables, the square root of the AVE value of the latent variable, and the Heterotrait-monotrait ratio (HTMT) coefficient (Hair Jr et al., 2017).

With the criterion of the cross-load coefficient, a discriminant value is generated when the system of load coefficients of the observed variable on its research variable is higher than all of its coefficients of cross-loading on other research variables. Compared to the results in Table 3, we observe that they meet the technical requirements. Therefore, we can conclude that all variables had discriminant values.

Table 3. Cross loading factor

	CM	CP	CX	DT	EI	FP	IT	TA
CM1	0.840	0.362	0.416	0.426	0.413	0.439	0.552	0.471
CM2	0.847	0.536	0.558	0.544	0.529	0.550	0.601	0.641
CM3	0.864	0.410	0.357	0.422	0.352	0.468	0.624	0.474
CM4	0.878	0.470	0.415	0.498	0.399	0.526	0.656	0.518
CP1	0.401	0.798	0.535	0.551	0.526	0.480	0.361	0.583
CP2	0.520	0.908	0.653	0.635	0.628	0.551	0.433	0.717
CP3	0.441	0.892	0.572	0.581	0.549	0.533	0.456	0.668
CX1	0.446	0.609	0.825	0.586	0.538	0.530	0.334	0.598
CX2	0.458	0.587	0.861	0.642	0.516	0.577	0.386	0.595
CX3	0.403	0.493	0.820	0.624	0.549	0.536	0.335	0.557
CX4	0.418	0.558	0.842	0.647	0.523	0.603	0.360	0.595
CX5	0.426	0.576	0.808	0.629	0.570	0.566	0.303	0.588
DT1	0.469	0.601	0.635	0.797	0.547	0.576	0.382	0.616
DT2	0.482	0.491	0.597	0.830	0.482	0.603	0.401	0.542
DT3	0.422	0.468	0.605	0.796	0.462	0.557	0.349	0.471
DT4	0.367	0.559	0.515	0.702	0.845	0.462	0.293	0.567
EI1	0.444	0.522	0.542	0.585	0.819	0.491	0.324	0.586
EI2	0.451	0.556	0.588	0.590	0.850	0.468	0.320	0.631
EI3	0.414	0.554	0.523	0.607	0.828	0.496	0.298	0.594
EI4	0.434	0.564	0.559	0.645	0.856	0.538	0.394	0.597
EI5	0.367	0.559	0.515	0.702	0.845	0.462	0.293	0.567
FP1	0.496	0.524	0.590	0.623	0.509	0.870	0.376	0.616
FP2	0.508	0.537	0.599	0.623	0.511	0.870	0.388	0.618
FP3	0.489	0.478	0.546	0.591	0.467	0.821	0.430	0.566
FP4	0.495	0.517	0.587	0.577	0.519	0.838	0.383	0.599
FP5	0.467	0.475	0.524	0.542	0.450	0.806	0.440	0.542
IT1	0.615	0.442	0.408	0.421	0.371	0.458	0.920	0.483
IT2	0.669	0.441	0.383	0.397	0.331	0.412	0.927	0.462
IT3	0.683	0.448	0.354	0.441	0.367	0.449	0.918	0.473
TA1	0.474	0.645	0.536	0.574	0.551	0.539	0.425	0.825
TA2	0.537	0.651	0.599	0.616	0.607	0.626	0.447	0.882
TA3	0.558	0.663	0.638	0.628	0.651	0.624	0.448	0.875
TA4	0.555	0.637	0.639	0.591	0.605	0.601	0.435	0.837

The results in Table 4 show that the square root value of the AVE (located on the main diagonal) of each research variable is larger than the correlation value between them and the other research variables in the model. Therefore, the research variables achieved discriminant values.

Table 4. Fornell-Lacker

	CM	CP	CX	DT	EI	FP	IT	TA
CM	0.857							
CP	0.526	0.867						
CX	0.517	0.679	0.831					
DT	0.558	0.680	0.753	0.783				
EI	0.500	0.657	0.648	0.748	0.840			
FP	0.584	0.602	0.677	0.704	0.584	0.841		
IT	0.711	0.482	0.414	0.456	0.388	0.478	0.921	
TA	0.622	0.759	0.706	0.705	0.707	0.700	0.513	0.855

Using the HTMT criterion, as shown in Table 5, the HTMT indices were less than the threshold of 0.9. On the contrary, the HTMT confidence interval in Table 6 does not contain the value 1. As expected, this value was discriminatory when compared with this threshold.

Table 5. Heterotrait-monotrait ratio (HTMT)

	CM	CP	CX	DT	EI	FP	IT	TA
CM								
CP	0.602							
CX	0.576	0.787						
DT	0.661	0.836	0.899					
EI	0.558	0.758	0.729	0.886				
FP	0.651	0.696	0.758	0.836	0.652			
IT	0.793	0.552	0.459	0.537	0.428	0.529		
TA	0.698	0.886	0.799	0.845	0.798	0.788	0.574	

Table 6. Confidence interval HTMT

	Original sample	Sample mean	2.5%	97.5%
CP <-> CM	0.602	0.602	0.504	0.691
CX <-> CM	0.576	0.576	0.488	0.657
CX <-> CP	0.787	0.787	0.720	0.847
DT <-> CM	0.661	0.660	0.572	0.742
DT <-> CP	0.836	0.835	0.770	0.895
DT <-> CX	0.899	0.899	0.852	0.942
EI <-> CM	0.558	0.558	0.463	0.643
EI <-> CP	0.758	0.758	0.670	0.836
EI <-> CX	0.729	0.728	0.651	0.797
EI <-> DT	0.886	0.885	0.843	0.924
FP <-> CM	0.651	0.649	0.579	0.713
FP <-> CP	0.696	0.695	0.619	0.765
FP <-> CX	0.758	0.758	0.697	0.812
FP <-> DT	0.836	0.836	0.780	0.884
FP <-> EI	0.652	0.652	0.576	0.723
IT <-> CM	0.793	0.792	0.740	0.840
IT <-> CP	0.552	0.552	0.467	0.633
IT <-> CX	0.459	0.460	0.373	0.542
IT <-> DT	0.537	0.536	0.450	0.618
IT <-> EI	0.428	0.428	0.338	0.513
IT <-> FP	0.529	0.528	0.444	0.606
TA <-> CM	0.698	0.696	0.626	0.761
TA <-> CP	0.886	0.886	0.831	0.934
TA <-> CX	0.799	0.799	0.731	0.857
TA <-> DT	0.845	0.844	0.789	0.894
TA <-> EI	0.798	0.799	0.723	0.863
TA <-> FP	0.788	0.787	0.731	0.838
TA <-> IT	0.574	0.573	0.499	0.645

All the criteria for evaluating the measurement model were satisfied, providing evidence of the reliability and validity of the CM, CP, CX, DT, EI, FP, IT, and TA scales in the model.

STRUCTURAL MODEL ANALYSIS RESULTS

Test for Collinearity Through the VIF

One measure of collinearity is the variance inflation factor (VIF). In the context of PLS-SEM, a VIF value ≥ 5 indicates significant collinearity (Hair Jr et al., 2017).

Table 7 shows that the VIF coefficients of the research variables are less than 4, which indicates that the multicollinearity between the research variables in the model does not affect the testing of the research hypotheses.

Table 7. Inner VIF values

	CM	CP	CX	DT	EI	FP	IT	TA
CM				1.662				
CP				2.550				
CX						2.413		
DT			1.000			2.454		
EI				2.161				
FP								
IT						1.291		
TA				3.259				

Path Coefficient

Path coefficients—which indicate the degree and importance of interactions between variables—were looked at to probe the structural model. With a significant and positive path coefficient of 0.753, digital transformation (DT) had the largest effect on customer experience (CX). With a path coefficient of 0.392, DT also clearly revealed a positive influence on firm performance (FP), therefore confirming beliefs that DT enhances organizational and customer-centric results (Table 8).

Table 8. Path coefficient

	Original sample	Sample mean	Standard deviation	T value	P value
CM -> DT	0.134	0.135	0.039	3.458	0.001
CP -> DT	0.202	0.203	0.038	5.263	0.000
CX -> FP	0.334	0.337	0.056	6.006	0.000

continued on following page

Table 8. Continued

	Original sample	Sample mean	Standard deviation	T value	P value
DT -> CX	0.753	0.753	0.023	33.391	0.000
DT -> FP	0.392	0.392	0.053	7.351	0.000
EI -> DT	0.436	0.436	0.047	9.211	0.000
IT -> FP	0.163	0.160	0.041	4.004	0.000
TA -> DT	0.160	0.159	0.049	3.270	0.001

Furthermore, showing a positive association with a path coefficient of 0.334 (Table 10) were customer experience (CX) and firm performance (FP), meaning that enhancing CX helps businesses to flourish generally. The interaction between DT and IT innovation (IT) in effecting FP was also validated with a path coefficient of 0.065 (Table 10), thereby validating the idea that IT innovation may improve the link between DT and corporate success.

Table 10. The results of testing the hypotheses in the model

Hypotheses	Relationship	Path coefficients (â)	T value	P value	Hypotheses results
H_1	CP -> DT	0.202	5.263	0.000	Accepted
H_2	TA -> DT	0.160	3.270	0.001	Accepted
H_3	CM -> DT	0.134	3.458	0.001	Accepted
H_4	EI -> DT	0.436	9.211	0.000	Accepted
H_5	DT -> CX -> FP	0.252	5.742	0.000	Accepted
H_{5a}	DT -> CX	0.753	33.391	0.000	Accepted
H_{5b}	CX -> FP	0.334	6.006	0.000	Accepted
H_6	DT -> FP	0.392	7.351	0.000	Accepted
H_7	IT x DT -> FP	0.065	2.579	0.010	Accepted

Other studies showing a connection between performance results and DT and CX (Masoud & Basahel, 2023; Braojos et al., 2019) find these route coefficients fit Strong correlations between these factors suggest that expenses in DT and IT innovation not only raise consumer pleasure but also enhance company efficiency and competitive positioning in addition.

Coefficient of Determination (R^2)

The R^2 values represent the proportion of variance in dependent variables the model's independent variables can account for. With R2 for digital transformation (DT) of 0.649, elements defining cost reduction, technological advancement, competitive pressure, and environmental effect accounted for 64.9% of its variation. DT explains 56.7% of the variance in customer experience (CX) by means of an R2 value of 0.567. Drawing on the R2 value for firm performance (FP), which came out to be 0.572, At last, DT and CX blame 57.2% of the variation in firm performance. These numbers highlight the model's outstanding explaining ability as well as the crucial part DT plays in deciding organizational and user results (Table 9 and Figure 1).

Figure 1. Research model results

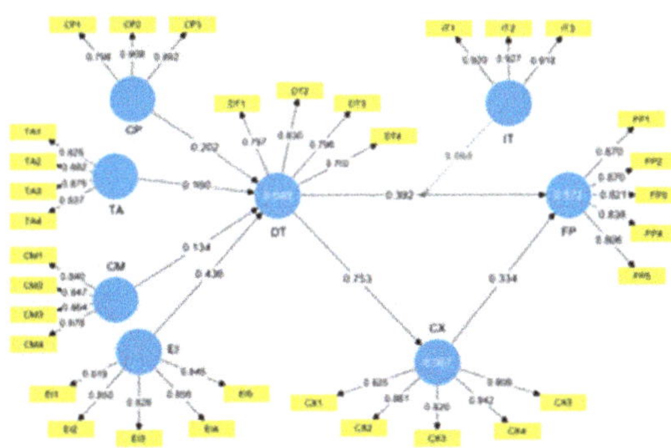

Table 9. Coefficient of determination R square adjusted (R^2Adj)

	R Square Adjusted (R^2_{Adj})
CX	0.567
DT	0.649
FP	0.572

Findings of Model Hypothesis Testing

The path coefficient value for the latent variable was used to evaluate the strength, weakness, and direction of the impact of each independent variable on the dependent variable. Statistical significance was set at a 5% level (p-values ≤ 0.05) using the bootstrapping procedure (Hair Jr et al., 2017).

Based on the path coefficient Beta – (β, normalized regression coefficient) to evaluate the importance of the independent variables in the influence on the dependent variable. The larger the beta coefficient, the stronger is the influence of the corresponding independent variable on the dependent variable. From Figure 1, it can be concluded that all hypotheses were accepted. The path coefficient from IT × DT to FP has the smallest value ($\beta = 0.065$; $p = 0.010$), and the path coefficient from DT to CX has the largest value ($\beta = 0.753$; $p = 0.000$).

Coefficient Effect f^2

The F-square is the change in the R-square when an exogenous variable is removed from the model. According to Cohen (1988), with f-square < 0.02: no effect; 0.02 ≤ f-square < 0.15: small effect; 0.15 ≤ f-square < 0.35: medium effect; and f-square ≥ 0.35: large effect.

Table 11 shows that the effect value of DT on CX was 1.311, indicating a very high effect. Similarly, EI to DT is 0.252, indicating a moderate effect. Next, DT to FP was 0.147, CX to FP was 0.109, IT to FP was 0.048, and CP to DT was 0.046, indicating a small effect. All effect coefficients were statistically significant at $p < 0.05$.

Table 11. Coefficient effect f^2

	Original sample	Sample mean	T statistic	P value
CM -> DT	0.031	0.034	1.632	0.103
CP -> DT	0.046	0.048	2.483	0.013
CX -> FP	0.109	0.114	2.937	0.003
DT -> CX	1.311	1.330	7.098	0.000
DT -> FP	0.147	0.151	3.275	0.001
EI -> DT	0.252	0.255	4.316	0.000
IT -> FP	0.048	0.050	1.966	0.049
TA -> DT	0.023	0.025	1.524	0.128

Therefore, two cases must be considered. The effect coefficient from CM to DT was 0.031 (p=0.103) and that from TA to DT was 0.023 (p=0.128). However, Table 10 shows that the path coefficients from CM and TA to DT are both positive and statistically significant ($p < 0.05$). Therefore, CM and TA have very small effects on the change in the R^2 value of the endogenous variable DT in the model. In summary, exogenous variables can explain endogenous variables at a very small to very large level.

Predictive Relevance Q^2

In addition to evaluating the R^2 values, researchers often measure the Q^2 value (Stone-Geisser test). This indicates the out-of-sample predictive power of the model. It was developed to evaluate the predictive value of exogenous latent variables and can be computed using a blindfolding procedure.

A $Q^2 > 0$ for an endogenous latent variable indicates the predictive relevance of the path model for the research variable. In contrast, $Q^2 \leq 0$ indicates a lack of predictive relevance (Geisser, 1974; Stone, 1974).

Table 12 shows that the Q^2 values of the three endogenous variables were all significantly higher than 0. Specifically, FP had the highest Q^2 value (0.402), followed by CX (0.388) and DT (0.384). The results show that all the Q^2 values are considerably above zero, thus supporting the model's predictive relevance regarding the endogenous latent variables.

Table 12. Q^2 value of endogenous variables

	SSO	SSE	Q^2 (=1-SSE/SSO)
CM	2724.000	2724.000	0.000
CP	2043.000	2043.000	0.000
CX	3405.000	2084.311	0.388
DT	2724.000	1677.242	0.384
EI	3405.000	3405.000	0.000
FP	3405.000	2035.589	0.402
IT	2043.000	2043.000	0.000
TA	2724.000	2724.000	0.000

Model Fit

The model fit was assessed with reference to the standardized root mean square residual (SRMR) value, 0.059. With this below the 0.08 level, Hair et al., 2017 show a reasonable model fit. Good alignment of the research model with the actual data increases its durability and reliability of the outcomes (Table 13).

Table 13. Model fit

	Saturated model	Estimated model
SRMR	0.059	0.076

Mediation Analysis

A mediation effect was produced when a third or research variable interfered with two related research variables. The role of the mediator variable was to clarify or explain the relationship between the two main research variables. The indirect effects involve a series of relationships with at least one relevant intervention research variable (Hair Jr et al., 2017).

> **H_5:** Customer experience positively mediates the relationship between DT and firm performance.

With t > 1.96 and p-value < 0.05, Table 14 will show that there is an indirect effect on the relationship between the independent and dependent variables. The variance accounted for (VAF) and confidence interval (CI) are two criteria that must be considered (Hair Jr et al., 2017). The first is the AVF index. A VAF value <20% indicates a very small mediating effect, which is considered as no mediating effect. When 20% < VAF < 80%, a partially mediated effect is observed. A VAF > 80% indicates a completely mediated effect. Second, CI was used. The model has a mediating effect if the CI (Total indirect effects) does not contain the value 0. In this study, CX mediated the relationship between DT and FP. Table 14 shows that the indirect effect coefficient was 0.252, reaching statistical significance at t = 5.742 and p=0.000. The calculated VAF value in the presence of the mediator variable is 0.392 (39.2%). In addition, the CI was in the range of 0.168–0.339 and did not contain the value 0. Therefore, by comparing the two criteria mentioned above, Hypothesis H_5 is accepted. This again confirms the support of Hypothesis H_5 that the mediator variable is statistically significant (as shown in Table 10). More specifically, firm performance partially mediates this relationship regarding the influence of DT on CX.

Table 14. The effect of mediator variable

Independent variable	Mediator variable	Dependent variable	Total effects	Indirect effects	VAF	CI	Hypotheses results
DT	CX	FP	0.643 (t=20.122 p=0.000)	0.252 (t=5.742 p=0.000)	39.2%	0.168 ÷ 0.339	Accepted

Moderation Analysis

The impact of IT innovation on the DT-firm performance connection was investigated in the moderation study More IT innovation increases the beneficial moderating influence of DT on corporate performance, according to the results ($\beta = 0.065$, $t = 2.579$, $p = 0.010$)(Table 15). Although it does not define DT's success by itself, the small but important impact size (f2 = 0.015) shows that IT innovation improves the degree to which DT yields higher performance (Table 16). This reinforces the corpus of research underlining how IT aids digital projects (Nambisan et al., 2019; Scuotto et al., 2017).

Table 15. The results of testing the moderated variable role in the model

Hypotheses	Relationship	Path coefficients (â)	T value	P value	Hypotheses results
H_7	IT x DT -> FP	0.065	2.579	0.010	Accepted

Table 16. Coefficient effect f^2 of moderator variable

	Original sample	Sample mean	T statistic	P value
IT x DT -> FP	0.015	0.017	1.248	0.212

At last, with IT innovation acting as a main mediator, the findings of this study offer significant proof of DT's favorable impact on firm success and customer experience. These results highlight the need of IT innovation in maximizing the advantages of DT top priority as well as the requirement of providing digital initiatives enhancing customer experience top priority. Strong statistical evidence validates the notion by showing that DT determines organizational success in Vietnamese SMEs predominantly.

DISCUSSION

The findings of the study offer important light on how customer experience (CX), IT innovation, and digital transformation (DT) influence operations of small and medium-sized enterprises (SMEs) in Vietnam. The results show that, with IT innovation acting as a moderating factor, DT is rather essential for improving company performance and customer experience. Emphasizing the implications for companies looking for DT for operational efficiency and competitive advantage, this talk will examine these outcomes in perspective of the corpus of present research. Simultaneously, the study examines the direct influence of DT on firm performance and its effect on CX.

The analysis indicates a strong and significant positive impact of DT on CX. This shows that DT enhances CX, which is consistent with the findings of previous studies (Immonen & Sintonen, 2015; Masoud & Basahel, 2023; Sasmoko, Mihardjo, Alamsjah, Elidjen, & Tarofder, 2019; Saunila et al., 2019). Successful DT not only improves CX but also enhances customer satisfaction, which fosters customer loyalty. Furthermore, it promotes overall organizational innovativeness, leading to increased productivity and reduced costs (Ferreira et al., 2019). Consequently, this leads to competition for products and services, increased customer retention rates, a larger market share, and improved profitability, ultimately improving firm performance.

Moreover, the analysis showed that CX has a significantly positive effect on firm performance. This finding aligns with those of the previous studies (Grønholdt et al., 2015; Masoud & Basahel, 2023; Mbama & Ezepue, 2018; Sharples, 2019). Enhanced CX leads to positive outcomes, such as favorable word-of-mouth and increased purchase intentions, customer satisfaction, and loyalty, leading to improved performance, such as retention rate, profitability, and market share (C. G. Lee & How, 2019).

Furthermore, a significant positive relationship exists when examining the impact of DT on firm performance. In other words, successful implementation of DT will lead to enhanced firm performance for SMEs. This outcome aligns with most studies exploring the effects of DT and technology adoption (Braojos et al., 2019; Y.-Y. K. Chen, Jaw, & Wu, 2016; Masoud & Basahel, 2023; Scott et al., 2017).

The results also indicate that IT innovation significantly and positively affects the relationship between DT and firm performance, indicating that effective IT innovativeness can enhance firm performance. DT involves the adoption of technology, and IT innovation is any innovation facilitated by technology (Nambisan, Lyytinen, Majchrzak, & Song, 2017). This result is consistent with the studies showing that technology adoption and IT infrastructure positively affect firm performance (Bharati & Chaudhury, 2019; Masoud & Basahel, 2023; Nambisan et al., 2019; Scuotto et al., 2017).

Chae, Koh, and Prybutok (2014) found that IT capabilities do not affect firm performance in terms of profitability or cost. They justified this by the fact that IT capability has lost importance because of reduced IT costs, outsourcing, and the availability of ready-made packages. However, this assertion is not true in the context of DT because it is a strategic process that is more than just implementing a technology.

Furthermore, the results demonstrate the significant positive impact of IT innovation on firm performance. Consequently, enhancing IT innovation improves firm performance, which aligns with the findings of previous research (Coad et al., 2016; Datta & Roumani, 2015; Masoud & Basahel, 2023; Turulja & Bajgoric, 2019).

Practical Implications for Vietnamese SMEs

These results underscore the importance of a thorough strategy for DT integrating IT creativity and customer experience for Vietnamese SMEs. SMEs can increase performance and market placement by including DT as a technical endeavor as well as a tool for motivating customer involvement and satisfaction. The study also urges SMEs to think about funding IT infrastructure and training to let their employees make maximum use of digital tools. Given the limited resources many SMEs have, this can entail implementing small improvements concentrated on key areas like enhancing online user interfaces and adopting cloud-based technologies for increased operational efficiency.

Theoretical Contributions

This study shows the complex interactions among DT, CX, IT innovation, and economic performance inside the framework of Vietnamese SMEs, therefore augmenting the corpus of knowledge. This study stresses the requirement of IT innovation and customer-centric design (CX) as basic components of successful digital strategy even while DT is often addressed independently. It advances present concepts by demonstrating, mediated by CX and controlled by IT innovation, that DT improves performance via both direct and indirect paths. This combined approach offers a more whole awareness of how DT could be optimized in different organizational environments—especially in developing economies.

Research Contributions

This effort generates several pragmatics as well as theoretical breakthroughs. By underlining how, especially in SMEs, IT innovation and customer experience may boost company performance, theoretically it adds to the body of research on

DT. This paper emphasizes the need of customer-centered digital strategy since it reveals that CX moderates the link between DT and company performance. The moderating influence of IT innovation highlights even more the necessity of companies to acquire IT competencies providing continuous adaptation and improvement of digital operations.

Practically speaking, the results provide SMEs with useful information, particularly in developing nations like Vietnam where resource limits call for strategic prioritizing. Seeing DT as a diverse project needing investment in IT innovation and customer experience in order to achieve performance advantages will help Vietnamese SMEs as well as Legislators and industry actors can use this data to enhance DT systems and provide SMEs focused support including digital infrastructure, technology adoption, and training using digital tools.

CONCLUSIONS

The findings of the research underscore the need of DT, CX, and IT innovation in enhancing corporate performance in Vietnamese SMEs. By proving that DT affects both CX and firm performance, therefore serving as a moderator and CX aids IT innovation to be understood. This gives a complete knowledge of the ways in which DT supports corporate success. These results highlight that DT's efficacy combines an emphasis on customer experience and flexibility through IT innovation, therefore transcending the acceptance of new technologies and in line with past studies.

This paper proposes that Vietnamese SMEs may become stronger and more competitive by investing in DT in a balanced manner and emphasizing on IT and CX innovation. Should Vietnamese SMEs rank these top priorities as the digital economy grows, they will be more fitted to control market fluctuations, satisfy consumer wants, and achieve sustainable development.

LIMITATIONS AND FUTURE RESEARCH

Future research should fix the shortcomings in this one. First, the sample included only of Vietnamese SMEs, therefore limiting the range of application for the results to other industries or bigger corporations. Wider relevance of DT, CX, and IT innovation will be established by future studies on how they affect company success across different sectors and organizational sizes.

Second, the cross-sectional character of the research restricts the capacity to prove causality. One could better understand how DT, CX, and IT innovation interact over time to affect company success by means of longitudinal study. Monitoring these components over time will enable researchers to see how corporate performance evolves as companies extend their DT projects and adjust to new digital tools.

At last, applying a limited set of criteria to assess corporate performance, this study mostly focused on operational and financial measures. More complex performance indicators including results of innovation, staff happiness, and market share expansion should be added in next studies. Moreover, considering other possible moderators and mediators—including organizational culture or leadership style—may enable one to better grasp the elements enabling effective digital transformation in various corporate settings. These recommendations would enable the DT literature to flourish and give legislators and professionals more sensitive understanding.

REFERENCES

Azam, M. S., & Quaddus, M. (2012). Adoption of e-commerce by the SMEs in Bangladesh: The effects of innovation characteristics and perceived risk. **Proceedings of the 15th International Business Research Conference**, 1–15.

Appio, F. P., Frattini, F., Petruzzelli, A. M., & Neirotti, P. (2021). Digital transformation and innovation management: A synthesis of existing research and an agenda for future studies. *. *Journal of Product Innovation Management, 38*(1), 4–20. DOI: 10.1111/jpim.12526

Bharati, P., & Chaudhury, A. (2019). SMEs and competitiveness: The role of information systems. *. *International Journal of Management and Enterprise Development, 1*(1), 1–19.

Braojos, J., Benitez, J., & Llorens, J. (2019). How do social commerce-IT capabilities influence firm performance? Theory and empirical evidence. *. *Information & Management, 56*(2), 155–171. DOI: 10.1016/j.im.2018.04.006

Cao, Q., & Zhang, F. (2011). The impact of digital transformation on business performance: Evidence from financial firms. *. *Management Information Systems Quarterly, 35*(4), 809–822. DOI: 10.2307/23043418

Datta, A., & Roumani, Y. (2015). Knowledge sharing in virtual communities: A study of citizenship behavior and social learning. *. *The Journal of Strategic Information Systems, 24*(1), 57–74.

Dahlman, C. (2018). *Technology, globalization, and international competitiveness: Challenges for developing countries*. United Nations Publications.

Goerzig, D., & Bauernhansl, T. (2018). Industry 4.0: How it changes the society and the economy. In *Digital Economy* (pp. 89–102). Springer.

Hair, J. F., Hult, G. T. M., Ringle, C., & Sarstedt, M. (2017). *A Primer on Partial Least Squares Structural Equation Modeling (PLS-SEM)* (2nd ed.). Sage Publications.

Hess, T., Matt, C., Benlian, A., & Wiesböck, F. (2016). Options for formulating a digital transformation strategy. *. *MIS Quarterly Executive, 15*(2), 123–139.

Immonen, M., & Sintonen, S. (2015). Customer experience and digital transformation: A framework for understanding the role of technology in customer interaction. *. *International Journal of Information Management, 35*(2), 215–224.

Kaperonis, S. (2024). How artificial intelligence (AI) is transforming the user experience in digital marketing. In *The use of artificial intelligence in digital marketing: Competitive strategies and tactics* (pp. 117–141). IGI Global.

Kotler, P., & Keller, K. L. (2016). *Marketing Management* (15th ed.). Pearson Education Limited.

Lemon, K. N., & Verhoef, P. C. (2016). Understanding customer experience throughout the customer journey. *. *Journal of Marketing*, *80*(6), 69–96. DOI: 10.1509/jm.15.0420

Liere-Netheler, K., Vogelsang, K., Packmohr, S., & Hoppe, U. (2018). A framework for customer-centric digital transformation. *. *Journal of Business Research*, *85*, 164–173. DOI: 10.1016/j.jbusres.2017.12.038

Martínez-Caro, E., Cegarra-Navarro, J. G., & Alfonso-Ruiz, F. J. (2020). Digital transformation and customer experience in SMEs: A conceptual framework. *. *Management Decision*, *58*(8), 1602–1624. DOI: 10.1108/MD-10-2018-1160

Nambisan, S., Wright, M., & Feldman, M. (2019). The digital transformation of innovation and entrepreneurship: Progress, challenges and key themes. *. *Research Policy*, *48*(8), 103773. DOI: 10.1016/j.respol.2019.03.018

Nunnally, J. C., & Bernstein, I. H. (1994). *Psychometric Theory* (3rd ed.). McGraw-Hill.

Porter, M. E., & Kramer, M. R. (2006). Strategy and society: The link between competitive advantage and corporate social responsibility. *. *Harvard Business Review*, *84*(12), 78–92. PMID: 17183795

Scott, S. V., Van Reenen, J., & Zachariadis, M. (2017). The long-term economic impact of digital transformation. *. *Economic Journal (London)*, *127*(603), 1203–1236. DOI: 10.1111/ecoj.12459

Tornatzky, L. G., & Fleischer, M. (1990). *The Processes of Technological Innovation*. Lexington Books.

Turulja, L., & Bajgoric, N. (2019). Information technology, knowledge management and their role in the innovation process: A review and research framework. *. *Journal of Knowledge Management*, *23*(3), 548–563.

Verhoef, P. C., Lemon, K. N., Parasuraman, A., Roggeveen, A., Tsiros, M., & Schlesinger, L. A. (2009). Customer experience creation: Determinants, dynamics and management strategies. *. *Journal of Retailing*, *85*(1), 31–41.

Westerman, G., Bonnet, D., & McAfee, A. (2014). *Leading Digital: Turning Technology into Business Transformation*. Harvard Business Review Press.

Zhai, L., Yang, L., & Chan, H. (2022). The role of digital transformation in customer experience and business performance. *. *Journal of Business Research*, *141*, 569–582. DOI: 10.1016/j.jbusres.2022.06.019

Chapter 8
Digital Fashion and Social Identity

Thomai Voulgari
https://orcid.org/0009-0003-0852-9013
Panteion University of Social and Political Sciences, Greece

Ilia Evangelia Sybil Sdralli
Panteion University, Greece

Yannis Skarpelos
https://orcid.org/0000-0002-8460-1676
Panteion University of Social and Political Sciences, Greece

ABSTRACT

The global fashion system changes decisively due to the impact of the growing technological innovation and social changes. Fashion is a method of self-expression that gives people the opportunity to be their true selves.The purpose of this study is to present a new perspective about the combination of fashion and gaming. The connection between the two seemingly different "worlds" is more relevant than ever considering that over three billion people worldwide are actively participate in-game experience. Further research seems necessary to explore how virtual fashion is used for new forms of self-identity. To develop a research model for approaching this aim, the study employs four theoretical approaches. So, this systematic literature review investigates the evolution of avatar fashion, the innovative ways that virtual environment used for forms of social identity, the collaboration between fashion-gaming industries and the Metaverse as a digital universe of avatars.

DOI: 10.4018/979-8-3693-8729-0.ch008

INTRODUCTION

The Emergence of Virtual Fashion

The global fashion system changes decisively due to the impact of the growing technological innovation and social changes. Fashion is a method of self-expression that gives people the opportunity to be their true selves. According to Bertola & Teunissen (2018), fashion industry is a dynamic field that enters a new era of interaction through the digital environment. Social identity is influenced by intergroup relationships. More specifically, social identity theory gives an explanation for the in-group bias, and also offers "a broader statement of how relationships between real-world groups relate to social identity" (Tajfel and Turner, 1979).

Nowadays users display aspects of their personal identity by representing themselves in social media communities and gaming through dressing up their digital avatars (Brunskill, 2013). The outfit in the virtual environment acts as a tool for creating a new self and fostering relationships between players. Users communicate and interact through the visual language of fashion (Lee & Park, 2011). Gamers spend significant amount of time customizing their characters in order to express their preferences and creativity (Karagöz, 2024). On the other hand, Takano & Taka (2022) claimed that communication between avatars in the game and emotional reactions reflect the lack of contact and social isolation in real-life.

Fashion brands are gradually expanding and joining Metaverse to gain popularity, reach new users and promote clothing which are often limited edition and worn exclusively in the digital world. Kaiser and Green (2022) asserted that through the way of dressing, individuals express visually what is often difficult to express in words. Fashion is inspired by social, artistic and cultural movements (Aswal, 2023). Kolia (2023) has found that young people tend to adopt any outrageous trend. It influences individual behavior especially Gen Z-ers who build their identity and perceive fashion culture through digital landscape. For instance, Tik Tok, Instagram and Snapchat trends are spreading at an unprecedented rate. It could be argued at this point that this has partly contributed to the democratization of fashion and the popularity of pop culture in the digital age (Patient, 2023).

However, while digital platforms encourage creative autonomy, they also blur the line between self-expression and social performance, placing users in a constant feedback loop of visibility and validation(Gonzales & Hancock, 2011).According to their conclusions, selective self-presentation in digital media, can lead to intensified relationship formation, and greatly a influence the users' impressions of the self.

The purpose of this study is to present a new perspective about the combination of fashion and gaming. The connection between the two seemingly different "worlds" is more relevant than ever considering that over three billion people worldwide are actively participate in-game experience. According to Statista, 3.22 billion gamers were recorded worldwide in 2023, a number that is expected to reach 3.32 billion by the end of 2024 (Statista, 2024). It also aims to investigate the effect of fashion and how clothing and colors combinations' chosen by the players contribute to the representation of their digital identity. As players use their profiles to portray a real or an ideal personality, outfit color combinations play a crucial role as it is a valuable medium of communication and one of the ways in which we define ourselves (Stanton, 2023). Koma (2023) argued that digital media facilitate the construction and representation of our digital selves.

The current study also touches on how virtual identity is not only curated but often multi-fragmented and strategically constructed, echoing Turkle's (1995) concept of the "multiple self" in online environments. Such multiplicity—though empowering and self expressive—can set ethical questions on authenticity and social comparison, especially among younger users navigating selfhood in hyper-visual virtual spaces.

Further research seems necessary to explore how virtual fashion is used for new forms of self-identity. To develop a research model for approaching this aim, the study employs four theoretical approaches. By grounding its analysis in theoretical frameworks such as social identity theory, postmodern identity (Bauman, 2000), and self-presentation theory (Goffman, 1959), this systematic review investigates the evolution of avatar fashion, the innovative ways that virtual environment used for forms of social identity, the collaboration between fashion-gaming industries and the Metaverse as a digital universe of avatars. The transition from social media aesthetics and user expression to immersive environments like the Metaverse marks a key thematic shift in this paper, demonstrating how identity is negotiated across layered and increasingly commercial digital terrains.

Virtual fashion operates as a bridge between the tangible and intangible. Most digital platforms like Roblox, Fortnite, and Animal Crossing have catalyzed this shift, providing interactive arenas where users craft unique identities through customizable skins and accessories. Cross brand collaborations, such as Balenciaga's partnership with Fortnite (Figure 1), Nike within the Roblox ecosystem with Nikeland (Figure 2), as well as Louis Vuitton's collaborate with Riot Games "League of Legends" (Figure 3), further illustrate the expanding blending of luxury fashion with digital innovation, signifying a broader cultural shift towards valuing digital aesthetics as a way of expressing themselves.

Figure 1. Balenciaga X Fortnite (Fortnite)

Figure 2. Roblox X Nikeland (Youtube)

Figure 3. Louis Vuitton X League of Legends

Note: As part of the League of Legends collaboration between K-pop band K/DA and luxury clothing brand Louis Vuitton.

Virtual fashion's accessibility democratizes style by offering individuals from diverse backgrounds opportunities to engage with fashion previously limited by geographic or financial barriers. Platforms like The Fabricant and DressX collaborate with high-end brands to make designer fashion available in the digital realm, often at lower costs than their physical counterparts (Lahti, 2023). While this evolution broadens fashion's reach, it also raises ethical concerns regarding cultural appropriation, as users may adopt styles without understanding their cultural significance (Anderson, 2020). This duality reflects virtual fashion's potential to both foster cultural exchange and risk superficial representation. Victoria McDermott in Digital Fashion and the New Social Identity (2021), describes virtual fashion as a tool for boundary-pushing self-expression.

More recently, virtual fashion has emerged as a compelling intersection between fashion and digital culture, transforming the ways individuals engage with clothing, self-representation, and identity (Mesjar, 2023). Virtual fashion in gaming environments challenges the traditional, physical constraints of fashion, enabling new forms of self-expression and identity creation. There is a very distinctive role fashion plays in gaming, with multiple implications for identity, self-perception, and cultural representation. Virtual fashion within gaming environments has grown significantly, as digital platforms allow users to purchase, design, and customize virtual clothing. As Balch (2022) explains, virtual fashion allows gamers to express themselves without the material limitations of real-world fashion, making it an accessible platform for experimentation and creativity. This shift aligns with theories on postmodern identity formation, where individuals are increasingly defined by their digital interactions and the curated images they project online (Turkle, 1995). Virtual fashion, therefore, is more than just an accessory within gaming; it has become an integral part of the gaming experience, allowing players to convey their personalities and connect with others through shared aesthetic language.

Aside from that, the evolution of avatar fashion has resulted in digital clothing becoming even more complex and realistic in design. Players are no longer restricted, they can express themselves freely and connect with their digital character as several innovative and personalized options are available (1883magazine.com). Two completely different worlds (fashion and gaming) are connected as a result of technological innovations. Samuel Jordan is a digital fashion designer who claimed that the most popular items in clothing and accessories are not like those that are perfectly designed, but they have the ability to express emotions (Segran, 2022). He creates accessories for avatars on Roblox and found out that clothing reveals significant information about a person's personality based on how they feel. Takano and Taka (2022) argue that the emotional reactions resulting from avatar communication compensate for the lack of meaningful social relationships in real life.

Many fashion brands design their collections not just for the catwalk, but also in order to join exclusively the game community. For example, Gucci company first created the "Sneaker Garage Gucci", digital products without a physical copy. Users have the ability to design their own sneakers by choosing design and colors of each element. By this way, they feel closer to the brand and have a unique experience by co-creating with Gucci (Segran, 2022). However, it raises questions about what digital fashion is and what it could be in the future. Schreder (2023) claims that: "When fashion is digital, more people can participate". In fact, in many cases the cost of virtual outfits, although intangible, is not much different from real luxury clothing. So, is virtual fashion likely to perpetuate the same inequalities and the gradual exclusion of people who cannot afford it? The new world of digital fashion and virtual clothing raises questions about what fashion is and what could it be. Actually, people feel closer to the brand, live a unique experience of sharing ideas and trends with other users and a sense of co-creating with brands. In the digital environment, users create a representation of self through fashion and a unique personal style by gathering information through their personal engagement, religion and cultural beliefs. They make an effort to bring out their ideal self by reflecting what they look like and often the digital person that have always aspired to be. Other researchers such as Mancini and Sibilla (2017) have shown that players may sometimes develop personalities that are totally out of touch with themselves offline.

Fashion brands and influencers are linked to the fashion industry and have particular appeal to Generation Z audiences because they have grown up in a digital world and view avatars as an extension of themselves in an authentic way. For example,

Levi's released clothes through social media platform Snapchat registering a high impact (Cambosa, 2020). There are certain factors that increase and influence the degree of identification of players with their avatar. Identification is enhanced by avatar adaptation with variables that have a relationship and connection with elements common to their real life, such as hairstyle and accessories. Several questions need to be answered. To what extent does dressing in the digital world influence the way we dress in the real world? Clothes are no longer designed to be worn exclusively on the catwalk. They are customizable to avatars and integrated into the game. However, "The Fabricant" is the first digital fashion house in the world creating objects that appear to exist only within the digital environment. A digital dress sold for $9,500 dress on the blockchain. It is sewn by the digital designer, Joanna Jaskowska, and this has the advantage of making it as valuable as cryptocurrency (Fingas, 2019).

FASHION BRANDS AND GAMING

Gaming industries significantly influence virtual fashion trends by establishing digital marketplaces and facilitating collaborations with luxury brands. Partnerships like Fortnite's collaboration with Balenciaga[1] illustrate how gaming platforms serve as virtual runways, enabling users to perform fashion identities in a community setting. However, this commercialization also introduces concerns about the commodification of self-expression, as players may feel social pressure to invest in digital fashion (Smith, 2021). Thorsten Busch and Dimitri Götz argue in Virtual Economies and Digital Fashion (2020) that the integration of fashion into gaming transforms it into symbolic representation, allowing users to construct identities that transcend physical appearance. This evolution underscores the potential for gaming environments to reshape notions of identity and consumption in the digital age.

A number of users buy clothes in video games just to impress friends in the digital environment. Illusion collapses since one cannot see the items in real life. Although it is something intangible, the cost is not far from the value of luxury clothing (Fingas, 2019). However, fashion in video games is not limited to clothing fabrics and textiles.

Avatars are evolving in graphic design, but also in the fashion aspect due to technological advancements. For instance, in "The Sims" series, the evolution of technology is related to fashion evolution in gaming industry[2].

Gucci designs virtual sneakers that can only be worn in the virtual environment using augmented reality. A player allocated $2,400 for a pair of virtual sneakers through the Aglet (aglet.app), while others have allocated high amounts for outfits available only on Instagram. Animal Crossing is developing partnerships with brands such as Net-A-Porter and has created a fashion island. Users dress the avatars in designer clothes from the new season. In 2007, The Sims2 released the ''H&M Stuff Pack'' which contained branded clothing and in this way users were able to dress the Sims as virtual versions of themselves (Anzu, 2023). However, fashion was passing quickly and they had to adapt to new trends. Louis Vuitton (LVMH) has created virtual dressing rooms, digital fashion shows and video game storytelling based on augmented reality. Digital characters **are** increasingly more realistic (Monteros, 2024).

The rapid integration of fashion brands into the virtual realm demonstrates their strategic adaptability and recognition of a burgeoning market. Fashion brands have translated their expertise in physical design into the digital sphere, offering consumers customizable and trend-conscious avatar wardrobes. Fashion collections for the metaverse often feature exclusively digital designs, capitalizing on the unique possibilities of virtual expression. A pivotal example of this intersection is the collaboration between Fortnite and Balenciaga[3], which underscored the synergy between haute couture and gaming. Drawing inspiration from Balenciaga's storied archives, the collaboration introduced the Balenciaga Fit Set, including standout virtual designs such as the Unchained Ramirez Outfit, Shady Doggo Outfit, Fashion Banshee Outfit, and Game Knight Outfit. Accompanying accessories, such as Fit Back Blings, the Speed Sneaker Pickaxe, and the Parasail Purse Glider, allowed players to further personalize their digital personas. According to the gaming platform Iion[4], these offerings provide players with unparalleled opportunities to assert individuality within the virtual world.

Similarly, the luxury giant Louis Vuitton emerged as a trailblazer among luxury houses entering the gaming sector. In 2019, the brand collaborated with League of Legends to create a dual collection of tangible fashion pieces and in-game digital skins. This initiative, crafted in partnership with Riot Games, achieved remarkable success, with the physical collection selling out in the European market within an hour of its release. Meanwhile, platforms like Roblox enable devoted fans to reimagine runway looks for their avatars using innovative design tools, demonstrating the creative potential of these digital ecosystems. In this context, pioneering designers such as Kunihiko Morinaga of Anrealage, are leveraging "cutting-edge" technological advancements to enhance the interactive potential of fashion[5].

The fusion of gaming and fashion extends beyond mere aesthetic appeal, embodying shared values of creativity, self-expression, and personalization. As gamification pioneer Fabio Viola notes, this intersection reflects a shared ethos of imaginative engagement. Psychological studies further affirm this connection, highlighting how both gaming and luxury fashion leverage the "dream factor" that shapes consumer desires. The dedication gamers exhibit in progressing through levels mirrors the emotional journey associated with luxury experiences, where both realms offer a balance of risk, reward, and tangible merit. The coveted "prize"—whether a virtual achievement or a luxury item—symbolizes the alignment of gaming and experiential luxury as avenues for self-expression and value creation. Digital platforms such as Roblox and Discord within the Web 3 landscape enable digital natives to interact directly with their favorite brands, underscoring the significance of virtual communities for contemporary commerce. Fashion brands, initially cautious about entering the gaming domain; have been compelled to respond to its explosive growth, particularly during the pandemic era of 2020–2021, which drove individuals toward online entertainment and social interaction. According to Statista, the global esports audience will exceed 318 million by 2025, signaling the growing importance of gaming as a cultural and economic force (Statista, 2024).

The immersive nature of gaming aligns seamlessly with the expressive possibilities of fashion, paving the way for highly personalized digital experiences. This synergy has catapulted gaming from fashion's periphery to a central platform for innovation, establishing it as a direct communication channel to Gen Z audiences, much as social media was for millennials. The rise of digital fashion and beauty within gaming environments has unlocked new revenue streams, serving as a precursor to the metaverse's potential. Marketing theories, such as gamification, have quickly adapted to harness this phenomenon. Gamification employs game elements—badges, trophies, or interactive challenges—within non-gaming contexts to boost engagement, customer loyalty, and brand recognition. By leveraging these methods, brands extend their physical presence into digital marketplaces, offering immersive shopping solutions. Luxury brands have embraced gamification with great success.

In 2024, several new and established fashion brands are innovating in the gaming space, leveraging virtual platforms to create unique skins and collaborations. Burberry and Fendi are actively exploring virtual commerce through platforms like Unmatereality[6]. They create digital twins of physical collections, providing virtual try-ons and interactive experiences for gamers and shoppers. This approach also supports sustainability by reducing physical samples. Ralph Lauren has continued to grow its presence in gaming, leveraging platforms like EA Sports FC and Fortnite to connect with younger audiences. These campaigns aim to build long-term brand loyalty among gamers who may become customers in the future.

SELF-EXPRESSION AND IDENTITY IN VIRTUAL SPACES

Technology has transformed our daily lives and it has lead to social change. A part of society feels that it expresses itself in a virtual environment in a different way from the physical world. What is more, virtual spaces can be used as influential social media platforms[7]. Could this be a way out of reality and a form of self-fashioning or a chance of representing different possible selves? Markus & Nurius (1986) argued that identity can be variable and adaptable. In the digital environment there is a wide range of forms and styles available equipped with technological platforms. At this point some questions arise. Does fashion in gaming tend to be a step forward in the freedom of clothing? Could it be a passage to freedom of self-expression through clothing?

Firstly, it is imperative to understand the relationship between fashion and society as it is a sociological phenomenon (Kawamura, 2005) as well as whether and how it enables or restricts self-expression and identity. In the book Fashion-ology, clothes are described as "more than just a piece of fabric" (Kawamura, 2005). Fashion is created and evolved within the socio-economic and political context. Roland Barthes (1967) in the book "The Fashion System'' used an in-depth semiotic analysis through the media's discourse. Barthes (1967:45) showed that: ''We have seen that a fashion utterance involves at least two systems of information: a specifically linguistic system, which is a language and a 'vestimentary' system, according to which the garment signifies the world or fashion". Kaiser & Green (2022) in the book "Fashion & Cultural Studies", claim that fashion choices represent our personal identity and inner world. And this can be interpreted as we tend to express ourselves more with visual and non-verbal communication. Entwistle (2001) noted that "Dress cannot be understood without reference to the body … and the body has always and everywhere been dressed". Kawamura (2005) also emphasized the association of fashion and dress. Rath et al. (1994), affirm that fashion is the style that is popular at a particular point in time. According to Williams (2021), clothing choices are an integral part of our identity, both online and in the physical environment. Luxury has always been an iconic tool through which we can enhance ourselves. For Bugg (2010), it is impossible to define fashion. For this reason, an interdisciplinary approach is crucial for a more holistic understanding of what fashion can be. Another theory suggest that fashion is considered as a way of hiding or revealing something, known as "camouflage", while for others is a way of highlighting multiple perspectives and aspects of the self (Efremov, Kertakova, & Kuzmanovska, 2021).

The act of gaming is deeply ingrained in human nature, offering excitement, adventure, and the triumph of victory—a phenomenon supported by scientific inquiry. Zichermann and Linder authors of "The Gamification Revolution", accurately articulate this dynamic, stating, "games are the only force in the known universe that can get people to take actions against their self-interest, predictably, without applying force" (Zichermann & Linder, 2013). And yet, games alone aren't the whole experience; the process of avatar customization is inherently key, a prelude that signals the start of the gaming journey. In contemporary gaming culture, in-game apparel—commonly referred to as "skins"—has evolved into a potent medium for self-expression, akin to the role of physical clothing in the material world.

The question arises and it needs to be asked is whether we dress our avatars considering them as our second selves. As Schwartz (2008) noted in the book "Virtual Worlds Research: Consumer Behavior in Virtual Worlds", the world "avatar" derives from the Sanskrit word "avatara," which means "incarnation''. In the context of on-line virtual worlds, the term denotes a metaphorical "incarnation", a graphical object corresponding to the user's virtual body in the world. Neal Stephenson made this use of the term "avatar" popular in his novel Snow Crash (Stephenson, 1992). Considering this, the choices that players make in dressing their avatars', serve as reflections of personal identity and aspirations, bridging the tangible and virtual realms into intersecting dualities.

Since the avatar is a proxy of a person in the virtual world – an "entire self-representation" (Yee & Bailenson, 2007) – designing the avatar appearance reflects how an individual projects his or her self-concept or self-image. (Messinger, et al., 2008). It is why the use of fashion, as a communicative tool, assumes heightened significance in digital environments-no matter the environment, it enables individuals to navigate, explore and express their identities. This burgeoning phenomenon underscores the integration of fashion and gaming, marking the rise of sophisticated virtual fashion trends within gaming communities. As players construct digital personas, the process becomes a meaningful exercise in self-representation or "self-enhancement", a fundamental human tendency to "propel the ego upward" (Koffka, 1935). Research supports this notion; a report by Roblox on Generation Z highlights that 47% of respondents view avatars as vital forms of self-expression, while 43% associate avatar customization with enhanced self-esteem (Roblox 2023, Digital Expression, Fashion & Beauty Trends Report). Furthermore, 94% of users have personalized their avatars, with 53% of Gen Z gamers dedicating significant effort to crafting unique digital identities (Roblox 2023, Digital Expression, Fashion & Beauty Trends Report). Notably, older members of this demographic, particularly females, exhibit heightened engagement in this form of customization, signaling the emergence of a distinct virtual fashion runway.

As mentioned above, clothing can function as a means of expressing identity and through the process of observation conclusions can be drawn regarding the identity of the individual, related to factors such as ethnicity, religion, age, education and social class (Akdemir, N. 2018). It reveals elements of cultural and social identity. According to Manuel Castells' theory of network society (2011), a sociological perspective suggests that all identities are constructed. He also claimed that: "The construction of identities uses building blocks from history, geography, biology, productive and reproductive institutions, collective memory and personal fantasies, power mechanisms and religious revelations" (Castell, 2010:7).

The rise of virtual worlds has transformed the concept of self, with fashion serving as a vital medium for identity construction. Research indicates that players often use virtual avatars to explore idealized or experimental identities, distinct from their physical selves (Vasalou & Joinson, 2009). These digital environments act as safe spaces for self-exploration, fostering fluid and dynamic identity construction through virtual clothing and customization. Vesa et al. (2021) highlight how virtual fashion empowers users to express marginalized or subcultural identities in ways not always possible offline, democratizing fashion while enabling diverse representations. Pioneering theorist Sherry Turkle emphasizes in "Life on the Screen" that virtual spaces facilitate the exploration of fragmented, decentralized identities, offering users the freedom to embody multiple selves (Turkle, 1997). This fluidity challenges traditional notions of identity by reframing it as an adaptable construct shaped by digital experiences. Similarly, Lisa Nakamura explores how fashion choices in gaming reflect broader narratives of self-identity and belonging, enabling players to experiment with cultural expression and gender (Nakamura, 2002).

FASHION GAMES-"DRESS TO IMPRESS"

In some cases, it has been observed that video game developers create games exclusively in relation to fashion. Examples include the following: Tommy Hilfiger's FashionVerse for android mobiles, Magic Tavern's Project Makeover, CrowdStar's Covet Fashion, or Super Stylist Fashion Makeover, accordingly Lady Popular designed by XS software). Some popular dress-up and fashion games are: fashion dreamer (Nintendo switch), love Nikki on mobile, style savvy, life makeover, shining Nikki, time princess, fashion designer World Tour and fashion designer New York.

Roblox's gaming platform is highly popular among Generation Z, and is a leader in the video game space with more than 40 million daily active users; it is the main enabler of fashion in games. "Dress To Impress" is the most popular fashion game in Roblox and globally with almost 2.4 billion visits (Seymore, 2024). She describes it as "a pop culture phenomenon".

Based on a random theme, players are asked to create an outfit and have five minutes to navigate a dressing room full of customizable clothing, accessories, make-up and hair options. The digital avatars then walk down a catwalk and other players rate each other from one to five stars. The user with the most stars wins (Figure 4).

Figure 4. Selection of the winners (Roblox / Eurogamer)

Recently the game has build a better presence on social media networks, TikTok and X, encouraging gamers to quickly boost their content, interact with other players and reach to a wider audience.

Main characteristics of the game:

- One of the main characteristics is that it has a simple and easy to understand structure.
- It uses a one to five-star rating system.
- There is a competitive environment among players of different skill levels (either beginners, intermediate or advanced).
- The five-minute time limit causes stress and pressure but keeps players alert.
- The game's UI is difficult to navigate. This too causes panic given the time constraint.
- It is inspired by Bratz fashion dolls[8]. This characteristic differentiates the game from competition by focusing on the needs of Gen Z grew up with similar beauty standards. Most dress-up games use the platform's Blocky default characters (Kent,2024).
- A wide range of different topics are available based on a mix of fashion trends that connect the past and the present such as: Halloween (Figure 5), Academia (Figure 6), Y2K fashion style with bright colors and patterns, pop-culture, grunge, coquette, Regency Era (Figure 7), Steampunk (Figure 8), Retro Glamour, Elegant, Gothic Romance (Figure 9&10), Vampire (Figure 11&12), Rock star (Figure 13), Softie coquette (Figure 14), 1950's (Figure 15), Art Teacher (Figure 16), Heavy Metal, Rock and Roll (Figure 17), Glam

goth (Figure 18), A Day in Paris, Folklore, Witch, Pirate (Figure 19), Ghost/ Horror (Figure 20), Celebrity, Barbie (Koepp, et al. 2024).

Figure 5. Halloween (Dress To Impress (@_Dress2Impress) / X)

Figure 6. Academia (Official X Page/ Dress To Impress (@_Dress2Impress) / X)

Figure 7. Regency Era (IGN/ https://www.ign.com/wikis/dress-to-impress/Dress_to_Impress_Outfit_Ideas_and_Themes_(DTI))

Figure 8. Steampunk (IGN/ https://www.ign.com/wikis/dress-to-impress/Dress_to_Impress_Outfit_Ideas_and_Themes_(DTI))

Figure 9. Gothic romance (Pinterest)

Figure 10. Gothic romance (Pinterest)

Figure 11. Vampire (Pinterest)

Figure 12. Vampire (Pinterest)

Figure 13. Rock star (Official TikTok Page)

Figure 14. Softie coquette (Official TikTok Page)

Figure 15. 1950's (Official TikTok Page)

Figure 16. Art teacher (Gals that game)

Figure 17. Rock N' Roll (Pinterest)

Figure 18. Glam goth (Pinterest)

Figure 19. Pirate (Pinterest)

Figure 20. Ghost/Horror (Pinterest)

- Another characteristic is the use of pink and black color in various shades from soft to deep.
- It has design elements of classic video games.
- Every time players can use up to 18 items for their avatar. This allows more creative freedom than the classic dress-up games.

"Dress To Impress" has a variety of preferences and the game is played in real time with thousands of players. It features many different quirky outfits that many users are hesitant to wear in real life. As mentioned above, players are given a random theme to create a complete outfit having five minutes to navigate a virtual dressing room and customize accessories, makeup, hair. Then they walk down a catwalk and other users rate from one to five stars. When a player becomes a VIP member then the clothing options increase significantly and the player gets a private room.

CHARLI XCX, started a collaboration with the game and this has increased its popularity (Adhikary, 2024) . It even seems to have become entrenched in the cultural spirit of Gen Z. Is it just an entertaining fashion game or is it now emerging as a cultural phenomenon through the game and its promotion in the MSM? Will its popularity establish itself over time or will it fade away like, fashion-famous game? Given the number of fashion-themed games, none have achieved the same level of popularity in the past and therefore deserve further investigation. Interpret the reasons that contributed to the game's upward trajectory. It is extremely popular with high-profile influencers, streamers and youtubers developing a partnership in the game.

"Layering" is a viable option in the game. In themes about ghosts; zombies; vampires, there is the illusion of blood. Some poses are extreme as players do butt stunts and glide across the stage. Pose "28" contains the comic element. There are YouTube channels specifically dedicated to covering the game, as well as Instagram and TikTok pages.

Recently, the game announced the introduction of models to fit each body type. Plus size bodies will be available in 2025 (Figure 21).

Figure 21. Size Body type to "Dress to Impress" (Instagram/ dtiroblox)

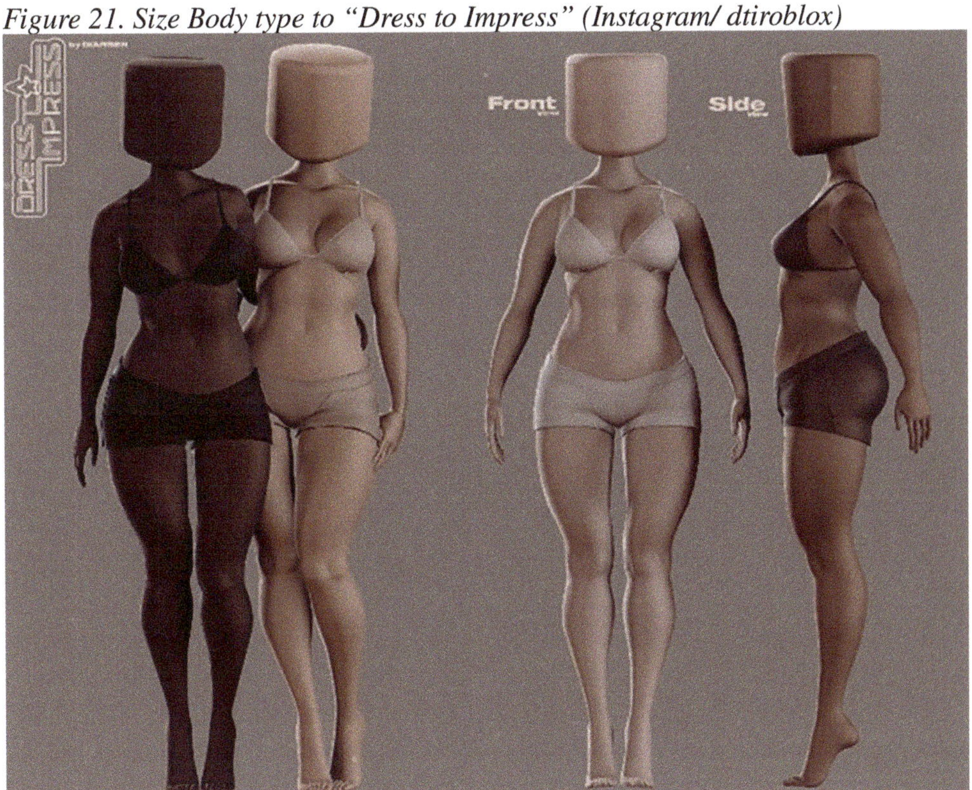

A future research that can extend or complement the current study may be focus to the following questions:

- How does "Dress to impress" affect the self-identity of users and which is the role of pink and black color?
- Do players dress up just for the game or they express their personality and creativity?

THE METAVERSE: A NEW REALM OF INTERACTION AND SELF-DISCOVERY

According to Hwang & Chien (2022), the transition to the virtual world, also referred to as the meta-universe, constitutes the upcoming generation of social connectivity. The American technological research Gartner predicted that, 25% of people are expected to spend at least one hour a day in the Metaverse by 2026 (Gartner, 2022). Numerous applications can be used in the virtual environment in the the Metaverse, such as the image-based social media networks (TikTok, Pinterest, Snapchat and Instagram), however, the most common is the use of digital games (Park & Kim, 2022).

Neal Stephenson, Snow Crash (1992) addressed the Metaverse as "a collective virtual shared space, created by the convergence of virtually enhanced physical reality and physically persistent virtual reality, including the sum of all virtual worlds, augmented reality, and the internet". Although this definition is often adapted, Stephenson's novel introduced the term "Metaverse" as a digital universe where users exist as avatars. The concept of the Metaverse represents an immersive digital reality that integrates social media, online gaming, augmented reality (AR), virtual reality (VR), and blockchain-based technologies, creating shared virtual spaces for interaction (Folger, 2022; Bobier, 2022). This rapidly expanding domain is anticipated to profoundly impact daily life, with Gartner (2022) predicting that by 2026, 25% of individuals will dedicate at least one hour per day to the Metaverse.

Metaverse-based games, a pivotal facet of this digital ecosystem, provide highly immersive experiences, fostering a global community of users (Shin, 2022). The gaming industry is instrumental in building the infrastructure for the Metaverse, enhancing its accessibility and appeal. As Meta CEO Mark Zuckerberg articulated in his 2021 Founder's Letter, the Metaverse holds unparalleled potential for self-expression: "In the Metaverse, you'll be able to do almost anything you can imagine... including expressing yourself in entirely new forms, building new experiences, and creating environments that feel real."

Or paraphrasing William Gibson inspired by his exploration of virtual realities in Neuromancer: “The distinction between reality and virtual reality is going to blur. In the Metaverse, people are not escaping reality—they are redefining it”. Within this digital frontier, self-expression transcends conventional boundaries. The Metaverse invites individuals to explore and redefine their identities, transforming play into a profound exploration of the self. This convergence of technology, fashion, and gaming signals the dawn of a new cultural paradigm, where virtual and physical realities intertwine to reshape human interaction and creativity.

VIRTUAL FASHION AND THE AUGMENTED SELF

Over 150 years ago, Hans Christian Andersen introduced a concept remarkably prescient of contemporary trends. In his tale The Emperor's New Clothes, two weavers deceitfully claim to craft an extraordinary garment visible only to those deemed competent and worthy of their roles. This narrative, rooted in allegory, subtly anticipates the emergence of a burgeoning phenomenon—clothing that exists beyond the physical realm. The concept of the augmented self, which exists across physical and digital domains, exemplifies the broader cultural shift driven by virtual fashion. Drawing on Goffman’s (1959) theories of self-presentation, this augmented self navigates both online and offline spaces, with virtual fashion bridging these realms. Ward (2023) describes virtual fashion as embodying a “digitally augmented identity,” where digital and physical selves interact symbiotically. Jean Baudrillard’s reflections in Simulacra and Simulation highlight the postmodern fragmentation of identity in digital spaces, where the self becomes an assemblage of signs and symbols detached from a single referential reality. In this context, T.L. Taylor asserts in Play Between Worlds that gaming environments offer unparalleled freedom for identity experimentation, enabling users to embody aspirational or fantastical selves. Indeed, according to Elizabeth Wissinger writer of This Year's Model: Fashion, Media, and the Making of Glamour: “In digital spaces, clothing transforms into a mode of affective labor, where its value lies not in materiality but in the ability to signify status, creativity, and personal branding.” As such, in a networked, multi-faceted society, personal identity is constructed through the interplay of real and virtual personas. Since virtual clothing, as part of this interchangeable dynamic, it radically allows for identity to be performed across contexts with a fluidity that is unprecedented in the real world.

WEB3: REDEFINING THE DIGITAL FRONTIER

Web3, often heralded as the future of the internet, is already reshaping the way individuals work, interact, and engage in leisure activities online. Although, The Web3 space needs time to develop further, so social and entertainment games provide opportunities for the growth of digital fashion, where creators exhibit their collections with a new community approach through digital events (Forbes, 2022).

Coined by Gavin Wood, co-founder of Ethereum, the term Web3 denotes a decentralized internet architecture underpinned by blockchain technology (Kharpal, 2022). Initially envisioned as a digital ecosystem characterized by cryptocurrencies, non-fungible tokens (NFTs), and decentralized autonomous organizations (DAOs), the Web3 paradigm signals a shift in power dynamics, transferring control from centralized corporations to user-driven networks. The vision was bold-according to–Chris Dixon, General Partner at Andreessen Horowit, as he claimed that Web3 will change the nature of digital ownership, enabling people to truly own their identities, data and digital assets across platforms (Dixon, 2024).

To further comprehend Web3, it is instructive to trace the evolution of the internet. The era of Web1, the foundational stage of the web, facilitated static information-sharing through open protocols like HTTP and FTP, with limited interactivity. The advent of Web2 in the early 2000s marked a transformative phase, enabling dynamic content creation and the rise of social media platforms. However, this era also consolidated power in the hands of private corporations such as Facebook, Google, and Amazon, centralizing control over user-generated content. The result is today's highly corporate owned online space and overall narrative. Web3 aspires to dismantle these hierarchies, fostering an ecosystem where users can actively own, shape and even monetize their digital identities. Decentralized networks are being envisioned as the tool to redefining the internet experience, creating new economic opportunities through DAOs (Law, 2021). This paradigm shift promises to transform online interactions, with human agency and creativity serving as its driving forces.

CONCLUSION

In conclusion, virtual fashion in gaming environments represents a paradigm shift in identity exploration, blending cultural exchange, technological innovation, and commercial dynamics. While this evolution democratizes fashion and empowers self-expression, it also reveals complexities around cultural representation and commodification. As digital and physical identities become increasingly intertwined, virtual fashion will remain a pivotal force shaping self-expression in the 21st century. Karagöz (2024) assures that fashion will keep having an important influence on the gaming, as it plays a crucial role allowing users to share their experiences with others while the gambling industry continue to grow rapidly over the next years.

The radically growing popularity of gaming and virtual spaces7 in general reveal this very profound evolution in the ways humans construct, express, and redefine identity through technological means. Central to this discourse is the avatar, a symbolic and functional extension of the self within the digital atmosphere-being also pivotal in the way users view and construct their identities online. As users customize avatars; they travel through a complex interplay of personal aspiration, cultural representation, and social interaction subconsciously aligning with broader theoretical frameworks on self-concept and identity formation. This process emphasizes the multifaceted nature of human self-expression and its enhancement inside the virtual zeitgeist. It is where virtual fashion, emerging as a pivotal medium within gaming and empowers users to experiment with identities that transcend traditional boundaries of gender, culture, and socioeconomic status.

Virtual spaces encourage self-exploration and experimentation that might not normally take place offline. They provide users with opportunities to perform idealized selves or engage with subcultural identities that might remain inaccessible in offline contexts. Sherry Turkle's theories on fragmented and decentralized identities highlight the transformative potential of virtual environments, where individuals embody multiple personas and challenge static notions of selfhood. Even otherwise marginalized groups engage with platforms to express their identities accessing a level of representation that transcends physical-world limitations.

The technological advancements shaping Web3, the Metaverse, and the rise in popularity of decentralized platforms provide a fertile ground for this reimagining of selfhood. Web3, with its emphasis on user agency and ownership, evangelizing the offering of greater control over individuals' digital identities (Cooper, 2024). As such, the Metaverse merges augmented and virtual realities to create immersive environments for interaction and self-discovery, a fresh experimentation on identity online that might radically influence identity perception across mediums. Fashion within these digital ecosystems operates as both a cultural artifact and a tool for self-assertion, enabling users to navigate and narrate their identities in ways that appear to be more dynamic and fluid than usual.

Virtual fashion allows users to self-identify as it reshapes broader cultural narratives. Digital identities are being positioned as integral components of modern selfhood. This very convergence of gaming, virtual fashion, and identity underscores a transformative era of human expression. By transcending material constraints we get access to an infinite canvas of opportunities for exploring the multifaceted nature of selfhood through our avatar selves. And while technology advances, the ways in which individuals construct and communicate their identities will continue to evolve, shaping a future where the boundaries between the real and the virtual blur.

However, the rise of virtual fashion and identity construction within those virtual spaces is not without cultural-some might say ethical- complexities. Today some research has already taken place regarding the commodification of self-expression, the social pressures surrounding digital consumption, and the risks of cultural appropriation. Virtual self-formation is seemly democratic but at the same time 'dictated' by the virtual means itself. And as virtual economies grow more prominent, the balance between fostering genuine creativity and navigating self-commodification and commercialization will become increasingly delicate. Truth is, we can't escape progress, and this fusion of gaming, fashion, and technology signals a broader cultural paradigm shift where digital and physical realities intertwine even further. As gaming companies and fashion brands continue to see this opportunity and collaborate in ever more intricate ways, the line between play, art, and commerce often blurs our effort for genuine self-representation online.

Gamers are a passionate and engaged audience so fashion industries need to find ways to connect with them in an authentic, genuine and innovative way. It is also very important virtual fashion include different body types giving users the choice to self-select and improving body representation. Many Roblox clothes are based on a "Layered Clothing" system (Larsson, 2022) designed to fit every body type- avatar, however this should be integrated into the wider gaming space and always taken into consideration. It could be extremely useful to create models with different body types. Theoretically, the fashion space has expanded and there are 'plus size' or 'petite' models. In fashion games or dress-up games should be created not only models with curves, but also these one with disabilities, plus size and mid-size. Future research could investigate factors that influencing digital ecosystem and limit people's fashion choices leading to feelings of oppression. Is technology finally the tool for freedom self- expression and the democratization of fashion? Undoubtedly, technology has allowed fashion to exist simultaneously in multiple time and space, as well as in the virtual digital world. However, on the opposite side are theories such as Busch's (2018), who considers the word democratic as an excuse for fashion by referring to exclusion, as he argues that there is a lack of inclusion in body types and people with disabilities.

Deeper knowledge is needed to determine the power of fashion in the representation of the self and the effect that its absence can have due to constraints. A case study that examined by many researchers and scholars is the school uniform. For some of the research community but also as Kwik (2022) found out, through school uniforms self-expression is covered and compliance is enforced. Are uniforms and dress codes methods and means of social control? We can claim that this question worthy of further investigation. Indeed, over time they have become more modern, less formal and more comfortable. However, it can be argued that despite the changes, the purpose remains unchanged.

Crewe (2013) explored the impact of digitally mediated communication on fashion consumers, finding that the power of the internet has transformed the world, allowing people to become more active users. However, the uncontrolled monitoring of fashion trends via the internet may contribute to the development of addictive behavior, and potentially renders the subject's relationship with fashion problematic and sometimes dependent when they are unable to control their online engagement time (Rose & Dhandayudham, 2014). At this point, it is worth investigating in the future if the hedonic motivation resulting from the use of gaming applications and possibly the fashion factor may intensify it and leads to impulse buying (Kim & Eastin, 2011).

REFERENCES

Adhikary, I. (2024). Dress to Impress Confirms 'Brat' Collab with Charli XCX. *Bee Bom.* Available at: https://beebom.com/dress-to-impress-charli-xcx-brat-collab-teased/

Akdemir, N. (2018). Visible Expression of Social Identity: The Clothing and Fashion. *Gaziantep University Journal of Social Sciences, 17*(4), 1389–1397. DOI: 10.21547/jss.411181

Anderson, M. (2020). Digital Fashion and the Ethics of Cultural Appropriation in Virtual Worlds. *Journal of Visual Culture, 3*(1), 25–38.

Anzu blog (2023). From Runway To Gameplay: The Rise of Fashion in Gaming. 17/01/2023. Available at: https://www.anzu.io/blog/fashion-brands-in-game-advertising

Aswal, P. (2023). The Impact Of Pop Culture On Fashion & Aesthetics. *Indian Institute of Art and Design.* Available at:https://www.iiad.edu.in/the-circle/the-impact-of-pop-culture-on-fashion-and-aesthetics/

Balch, K. (2022). Gaming and Fashion: Virtual Realities and the Democratization of Style. *Game Culture Review, 8*(2), 112–129.

Barthes, R. (1967). *Système de la mode*. Seuil.

Bertola, P., & Teunissen, J. (2018). Fashion 4.0. Innovating Fashion Industry through Digital Transformation. *Research Journal of Textile and Apparel, 22*, 352–369. DOI: 10.1108/RJTA-03-2018-0023

Brunskill, D. (2013). Introducing the Social Avatar. *The British Journal of Psychiatry, 203*, 386.

Bugg, J. (2010). The shifting focus: culture, fashion and identity. *24th International Costume Congress: Fashion, Culture and Identity*, (pp. 23-30). Yonsei University, Seoul, Korea.

Busch, T., & Götz, D. (2020). Virtual Economies and Digital Fashion. *Journal of Digital Culture Studies, 5*(3), 210–225.

Cambosa, T. (2020). Avatars' wardrobe on Bitmoji, Snapchat gets upgrade with new Levi's tie-up. Available at: https://marketech-apac.com/avatars-wardrobe-on-bitmoji-snapchat-gets-upgrade-with-new-levis-tie-up/

Castells, M. (2011a). Introduction to the Workshop: The Promise of Network theory. *International Journal of Communication, 2011*(5), 794–795.

Castells, M. (2011b). Introduction in International Seminar on Network Theory: Network Multidimensionality in the Digital Age. The Annenberg Networks Network (ANN), University of Southern California. YouTube video, https://www.youtube.com/watch?v=vDh7eGJwhwg

Cooper, C. (2024). Understanding Web3:The Decentralised Revolution of the Internet. *Medium.* Available at: https://medium.com/@colin-cooper/understanding-web3-the-decentralised-revolution-of-the-internet-2c867be44415

Dixon, C. (2024). *Read write own: building the next era of the Internet.* Random House.

Efremov, J., Kertakova, M., & Kuzmanovska, V. D. (2021). Expression of Personality Through Dressing. *Tekstilna Industrija*, *69*(1), 28–35.

Fingas, J. (2019). A digital 'dress' sold for $9,500. *Engagement.* Available at: https://www.engadget.com/2019-05-27-fabricant-blockchain-digital-dress.html?guccounter=1&guce_referrer=aHR0cHM6Ly93d3cuZ29vZ2xlLmNvbS8&guce_referrer_sig=AQAAAMfAOtfKtFW0tRw8QZFHD2n-uCj_FfP9d3sXDxSpPpQF80HY21MQDp_vABlFcrxw2KeJw0EToxjx7V5Db8YT_wyvDCN9yhha3SyFGuqtrod2CSWSdF72JQoz_SnMnXxP1QvGAYnt6VD9wb2xxMklNMBta6hax_AWaJwsM_kPLZrb

Goffman, E. (1959). *The Presentation of Self in Everyday Life.* Anchor Books.

Gonzales, A. L., & Hancock, J. T. (2011). Mirror, Mirror on My Facebook Wall: Effects of Exposure to Facebook on Self-Esteem. *Cyberpsychology, Behavior, and Social Networking*, *14*, 79–83. DOI: 10.1089/cyber.2009.0411

Hwang, G.-J., & Chien, S.-Y. (2022). Definition, roles, and potential research issues of the metaverse in education: An artificial intelligence perspective. *Comput. Educ. Artif. Intell.*, *3*, 100082. DOI: 10.1016/j.caeai.2022.100082

Kaiser, S. B., & Green, D. N. (2022). *Fashion and cultural studies.* Bloomsbury.

Karagöz, E. (2024). Fashion in the Gaming Industry. *iion.io.* Available at: https://www.iion.io/blog/fashion-in-the-gaming-industry

Kawamura, Y. (2005). *Fashion-ology: An Introduction to Fashion Studies.* Berg Publishers.

Kent, E. (2024). Viral sensation Dress to Impress has a secret ingredient: fabulous video game jank. *Eurogamer.* Available at: https://www.eurogamer.net/viral-sensation-dress-to-impress-has-a-secret-ingredient-fabulous-video-game-jank

Kharpal, A. (2022). What is 'Web3'? Here's the vision for the future of the internet from the man who coined the phrase. *Cnbc.* Available at: https://www.cnbc.com/2022/04/20/what-is-web3-gavin-wood-who-invented-the-word-gives-his-vision.html

Kim, S., & Eastin, M. S. (2011). Hedonic Tendencies and the Online Consumer: An Investigation of the Online Shopping Process. *Journal of Internet Commerce, 10*(1).

Koffka, K. (1935). *Principles of Gestalt psychology*. Harcourt, Brace.

Koma, K. (2023). Fashion and Identity in Virtual Spaces. In Gardetti, M. Á., & Larios-Francia, R. P. (Eds.), *Sustainability Challenges in the Fashion Industry. Sustainable Textiles: Production, Processing, Manufacturing & Chemistry*. Springer., DOI: 10.1007/978-981-99-0349-8_12

Kwik, H., & Gondoputranto, O. (2022). Constructing Self Identity through the Lens of Fashion. Folio, 3(2), Retrieved from: https://journal.uc.ac.id/index.php/FOLIO/article/view/3488

Lahti, T. (2023). The Digital Runway: Virtual Fashion, Luxury, and the New Consumer. *Fashion Futures Journal*, *5*(3), 89–104.

Larsson-Book, B. (2022). Layers of Genius Behind Layered Clothing. *corp.roblox.* Available at: https://corp.roblox.com/newsroom/2022/04/layers-genius-behind-layered-clothing

Law, A. W. (2021). The Rise of Decentralized Autonomous Organizations: Opportunities and Challenges. *Stanford Journal of Blockchain Law & Policy.* Retrieved from: https://stanford-jblp.pubpub.org/pub/rise-of-daos https://corp.roblox.com/newsroom/2022/04/layers-genius-behind-layered-clothing

Lee, J. J., & Park, S. M. (2011). *Fashion, Reflection of Culture*. Yekyong.

Mancini, T., & Sibilla, F. (2017). Offline personality and avatar customisation. discrepancy profiles and avatar identification in a sample of mmorpg players. *Computers in Human Behavior*, *69*, 275–283. DOI: 10.1016/j.chb.2016.12.031

Markus, H., & Nurius, P. (1986). Possible selves. *The American Psychologist*, *41*(9), 954–969. DOI: 10.1037/0003-066X.41.9.954

McDermott, V. (2021). Digital Fashion and the New Social Identity. *Journal of Virtual Fashion*, *7*(4), 97–110.

Mesjar, L., Cross, K., Jiang, Y., & Steed, J. (2023). The Intersection of Fashion, Immersive Technology, and Sustainability: A Literature Review. *Sustainability, 15*(4), 3761. DOI: 10.3390/su15043761

Messinger, P. R., Ge, X., Stroulia, E., Lyons, K., Smirnov, K., & Bone, M. (2008). On the Relationship between My Avatar and Myself. *Journal of Virtual Worlds Research, 1*(2), 1–17.

Monteros, M. (2024). Why fashion brands are inching into the gaming arena. *Modernretail.co.* Available at: https://www.modernretail.co/technology/why-fashion-brands-are-inching-into-the-gaming-arena/

Nakamura, L. (2002). *Cybertypes: Race, Ethnicity, and Identity on the Internet.* Routledge.

Park, S.-M., & Kim, Y.-G. (2022). A metaverse: Taxonomy, components, applications, and open challenges. *IEEE Access : Practical Innovations, Open Solutions, 10*, 4209–4251. DOI: 10.1109/access.2021.3140175

Patient, G. (2023). The Influence of Pop Culture on Fashion Trends. *Merchmates.co.uk.* Available at: https://www.merchmates.co.uk/blog/culture/influence-of-pop-culture-on-fashion/

Rath, M. P., Peterson, J., Greensley, P., & Gill, P. (1994). *Introduction to fashion merchandising*. Delmar Pub.

Schreder, S. (2023). Why the metaverse needs fashion to feel human. *Culture3.com.* Available at: https://www.culture3.com/posts/why-the-metaverse-needs-digital-fashion-to-feel-human

Schwartz, G. (2008). Virtual Worlds Research: Global X Local Agendas. In *Virtual Worlds Research: Consumer Behavior in Virtual Worlds,* 1(2), ISSN: 1941-8477.

Segran, E. (2022). Karlie Kloss wants to dress your avatar. *Fast Company.* Available at: https://www.fastcompany.com/90768159/karlie-kloss-wants-to-dress-your-avatar

Seymore, C. (2024). Dress to Impress offers new era of gaming, fashion fusion. *Thepostathens.com.* Available at: https://www.thepostathens.com/article/2024/09/dress-to-impress-roblox#google_vignette

Smith, R. (2021). Commodifying Self-Expression: The Economics of Digital Fashion in Gaming Environments. *Journal of Digital Economies, 14*(4), 400–418.

Stanton, A. (2023). How Does The Color Of Your Clothing Affect Your Mood? *The good trade*. Available at: https://www.thegoodtrade.com/features/clothing-color-psychology/

Statista (2024). Distribution of video game users in the United States in 2023, by gender. Available at: https://www.statista.com/forecasts/494867/distribution-of-gamers-by-gender-usa

Stephenson, N. (1992). *Snow crash*. Bantam Books.

Tajfel, H., & Turner, J. C. (1979). An integrative theory of intergroup conflict. In Austin, W. G., & Worchel, S. (Eds.), *The social psychology of intergroup relations* (pp. 33–37). Brooks/Cole.

Takano, M., & Taka, F. (2022). Fancy avatar identification and behaviors in the virtual world: Preceding avatar customization and succeeding communication. *Computers in Human Behavior Reports*, *6*, 100176.

Turkle, S. (1995). *Life on the Screen: Identity in the Age of the Internet*. Simon & Schuster.

Turkle, S. (1997). Computational Technologies and Images of the Self. *Social Research*, *64*(3), 1093–1111. https://www.jstor.org/stable/40971200

Vasalou, A., & Joinson, A. N. (2009). Me, Myself and I: Avatars and the Idealized Self in Virtual Worlds. *Psychology of Popular Media Culture*, *18*(3), 183–200.

Vesa, M.. (2021). Virtual Selfhood and Digital Identity Construction in Online Gaming. *New Media & Society*, *23*(6), 1045–1063.

Ward, C. (2023). Augmented Identities: Exploring the Intersection of Fashion and Selfhood in Digital Spaces. *Journal of Digital Studies*, *10*(1), 52–75.

Yee, N., & Bailenson, J. (2007). The Proteus Effect: The effect of transformed self-representation on behavior. *Human Communication Research*, *33*(3), 271–290. DOI: 10.1111/j.1468-2958.2007.00299.x

Zichermann, G., & Linder, J. (2013). *The Gamification Revolution: How Leaders Leverage Game Mechanics to Crush the Competition* (1st ed.). McGraw-Hill.

ENDNOTES

1 https://www.fortnite.com/news/high-digital-fashion-drops-into-fortnite-with-balenciaga?lang=en-US

2 https://www.edarabia.com/fashion-video-games-why-they-match/

3 https://www.epicgames.com/site/en-US/news/balenciaga-brings-high-fashion-to-fortnite

4 https://www.iion.io/

5 https://fashiontechnews.zozo.com/en/features/features000/kunihiko_morinaga?page=8

6 https://www.forbes.com/sites/katiebaron/2020/05/26/meet-ada-the-new-fashion-game-smashing-down-the-door-to-democratized-luxe/

7 https://seoma.com/the-evolution-of-virtual-worlds-from-games-to-social-ecosystems/

8 https://www.bratz.com/

Chapter 9
Visual Politics at the Time of Platform Society

Christian Ruggiero
https://orcid.org/0000-0001-9642-6257
Sapienza University of Rome, Italy

Mauro Bomba
https://orcid.org/0009-0003-1204-9290
Sapienza University of Rome, Italy

ABSTRACT

Visual Politics has its roots in the history of political communication, and it takes on particular relevance in the contemporary media scenario, together with the assumption of centrality of social media. The chapter begins with a brief history of the role of image in electoral campaigns. The contribution of Visual Politics properly so-called is presented through the evolution of the theoretical framework and exemplary research conducted over time, first in the television field then in the social media field. The social media environment is then identified as an exemplary place where Visual Politics can express its potential in the hybridization of research methods belonging to the sociology of communication and computer sciences. Some first evidence on the contribution of Computer Vision to the study of the 2024 European Elections, provided by two joint research programs in which the authors are involved, are presented.

INTRODUCTION

The role of visuals in analysing political communication has become increasingly important. However, it has come to be associated with two major moments of accel-

DOI: 10.4018/979-8-3693-8729-0.ch009

eration in the impact of communication technologies on society. Both moments are related to a key concept in political communication studies: disintermediation, which refers to the process by which technological advancements significantly reduce the distance between politics and citizens, thereby altering the delicate balance between politics, citizens, and the media system. These are highly ambiguous processes that redraw the roles of the actors involved, leading, according to some interpretations, to an impoverishment of the very concept of democracy; according to others, to its enlargement. Furthermore, these are processes that, depending on one's point of view, strongly emphasise or depress the role of the media system in the "triangle" described above. Slowly but surely, the concept of performance associated with the media is changing (McQuail, 1992), becoming less and less related to their role as mediators between a complex system such as politics and citizens, which is the basis of journalism and, more generally, of the contribution that the media make to the democratic system, and increasingly related to the feeding of a system based on economic exploitation.

The first historical moment to which we are alluding is the birth and rapid assumption of an almost hegemonic role in the media system by television. Seeing leaders on screen, not in static, solemn poses as in posters or newspaper photos, but amid a performative gesture, whether responding to an interview or participating in some form of entertainment, certainly had a powerful impact on citizens. Politicians suddenly seemed more human, which on the one hand gave voters an additional yardstick with which to evaluate their proposals, but on the other exposed them to the risk of being "too human", mercilessly exposed with their tics, hesitations, and gaffes (Thompson, 1995). Hence, the enthusiastic or apocalyptic interpretations of the anthropological change that would follow the centrality of the image, which was replacing the centrality of the written word.

The second moment in the reconstruction of the role of the visual in politics that we propose here is much more recent. However, it follows the same dynamic: the growth in importance of image- and video-based social media. The rich catalogue of new opportunities and risks brought about by the transition from print to television is repeated in the transition from text messages on Twitter to images on Facebook, Instagram, and then TikTok. Television viewing has been accused of trivialising and flattening debate, forcing interlocutors to adapt to the rules of "fast thinking" (Bourdieu, 1996); thirty years later, politics is grappling with the need to 'reduce' a message to the temporal dimensions and technical affordances of a TikTok post. Conversely, thirty years ago, a politician's appearance on television allowed an infinitely greater number of citizens, beyond his supporters, to learn about his ideas; today, TikTok's logic, which is no longer networked but clustered (Gerbaudo, 2024), revives and amplifies this opportunity.

The first part of this contribution will therefore attempt to reconstruct these two passages in the history of communication and its link with politics, and to identify the elements that remain unchanged in some way - the development of strategies for personalising and popularising the political message, the use of interdisciplinary frameworks to understand the impact of multimodal communication on its recipients, the need for leaders to adapt to languages and formats that are technically difficult to manoeuvre, often the preserve of the generation following their own.

To delve deeper into the second phase of the "visual turn" described above, we will then examine one of the major differences that distinguish it from the television era. Thanks to advances in research technologies and the increasingly interdisciplinary nature of the study of visual politics, the tools of political and social sciences are now complemented by those of computer sciences. To demonstrate the fruitful contribution that an interdisciplinary research framework can make, we will report some initial results of research conducted on visual politics materials published by candidates and parties during the Italian election campaign for the 2024 European elections.

THE ROLE OF IMAGE IN POLITICS

The iconic dimension of political communication becomes increasingly important over time. This reflects a progressive dominance of the image in the evolution of media, particularly in those that characterise election campaigns at different phases of their development. If we consider the tripartition elaborated by Pippa Norris (2000) and re-elaborated by Fritz and Gunda Plasser (2002), a pre-modern phase of political communication corresponds to a multiplicity of reference media: party press, posters, newspaper advertisements, and radio programmes. The iconographic dimension is central in only two of these media, and it finds immediate recognition in terms of political communication, primarily in posters. On the other hand, in a historical period in which the dominant style is identified in the party message, it is logical that its best transposition is achieved through media that, to put it *à la* Lippmann (1922), still require a certain intellectual effort to create the images they intend to convey in the minds of their audience.

The transition to the modern phase of election campaigns marks the rise of television and of a style of communication based primarily on soundbites and the construction of the politician's image. On the one hand, we find one of the first forms of disintermediation of politics, given the possibility of getting the core of one's message across through one's viva voce (Hallin, 1992). On the other hand, we find a crucial shift towards a politics of image, which animated the debate among scholars throughout the second half of the 20th century, aimed at "demonstrating" the ability of the image to privilege slogan over reasoning, dream over project, and marketing over politics. Marshall McLuhan (1964) based the distinction between "hot" and "cold" media on evidences such as the circumstance that during the 1960 election campaign between John F. Kennedy and Richard M. Nixon, voters who *listened to* the debates between the two presidential candidates tended to declare Nixon's victory, while those who *watched* them on television were convinced of Kennedy's superiority. Neil Postman (1985) went further in his critique of the anthropological mutation brought about by the shift from the socio-cultural primacy of the press to that of the audiovisual model, accusing television of reducing all content that passes across its screen to the logic of entertainment, to the point of irreparably compromising its viewers' capacity for critical thought. Along these lines is the Italian political scientist Giovanni Sartori (1997), who attributed to the circumstance that the world was now known first through images, especially television, and then through literacy, the loss of the conceptualisation and abstraction capacities necessary for every good citizen to understand the political message and translate it into conscious voting decisions.

The most important theories on the effects of mediatisation are developed based on this observation. On one side, there are the systemic effects resulting from the media's impact on the functioning of the political system. At the same time, on the other hand, there are the psychosocial effects, stemming from the media's influence on citizens' political attitudes, opinions, and behaviour (Mazzoleni, 1998). Among the former elements, the most relevant is certainly the spectacularization effect, which derives from the effectiveness of a political message modulated by the linguistic and narrative patterns of mass culture, which require attractive, entertaining, and sensationalist content (Nimmo & Combs, 1990). The aforementioned effect of fragmenting political discourse through the use of soundbites is also of considerable interest, which is part of the more general simplification of the political message to make it more attractive, compelling, and brilliant (Franklin, 1994). Prominent among the political effects is the personalisation of politics, which has led to the construction of the leader as a *person*, with their peculiarities, rather than as a representative of a party or ideology (Van Zoonen, 1998). Connected to this is the effect of leaderization, i.e., the personalisation of the political leader (Cavalli, 1992), and the logic of selecting political elites due to their ability to conform to the rules of "politics as a spectacle" (Statera, 1986). The latter effect marks a point of no return for the centrality of politics as ideology and parties as centre stage subjects, as the "mediality" of a leader replaces the cursus honorum that used to be built through a more or less long and difficult commitment to party life (Louw, 2010; Manin, 1995).

The concerns that characterised the modern phase of electoral campaigns have partly receded as we move into the post-modern phase: the dawn of the Internet again ensured a wider range of media available, making the centrality of the television medium less dramatic. At the same time, the style of political communication became more fragmented. The declination of the political message on different platforms and the "hunt" for an increasingly volatile electorate brought with it important questions as much about the control of the political message by parties and candidates as about the ability of the public of citizen-voters to select and manage their sources of political information (Towner & Dulio, 2012). It follows that, as Strömbäck (2008) suggests, the Internet poses the same questions concerning the equilibrium/dependency dynamics of the logic of politics and the logic of the media that scholars have posed concerning the dynamics of mediatisation.

In this context, moreover, television remains relevant from the perspective of a *hybrid media system* (Chadwick, 2013). There is, indeed, a constant interchange between the contents broadcast by legacy media, television *in primis*, and those posted at different levels of professionalism on the web, hence a hybridisation not only of contents and modes of fruition, but of the media logic itself (Altheide & Snow, 1979) underlying these practices. In such a composed media ecosystem, it becomes central to understanding political phenomena to have an unprecedented datum, i.e., "the sum of media available to people at a defined place and time, including the priorities of these media" (Prior, 2007, p. 9). The emphasis on preferences is linked to defining the environment in which the individual-voter operates as a *high-choice media environment*. In this environment, not only are the logics of prime-time comedy or soap operas mixed with those of political discourse, but within this now extremely rich basket, the attitude towards politics will be influenced both quantitatively and qualitatively by preferences for entertainment content.

The picture, as we shall see in this chapter, becomes even more complex with the entry into the field of social media, which features one-to-many modes of use, such as Twitter (now X), Facebook, Instagram, and TikTok, that soon assume a central role in *platform society*. On the one hand, the latter succeed in taking that logic of mass self-communication that Castells (2009) had attributed to the network to its extreme consequences, further breaking down barriers to entry and disproportionately expanding the global audience of the messages published on them. On the other hand, they constitute a new type of actor in the communication ecosystem. An actor that is capable of defining the characteristics of the channel, but without being involved in the construction of the physical infrastructure that makes it transmissible. An actor that influences the characteristics of the message, but without having an editorial responsibility for what is published (van Dijck, Poell, de Waal, 2018). From a political perspective, the consequences are extremely ambiguous. On the one hand, the process whereby communication operates as an organisational process is further strengthened, and as a resource that is not only symbolic for mobilisation, leadership, and collective action framing, which, however, according to Bennett and Segerberg (2012), should have resulted in the affirmation of a logic of connective action. On the other hand, the emergence of platform logic immediately stands as a counterweight to the "public values" that should nourish collective action, to the point of generating what, rather than collective and widespread forms of action, become widespread but ultimately marginal pockets of resistance (Bonini, Treré, 2024).

TAKING VISUAL SERIOUSLY

The call to take seriously the visual approach to communication, and specifically political communication, is significantly transversal to both the era of applying this approach to the flow of images transmitted by television and the period in which this flow has multiplied and fragmented in web and social environments.

Three elements will stand out in the transition between the two "eras" and the comparison between them. First: the structure of the communication system. The television era laid the foundations for the "mainstream" that Marxist critics accused of bringing the flow of meanings and values of a specific culture into American homes (Williams, 1974). The platform era has unified the operating logic of these increasingly widespread information tools. However, it has dismantled the very concept of mainstream in favour of a context of fragmentation and information overload. Second: the technological capacity of the research tools available to scholars. We will see how those who studied political performance on television were troubled by having to invest considerable time and personnel in the manual analysis of visual materials. Their prayers have been answered, as advances in automatic content detection and analysis techniques now make it possible to tackle once inaccessible data volumes. Third: the increasingly decisive interdisciplinarity. Political and social sciences soon had to turn to psychology to understand the implicit meaning and nonverbal dimensions of what appeared on the screen (Bull, 1987; Bull & Mayer, 1993). The challenge of the visual-based social network era is to integrate these approaches with those of computer science and attempt to bring the results back to a common theoretical framework.

The Television Era

Maria Elizabeth Grabe and Erik Page Bucy (2009) recall how, while newspapers have been lauded for promoting civic vitality, building a market for political controversy, and advantaging public discourse, television has been accused of contributing to promoting unreason, building an arena to deal with irrelevant issues, and advantaging the trivialisation of public discourses. However, the transition from a public sphere mobilised by written forms of communication to one whose driving forces follow the logic of the audiovisual represents a significant challenge for scholars. It constitutes an even greater challenge for the protagonists of political communication, whether they are the leaders who must convey a message, the media that must mediate it, or the public audiences that must interpret it. In the words of Doris Graber (1988, p. 174): “The television age demands a reconsideration of our print-age value structure, which routinely prizes abstractions conveyed through words more than the realities and feelings conveyed through pictures.”

Graber’s words send us back to two orders of reflections.

Firstly, there is a need not to ignore the importance of visual communication. Ignoring the visuals of a televised news report means disregarding the strategies employed by both players on the scene, on the one hand, politicians, who try to promote their desired qualities and favoured topics through the visual dimension of their performance, and to reinforce their policy positions. On the other hand, journalists apply camera and editing techniques to place their interviewees in a more or less favourable visual light. Furthermore, the potential impact of politicians’ non-verbal behaviour on influencing audience attitudes should not be underestimated. This is because the visual dimension of their performances adds several informational aspects to the public, evoking emotions and shaping impressions both on the politicians themselves and on the topics they are dealing with.

Secondly, the challenge that visual communication brings. Investigating images “requires methodological resourcefulness and an investment of time and human resources that goes well beyond computerised content analyses of news databases to identify visual patterns and trends” (Grabe & Bucy, 2009, p. 6). The latter statement should be read because, at the end of the first decade of the 2000s, “broad groupings of streamlined coding procedures do not provide the detail or subtlety necessary for isolating visual effects” (Grabe & Bucy, 2009, p. 7). This limitation was overcome a decade later by the new possibilities introduced by the combination of cognitive frameworks and investigative tools from the sociology of communication and computer science.

It is interesting to begin our examination of visual politics by considering how Grabe and Bucy apply the caveats formulated above to their research. The analysis conducted by the two authors on television news encompasses elements of behavioural biology and cognitive neuroscience (Somit & Peterson, 1998). This is an attempt to investigate the effects of campaign news coverage on voters’ opinions, attitudes, and behaviour with greater scientific rigour than that guaranteed by the spurious relationships that can be identified ex post between a certain type of coverage and a certain type of result. The new attention to the visual displays of political leaders lies in this search for greater scientificity of media analysis; it is an element that viewers would be able to process and remember with extreme ease, and which would consequently have a decisive weight in determining the rise in their perception of certain emotions and feelings of approval or disapproval. Hence, the focus is on the non-verbal dimension of televised political communication, including gestures and facial expressions that convey emotions while also revealing the speaker's intentions and the reactions expected from the audience. It is essential to consider both of these verbs, as Grabe and Bucy's research specifically aims to address the need to move beyond focusing on measuring soundbites and instead examine the impact of image bites in U.S. election campaigns between 1992 and 2004.

To study a total of 178 newscasts from the three examined broadcast networks—ABC, CBS, and NBC—the authors employed two units of analysis. Firstly, the campaign story, where presidential and vice-presidential candidates were central to the narrative. Secondly, the individual candidates for president and vice president. To emphasise the importance of the visual dimension of the research, the soundbites of the candidates were separated into three different categories: bites in which candidates were *shown speaking*, bites in which they were *heard but not shown*, and bites in which they were *shown but not heard*. The specific analysis of the image bites also revealed three further categories: the shown-but-not-heard presentation of candidates when they are not engaged in some form of public address, where the viewer therefore does not hear what they say as the primary sound, because the reporter's narration dominates the audio track. Then, the shown-but-not-heard appearances occur when the candidate is shown in a formal talking mode; the specificity of this category lies in the fact that while the reporter's voiceover is audible, the candidate is shown in speech mode. Finally, the "sound-up": that is, a brief bite of a candidate talking, occurring when the reporter's voiceover is paused to make the candidate's informal response audible, after which the reporter's narration continues.

The results were encouraging. In absolute terms, soundbites made up 14.28% of all election coverage, whereas image bites constituted 25.07%. The authors comment: "Even as soundbites continue to shrink, the amount of time candidates are appearing visually in newscasts is increasing in duration, gradually rising each election year. This suggests that viewers of the evening news are receiving more direct visual cues about candidates than verbal information. Because pictures are perceived as first-hand knowledge about political actors and events, including presidential candidates and campaign-related developments, news visuals are undoubtedly delivering more information to viewers than previously acknowledged" (Grabe & Bucy, 2009, p. 76).

The Platform Era

The second "visual turn" in political communication builds upon frameworks and research approaches similar to those of the first, but with more powerful research tools and increased access to data. Lilleker (2019) picks up both on the theoretical framework, which is inspired by the psychological importance of images as a tool for an in-depth understanding of 21st-century politics, and on Grabe and Bucy's frustration at the impossibility of accurately measuring the impact of each image in the everyday lives of individuals who come into contact with them. Thus, he chooses to deconstruct the power of the image from a psychological perspective through two tools: a discussion of processes of socialisation that create normative standards against which the real world is measured, and an assessment of the cognitive processes that occur at the point of exposure to visual political communication.

Since images are invariably part of the socialisation process, Lilleker emphasises how they can be used "to portray ideal roles, behaviours and the norms of a given society, for example, how a world leader should look or what makes a leader appear more authentic, honest, decisive or competent" (Lilleker, 2019, p. 39). At the same time, they can be constructed to create shock, to generate resonance through the breaking of conventions. Built around the fundamental opposition between right and wrong, images enable judgements to be made about what the individual experiences, including political action and the action of the politician. When one associates hope, unity, patriotism, economic stability, and foreign policy strength with a particular party in an election campaign, one constructs a schema, a cognitive categorisation around the candidate in question. The choices underlying the construction and maintenance of a certain image of a candidate are functional to make him appear more "authentic" or "in touch", more accessible or likeable, or even more trustworthy. For instance, concerning a survey of young voters conducted on the occasion of the 2017 British General Elections, "extensive exposure to images of Corbyn with ordinary people, rappers, people from underprivileged and ethnic minority backgrounds built a schema associating him as being caring and in touch" (Lilleker, 2019, p. 40).

Entrusting the construction of a judgment to an image, although it refers to a practice well rooted in the history of human nature, takes on the contours of a democratic emergency. This happens when there is "an overabundance of information, perhaps coupled with low interest in the nature of the task (scrolling through a social media feed) rather than the end goal (keeping entertained on a commute), errors of judgement can occur due to certain pieces of information standing out while others fade to the background" (Lilleker, 2019, p. 41). It is for this purpose that Joo et al. (2014) subjected a series of images of Obama widely circulated during the 2012 presidential election to computer analysis, to check the correct inference in terms of being favourable (i.e., appearing competent, trustworthy or compassionate) or unfavourable (i.e., appearing angry or fearful). At the end of their analysis, they "show that the resulting favourability judgements correlate in informative ways with independent measures of public opinion, proposing the contrasting media behaviours of agenda setting and mirroring" (Joo et al., 2014, p. 7).

Lilleker's conclusion is very firm: "What we see, what resonates with us and what we believe is a matter for the individual within this environment. Images are proven to be memorable and emotionally resonant; hence, they are likely to impact attitudes and behaviours. This chapter outlines the primary reasons why images may have a profound impact in politics, facilitating an understanding of how those with limited interest may seek a simple solution to complex and contested political questions. It places the visual as one important component of the process of finding cognitive shortcuts, and while words, slogans and headlines equally provide shortcuts, the visual may be the most memorable and resonant form of shortcut, and it is certainly one that is used frequently within political communication" (Lilleker, 2019, p. 48). In order to put this to the test, it is possible to cite some examples of studies that, in recent years, have consciously or unconsciously placed the image at the centre of political communication strategies via social media.

It is useful here to mention Ceccobelli's (2017) research, which aims to understand the articulations of the political sphere on a digital platform like Facebook. This research focuses particular attention on how this social media platform fosters the emergence of niche audiences and its ability to expose different types of users, ranging from those more engaged to those less engaged, to political content. A central element of Ceccobelli's research, which offers interesting insights into the evolution of the "politics of the image" on social media, is a new definition of the popularisation of political communication. Ceccobellli describes it as "a communicative action through which political actors try to communicate with citizens who are less interested in news, actors and dynamics inherent to the political sphere. It takes two distinct forms, namely popularisation as a style or as a type of communication, and can be achieved mainly through three types of strategies (i) resorting to people and the vicissitudes concerning their private lives; (ii) using and interacting with media environments, rhetorics, and the actors themselves belonging to the world of celebrities from television, music, sport and film; (iii) reproducing and interpreting the lifestyles and everyday life of citizens, or the dominant customs and practices of the popular culture of reference" (Ceccobellli, 2017, p. 42). Starting from the hypothesis that Facebook is the digital platform on which politicians suffer most from an incentive to popularise their political communication, Ceccobelli identifies examples of such popularisation through an analysis of the content of the Facebook pages of leading political leaders in Western Europe, North and South America, Australia and New Zealand. They fall into three types, corresponding to as many strategies known in the literature: intimate politics (Stanyer, 2013), celebrity politics (Street, 2004) and lifestyle politics (Bennet, 1998). Here, then, is the Swedish prime minister sharing a playful moment with his granddaughter, the former Australian prime minister posing with two national cricket team players, a selfie through which the former Argentine president immortalises a crowd bath. Although the selected unit of analysis is the post, and not the eventual photographic content, as expected, the results of Ceccobelli's analysis reveal the publication of photographic content with the image of the political actor as the first indicator of a case of direct personalisation. The aforementioned selfie tool is also the preferred mode of expression of the leader's closeness to his electorate, i.e., lifestyle politics, which Ceccobelli's results identify as the first mode of expression of popularisation in the analysed sample. Even the examples of strategic use of a personal Facebook page, not only to publish content that is not directly political but also to insert a possible "story" into the flow of the hybrid media system – the image of the Norwegian prime minister playing hide-and-seek with some children, taken up by major national newspapers – brings one back to the iconographic dimension. The way this dimension emerges consistently and transversally in an investigation that does not directly focus on the specifics of the image in political communication via social media gives an idea of how powerfully the visual element emerges in the analysis of communication flows through social media.

Why does Ceccobelli's research not fit into the visual politics groove, or does it not regard it as a recognised strand of study? Because, despite the growing importance of image- and video-based social media, placing the iconographic dimension at the centre of attention, scholars have struggled to recognise this 'visual turn' until very recently. As Dumitrescu (2016) argues, echoing Grabe and Bucy, only in the last decade or so have social scientists begun to take visuals seriously. This expression is quoted by Bucy himself in the introduction, co-written with Jungseock Joo, to an issue of *The International Journal of Press/Politics* devoted precisely to Visual Politics (Bucy & Joo, 2021). It is in this issue that the challenge of interdisciplinarity is emphasised, not only concerning the aforementioned cognitive frameworks belonging to political psychology, but concerning disciplinary traditions that on the one hand include the humanities and in particular the aesthetic disciplines, and on the other hand, the so-called "hard sciences" and in particular computer science. At the intersection of the views and tools made available by these disciplines, it becomes possible, according to the authors, to identify the much-desired paradigms of influence of visual, non-verbal, and iconographic categories on the opinions, attitudes, and behaviour of citizen-voters.

There are several possible fields of application, all identified by the polysemic quality of the images, which allows them to have more than one meaning or connotation, opening them up to different interpretations depending on the viewer's orientation. "For instance, differentially framed images of fracking (e.g., economic benefit, environmental risk) may be given preferred, negotiated, or oppositional readings that vary with the viewer's standing attitudes towards this controversial practice of oil and gas extraction" (Krause & Bucy, 2018). Analysis of open-ended responses to fracking images revealed that individuals who disagreed with the practice were more likely to express concern for the environment, regardless of the frame shown. In contrast, undecided individuals and supporters cited the impact on human health more frequently. A similar pattern of standing attitudes shaping perceptions of visual representations of politically consequential events was found in a study of protest imagery using 360-degree camera technology (Bucy, 2020).

Finally, one article in the aforementioned issue of *The International Journal of Press/Politics* is worth mentioning. The article by Xénia Farkas and Márton Bene (2021) aims to identify and compare the patterns and effects of Hungarian politicians' image-based communication on Facebook and Instagram during the 2018 parliamentary election campaign. The authors argue that, from a strategic point of view, visuals are strongly connected to personalisation – an assumption they draw from Enli and Skogerbø's (2013) and Hermans and Vergeer's (2012) studies – but they also link their research pattern to that used by Ceccobelli. Farkas and Bene find it particularly advantageous to study the dynamics of personalisation conveyed through visuals in the social media environment. As Ceccobelli did with popularisation, they "disassemble" the dimension of personalisation into two main branches, built on the studies of Van Aelst and Collghi (2012): "individualisation (i.e., focus on politicians' political work) and privatisation (i.e., focus on politicians' personality and personal background) dimensions of personalisation" (Farkas & Bene, 2021, p. 120).

With their research, Farkas and Bene intend to bridge three significant gaps in the field of visual studies. This is why their study is well suited to conclude this examination of the "adventure" of visual studies. The first gap is methodological, and again raises questions about the need to develop a coding scheme that is specific enough to investigate images as objects of interest in their own right and provide a detailed map of political actors' visual strategies. While operating according to a traditional content-analysis approach, the authors intend to refine this method through the use of "categories we rely on a comprehensive inductive visual analysis of the specific details of the images, such as their types or the persons depicted inside the frame, cultural and political references, the sentiment projected by the depiction, and the visual character of the image" (Farkas & Bene, 2021, pp. 121-122). The second gap concerns the environment in which the research takes place: the authors emphasise the tendency of previous research to focus on a single platform, and the innovativeness of applying a cross-platform approach. The latter is made possible not only by the common ownership of the two platforms (both Facebook and Instagram are part of the communications giant Meta) but also by the broad overlap in architectures and affordances between the two environments. These elements can significantly shape communication strategies (Bossetta, 2018). What makes the analysis interesting are the differences between the two environments. At the time of analysis, Facebook is significantly more popular in Hungary than Instagram.

On the other hand, Instagram's user base is much more specific and composed of young people, which explains at least in part why the networks of Facebook users are more offline-anchored than those of Instagram users. Facebook, by its very structure, provides for a more two-way type of communication, at least ideally, than Instagram. This difference is reduced when using the page tool on Facebook. Facebook's logic is based on the possibility of sharing content, which is impossible on Instagram. The use of Instagram is more widespread among celebrities and influencers than that of Facebook. The combination of these elements leads Farkas and Bene to hypothesise that "visual political communication on Instagram is more informal and thereby more focused on the privatisation dimension of personalisation than on Facebook (H2), while on Instagram images should be used to display more informal aspects of candidates' lives, relevant to the individualisation dimension of personalisation (H3)" (Farkas & Bene, 2021, p. 125). The third gap concerns the study of the effects of different images on user engagement, thus extending a type of analysis that is now widespread in the study of social media and politics (Bene, 2017; Heiss et al., 2019), but predominantly on textual elements. In terms of research questions, these elements translate into the focus on what types of visual tools trigger engagement (e.g., likes) and what cross-platform similarities and differences can be identified in the engaging responses to politicians' visual communication.

By analysing each image-based Facebook and Instagram post of candidates who were active on both platforms during the campaign (n = 51), Farkas and Bene find that the visual communication on both platforms is highly personalised. However, some differences can be found. For example, visual communication on Instagram is much more personalised than on Facebook, and conversely, each formal element is significantly more employed on Facebook than on Instagram. Facebook appears to be a platform more dedicated to campaign posts, whereas on Instagram, content featuring leaders in informal and familiar contexts takes up a larger share of space. "It seems that while individualisation has a strong presence on both platforms, it is dominant on Facebook, while on Instagram it is mingled with privatisation" (Farkas & Bene, 2021, p. 134). However, this difference does not automatically translate into greater or lesser engagement for certain types of content on either platform: "selfies, personal content, and references to popular cultural objects show no significant relationship with likes on either platform" (Farkas & Bene, 2021, p. 137). This leaves further room for investigation about what remains the central crux of visual politics: the possible effects that the visual content posted on social media by political leaders may have on the citizen-voters exposed to the communication flows of social media. By configuring a more or less voluntary exposure due to their algorithmic logic, this makes the crux even more complex to unravel. Hence, calling into play the contribution of the hard sciences, which have now reached a level of sophistication and flexibility of use that overcomes the doubts expressed back in 2009 by Grabe and Bucy.

THE CHALLENGE OF INTERDISCIPLINARITY

In 2012, Schill concluded one of the most influential reviews on the state of research in Visual Political Communication (VPC) with a warning: “When scholars only examine written or verbal texts, they are only seeing a small part of the political communication process” (Schill, 2012, p.133).

So far, we have extensively discussed the importance of visual communication, dating back to the era when television was the primary medium for disseminating and consuming political information, statements, and content. The digitalisation of political communication and the adaptation of political language to the norms of social media platforms were shown to have led to an exponential increase in the dissemination of visual political content. However, at the time of Schill's review, this process was still in its early stages of development. Within the field of political communication studies, as he notes, “The visual aspects of political communication remain one of the least studied and least understood areas” (Schill, 2012, p. 119).

Schill’s call to action emphasises the need to expand this field of research, acknowledging the central role of imagery in contemporary political communication.

A decade later, Farkas’s (2023) review demonstrates not only that Schill’s call to action was widely heeded but also that the integration of social networks into political communication channels has significantly fuelled scholarly interest. This development has driven substantial progress in the study of visual political communication and the advancement of visual data analysis techniques. Farkas's work offers a comprehensive review of the evolution of Visual Political Communication between 2012 and 2022, compiling all research outputs primarily focused on the visual dimensions of political communication. The most notable difference from Schill's study lies in the centrality of social networks. Among the 499 contributions published during the ten years analysed, those dedicated to visual political communication on social media more than double the number of studies focused on television (Farkas, 2023, p. 12).

The use of social media as a channel for political communication has revolutionised access to and consumption of political imagery, with effects extending beyond digital environments (Chadwick, 2013; Blumler, 2016; Lilleker, 2019). It has also opened new avenues for research in Visual Political Communication (VPC). The digital era has ushered in an unprecedented volume of visual information, compelling political scientists to increasingly recognise the need for advanced methodologies. These include leveraging computational techniques for the automated analysis of large datasets to address this growing body of data effectively. In her review, Farkas observes that, despite the extensive body of research on social media, few studies to date have employed advanced methods of visual analysis, such as those enabled by computer vision. Among the studies she surveyed, fewer than 3% rely on automated methods and computer-based techniques. This gap represents a significant limitation in the field, as methods such as machine learning and computer vision enable the systematic analysis of large volumes of visual data. Such techniques can overcome one of the key challenges in studying visual politics: the manual construction and analysis of visual corpora, which often restricts research to a limited number of items. This constraint diminishes the generalizability of findings and hampers the depth of analysis.

The adoption of computer vision methodologies for studying visual political communication, compared to manual coding, enables the analysis of larger and more complex datasets. It facilitates the identification of connections, structures, and features – even at a micro level – more comprehensively and reliably (Joo & Steinert-Threlkeld, 2018). With advancements in computer vision techniques, the scope of visual communication analysis has expanded significantly. Bucy (2023) highlights how computer vision technologies have transformed VPC, allowing researchers to examine images and videos with unprecedented precision. These tools automatically identify visual characteristics and non-verbal signals, enabling the study of how images convey political messages, represent identities, and influence public perception at scale. For this reason, Farkas – echoing the calls of other scholars (Joo & Steinert-Threlkeld, 2018; 2019; Bucy & Joo, 2021; Bucy, 2023) – concludes her review by urging researchers to invest more in these emerging technologies to explore new dimensions of VPC using an interdisciplinary approach, encouraging collaboration between political communication scholars and experts in computational sciences.

Nevertheless, within the nearly 3% of studies that integrate the perspective of social and political sciences with the methodologies and tools of automated computational analysis – a figure likely to have increased in the years following Farkas's study – lies the foundation of a burgeoning field. This emerging area has already demonstrated significant potential in its applications.

In this context, Joo and Steinert-Threlkeld (2018) draw an insightful parallel with developments in the study of political texts in a pioneering study on the most commonly used methodologies in Automated Visual Content Analysis for Political Science.

In response to the growing availability of textual data and the challenges of conducting in-depth analysis through manual methods, political scientists have developed techniques that leverage computers and specialised software to analyse large textual datasets. Over the past decade, the authors note, these automated content analysis methods have been widely adopted in political science, paving the way for methodologies now well-established in the social and political sciences, such as text mining, topic modelling, sentiment analysis, and opinion mining.

Joo and Steinert-Threlkeld argue that, in the study of visual data, we are at "a very similar turning point" (Joo & Steinert-Threlkeld, 2018, p. 2). The analysis of visual content involves handling vast amounts of data that make manual examination impractical without significantly reducing the scope and scale of the analysis. This challenge underscores the growing need to adopt innovative methodologies that offer deeper insights and scalable approaches to studying these phenomena.

Focusing on computer vision methodologies applicable to visual political communication, numerous scholars – among the most notable contributions are Krizhevsky et al. (2012), Bucy and Stewart (2018), Joo and Steinert-Threlkeld (2018; 2019), Peng (2021), Peng and Lu (2023), and Hoffmann et al. (2022) – have highlighted how certain deep learning models have become the standard for image classification. Specifically, Deep Neural Networks (DNNs) have gained prominence; they form an advanced type of artificial neural network that is organised in multiple layers. These networks are termed "deep" because they comprise several hidden layers between the input layer and the output layer. Each layer consists of computational units, or neurons, which process data from the preceding layer and pass it to the next. In simple terms, DNNs function through a hierarchical learning system. The initial layers identify basic features in the data (e.g., the outline of an image).

In contrast, subsequent layers combine this information to recognise more complex patterns (e.g., shapes within the outline). The connections between neurons, and thus between layers, are trained using algorithms that correct errors and enhance performance based on labelled datasets aligned with the variables of interest for analysis (Chollet, 2017). In practice, this involves constructing a large visual corpus (comprising static images or videos), coding a subset of these images with labels derived from the variables under study, and training computer classifiers using the information from the labelled subcorpus (Hoffmann, 2022; Bucy, 2023).

Among deep learning models, Convolutional Neural Networks (CNNs) are particularly suited to this type of approach for image analysis. CNNs are a specialised category of DNN that are designed to process grid-like structured data, such as images or videos. They are especially effective in recognising visual patterns due to their ability to automatically learn relevant features from raw data. This makes CNNs highly valuable for studying visual political communication.

Similar to DNNs, convolutional networks operate based on a layered structure and hierarchical learning. The initial layers, known as Convolutional Layers, apply filters (referred to as kernels or convolutional filters) to portions of the image to extract key features such as edges or colours. Subsequent layers (Pooling Layers) reduce the image size and the number of parameters to process, using the information derived from the previous layer. At the end of the network, Fully Connected Layers combine the features learned throughout the process to produce the final classification.

Compared to DNNs, CNNs identify patterns directly from the pixels of the image. These patterns are translation-invariant, meaning they can recognise a feature regardless of its position within the image, and involve fewer parameters compared to traditional neural networks due to their ability to leverage the spatial structure of images. This characteristic makes CNNs particularly efficient in handling high-dimensional data (Chollet, 2017).

Peng (2021) provides a comprehensive literature review aimed at identifying the primary deep-learning-based methodological approaches used for visual analysis in the social and political sciences. Peng identifies four main approaches. The first relies on open-source computer vision libraries (such as OpenFace, ImageNet) and commercial APIs (such as Face++), which facilitate standardised analyses like facial recognition, object or person detection within images, and Optical Character Recognition (OCR) techniques for extracting text from images. This approach is considered unsupervised because it utilises prebuilt models, thereby limiting the analysis to the predefined attributes available on these platforms (Peng, 2018).

The second approach is referred to as supervised learning. In this model, researchers can train neural networks to identify specific, customised attributes based on datasets that have been manually labelled according to the variables or features of interest. However, the primary limitation of this approach is its reliance on very large datasets – often comprising millions of images – for effective training.

The third approach, i.e., Transfer Learning, is arguably the most practical for the social sciences, which frequently deal with much smaller datasets. This model is compatible with both supervised and unsupervised learning. It involves adapting networks that were initially trained on large datasets to work with smaller ones. In supervised learning, features learned during the initial training phase can be repurposed and tailored to fit the new dataset. In unsupervised learning, the model can identify hidden patterns within the data without relying on predefined labels.

The final approach described by Peng (2021) pertains to the aesthetic qualities of images. Computer vision tools can be used to detect visual attributes such as brightness, colour, or composition. In the field of visual political communication studies, this type of analysis can be particularly useful for assessing an image's effectiveness based on audience reactions.

The practical applications of these techniques span a range of tasks, including *Image Classification*, which analyses images based on their features and assigns them a set of labels. These labels are derived either from a predefined set of categories (in the case of unsupervised learning) or from user-defined categories that have been trained into the model (in the case of supervised learning). Another key application is *Object Detection*, which identifies and classifies objects within an image while specifying their location.

Additionally, these techniques are extensively applied to the analysis of individuals depicted in images. This type of analysis primarily focuses on the presence of people and groups, as well as the examination of faces and facial expressions in photographs. Research in this domain is generally divided into three main areas: *Face Detection, Face Recognition*, and *Face Attribute Classification* (Peng, 2018; 2021). Specifically, Face Detection involves analysing images to determine whether faces are present. Face Recognition extends this process by verifying an individual's identity by comparing the image to those in the model (either supervised or unsupervised). Finally, Face Attribute Classification analyses faces in images to extract various attributes, including gender, age, posture, expressions, and emotions. Models designed to identify individuals are particularly applied in public contexts, especially when images feature crowds or large groups of people. As will be discussed further, these techniques have been widely employed in studies of political events and protests (Sobolev, 2020; Bucy, 2023).

In the field of political studies, these applications have given rise to various analytical approaches. Joo and Steinert-Threlkeld (2018) identify five main areas where visual data is employed to study political communication:

1. **Political Behaviour:** Automated methods analyse demographic information from images, such as gender, ethnicity, and age group, to identify correlations between these characteristics and political participation. For instance, Wang, Li, and Luo (2016) used Twitter profile pictures to infer the demographic composition of followers of political candidates during the 2016 U.S. presidential elections. Similarly, these techniques were employed to study the demographics of protesters at rallies and compare them with media representations (Sobolev, 2020) or to infer the relationship between the composition of a protest crowd and political changes (Steinert-Threlkeld et al., 2022).
2. **Political Communication:** Computer vision models facilitate the analysis of politically relevant attributes of actors or events depicted in extensive image datasets. Joo et al. (2014) used these techniques to infer the communicative intentions hidden in photographs of politicians, analysing personal traits such as facial expressions or hand gestures to persuade the public visually. Automated models have also been employed to identify facial characteristics of winning and losing candidates (Joo, Steen, & Zhu, 2015) and to extract visual features from facial images to predict subjective judgments on traits such as intelligence or trustworthiness (Rojas et al., 2011; Vernon et al., 2014).
3. **Development:** Computer vision techniques enhance development initiatives by leveraging remote sensing data. These data can detect land use and urban characteristics, providing census-like insights (Tapiador et al., 2011). Additionally, light emissions measured via remote sensing can serve as indicators of wealth

in impoverished regions (Weidmann & Schutte, 2017). Deep learning models applied to satellite imagery have been used to estimate consumption and household wealth (Jean et al., 2016). Google Street View data has been leveraged to estimate income and socioeconomic characteristics by analysing vehicles or neighbourhoods (Glaeser et al., 2018). Additionally, pre- and post-disaster images aid in assessing damage and recovery efforts, shedding light on the role of institutions (Hsiang et al., 2013).

4. **Sub-National Conflict:** Deep learning models applied to images are also employed to study protests and state responses. Joo and Steinert-Threlkeld refer to research on the "repression-dissent puzzle", highlighting its mixed findings: repression may have no effect (Ritter & Conrad, 2016); it could escalate protest intensity (Steinert-Threlkeld, 2017), reduce it, or produce time-dependent outcomes. These techniques enable the collection of data on protester numbers, the progression of violent events (Steinert-Threlkeld, Won, & Joo, 2018), and even the motivations behind protests (Hunziker & Cederman, 2017; Weidmann & Schutte, 2017).
5. **Protest Analysis:** Computer vision models applied to images of demonstrations allow researchers to estimate crowd size, detect the presence and behaviour of police forces, identify acts of violence or protest banners, and analyse the demographic composition of participants (such as gender, ethnicity, and the presence of children). These tools have been utilised, for example, in the analysis of the 2016–2017 South Korean protests, allowing for daily-level assessments of shifts in protester profiles and the intensity of conflict (Won, Steinert-Threlkeld, & Joo, 2017).

Joo and Steinert-Threlkeld's review of these contributions represents a significant step in defining the field of visual political communication studies using computer vision techniques. However, it is also reflective of the time in which it was published, when the discipline was still in its formative stages. As a result, many of the applications highlighted by the author are related to the study of political communication, often in a cross-cutting manner.

In the following years, with the growing interest in these methodologies within the field of social and political studies, Bucy (2023) provides an updated overview of the potential applications of computer vision to the study of visual political communication.

The author identifies six key areas in which computer vision techniques are effectively used:

1. **Political Protests:** Thanks to the increasing availability of visual materials from protests and demonstrations, these have become the subject of numerous studies

using computational techniques. The previously mentioned study by Sobolev et al. (2020) provides an interesting example in this regard. By analysing protest images with a classifier and a facial detector, the researchers found that the participation estimates provided by the media and politics regarding certain protests were accurate. Another noteworthy study is that of Steinert-Threlkeld et al. (2022), who analysed a global sample of geolocated protest images and explored how the violence of protesters negatively influences participation in protests, and how, conversely, state repression can positively or negatively affect protest participation depending on the intensity of the violence.

2. **Candidate Self-Presentation Style:** Computer vision techniques allow for the analysis of promotional images of candidates, revealing how their visual presentations influence public perception. A notable example is the study by Yoon et al. (2020), which demonstrated how automatic facial feature classification can accurately identify leaders and their corresponding social roles. In the 2020 U.S. elections, Bossetta and Schmøkel (2023) employed computer vision and clustering techniques to compare candidate images on Facebook and Instagram, revealing that Instagram users reacted more favourably to emotional displays than Facebook users; for instance, expressions of anger received more interactions, but expressions of happiness generated more “likes”. Similarly, Haim and Jungblut (2021) highlighted that candidates’ self-representations tend to be more positive than media portrayals, which often appear more negative. Peng (2021) also demonstrated that the size of the face and emotional content in Instagram posts significantly influence positive engagement.
3. **Party Symbolism and Affective Polarisation:** Computer vision is also used to study symbolism in political materials and to analyse affective polarisation between political groups. Computational techniques enable the examination of how images influence emotions and public reactions. Ash and Caliskan (2022), for instance, developed an emotional classification model to analyse the non-verbal responses of news anchors during partisan conflicts on American cable news networks, finding that fear is the primary emotion expressed by the anchors when political adversaries are mentioned. Xi et al. (2020) examined the symbolism in Facebook posts by U.S. Congress members, showing that Republicans use conservative clothing and patriotic and military symbols to express their ideology, while Democrats tend to represent diversity and address issues of inequality. An emerging strand of research in this area also highlights the disparities with which different political factions represent individuals, groups, and issues depending on their proximity or distance from them (Thomas & Kovashka, 2021).
4. **Gender Stereotypes:** The study of gender stereotypes in visual politics has also benefited from the automation of visual political studies, transitioning

from small-scale analyses to broader research using computational methods. Boussalis et al. (2021) demonstrate that women in Congress post more images of happiness than men, but their verbal tone in these posts tends to be more negative. Thus, women are said to perform more "emotional labour" on social media during election campaigns compared to their male counterparts. Chen et al. (2020) find that masculine traits are associated with electoral success, while feminine traits are not, using CNNs to infer political misogyny in social media images as a predictor of electoral victories.

5. **Debate Performance and Voter Impression Formation:** Computational analyses of debates reported by Bucy (2023) demonstrate that displays of anger and defiance are just as effective in mobilising supporters as expressions of happiness, particularly in polarised contexts such as U.S. presidential primaries or elections. Anger and other strong emotions unite political supporters against perceived threats from external groups (Bucy & Grabe, 2008), while happiness and reassuring expressions encourage participation by suggesting affinity.
6. **Public Policy and Electoral Fraud:** Some of the most interesting examples cited by the author relate to the use of surveillance camera images in the Netherlands to measure people's adherence to social distancing recommendations during the COVID-19 pandemic (Bernasco et al., 2022). Similarly, Twitter posts with images of face masks in the United States were correlated with national survey data to estimate large-scale mask usage (Lin et al., 2022). Image analysis has also been applied to detect electoral fraud: Cantu (2019) used CNNs to identify alterations in ballots during the 1988 Mexican elections, suggesting the occurrence of electoral fraud.

LIMITS, BIAS AND ETHICAL IMPLICATIONS OF MACHINE LEARNING

The examples presented by Bucy (2023) illustrate how the application of computational techniques in the study of visual political communication has expanded, on the one hand. On the other hand, such techniques are increasingly defined both as a specific field of study and through the use of established methodologies that have gained consensus within the scientific community. The path toward defining a true interdisciplinary field of study is still in its early stages; however, the potential demonstrated in the studies discussed here highlights the scientific community's growing interest in these applications and the increasing investment in this direction.

Nevertheless, Bucy concludes his review by pointing out several concerns and cautions. He emphasises the risks associated with the use of facial recognition techniques, which raise concerns about privacy and potential implications for the repression of freedom of expression and the right to protest. The analysis of massive image datasets poses significant privacy challenges, especially when images are collected without consent at public events. Bucy stresses that it is essential to uphold individuals' privacy rights, particularly under authoritarian regimes where these tools could be deployed for repressive surveillance. Researchers must therefore adopt protocols that protect subjects' rights—employing anonymisation measures when necessary—and navigate complex data-protection frameworks such as the GDPR, which restricts the processing of sensitive attributes like ethnicity or gender.

The use of algorithms to analyse political images can still lead to distorted results or interpretations, as the models upon which the analyses are based may overrepresent or underrepresent certain ethnic groups, genders, or characteristics. Machine learning tools are far from neutral and primarily reflect the data on which they were trained. This training phase is crucial for defining a model's classification criteria, and the most significant risk lies in using datasets that contain implicit biases. Algorithmic bias can lead machine learning models to produce skewed or discriminatory outcomes. Laranjeira and colleagues (2024) identified the most worrisome bias types in computer vision:

1. **Selection Bias,** from datasets skewed toward particular genders or ethnicities (such as Caucasian males), causing models to "specialise" in overrepresented traits and lose accuracy on underrepresented groups, often compounded by a binary gender framework (Scheuerman et al., 2019);
2. **Annotation Bias,** from vague labelling guidelines that lead coders to misinterpret culturally normative postures as threats;
3. **Measurement and Recall Bias,** when models tuned to light skin tones capture fewer details on darker skin, resulting in lower recognition rates;
4. **Evaluation Bias** when aggregate metrics mask poor performance on minority populations; and
5. **Overgeneralization Bias** occurs when a model's capacity to scale is overestimated, ignoring data quality disparities. These risks are particularly evident in racial bias within facial recognition systems and in gender bias within classification systems. The former often arises from training datasets that are skewed toward Caucasians, use non-representative metrics, or have historical prejudices embedded in the data. The authors cite the example of Google Cloud Vision labelling the photo of a dark-skinned hand holding a thermometer as "gun," while the same image with lighter skin was labelled "monocular" (AlgorithmWatch, 2020). They mention the well-known case of the COMPAS software used by the U.S. justice system, which embedded racial prejudice by assigning African Americans a "high" recidivism risk more than twice that of Whites despite similar actual reoffense rates (Dieterich et al., 2016). In the latter case, models

perpetuate gender stereotypes or underrepresent segments of the population. Prabhu and Birhane (2020), for example, showed how ImageNet - one of the most widely used image databases for model training - paired misogynistic labels (e.g., "bitch," "slut") with female images and even included non-consensual "upskirt shots." Otterbacher et al. (2017) similarly demonstrated that queries like "CEO" predominantly return men in positions of power, while "nurse" more often return women in caregiving roles. Additionally, they found that adjectives such as "competent" associate with men, whereas "warm" associates with women. In the same vein, Wang and Kosinski's (2018) "sexual orientation detector," built from dating-site datasets, relied on grooming cues (make-up, eyewear) rather than facial features; its "average faces" depict heterosexual women wearing make-up and lesbians without, highlighting selection bias and clear spurious correlations.

These differential errors are not mere technical glitches but translate into decisions about bail, sentencing, and civil liberties, amplifying structural injustices and necessitating rigorous data disaggregation and the adoption of group-specific fairness metrics (Risser et al., 2022).

Moreover, the use of these techniques requires significant investment in both computing equipment and expertise, as well as human resources. This could result in inequalities that limit access to this type of research for scholars with fewer resources. For this reason, Bucy (2023) concludes with a call, which we also endorse, for collaboration and interdisciplinarity with computer scientists to make the potential of computer vision techniques for the study of visual politics more democratic and accessible.

THE CONTRIBUTION OF COMPUTER VISION TO THE STUDY OF THE 2024 EUROPEAN ELECTIONS

Methodology and Objective

We conclude this chapter by presenting the results of a preliminary study conducted in collaboration with the Department of Computer Science at Sapienza University of Rome. In this study, we tested various deep learning techniques to analyse visual content produced by politicians during the 2024 electoral campaign. This research addresses two main objectives: the first is to fill a gap in Italian academic literature, as highlighted in Farkas's (2023) review, which lacks recent contributions that systematically examine the role and functions of visual politics in Italy. The second objective stems directly from the first. It involves testing these computational techniques in a political context where they have not yet been applied on a large scale for research purposes.

The research relies on two corpora of images collected from posts published by the main accounts of Italian political leaders and parties during the 2019 and 2024 European elections. The 2019 corpus was manually coded by a team of analysts using labels designed to capture various dimensions: the type of image (photographic, textual, or mixed); the production techniques involved in its creation; the presence of politicians, their attire, primary facial expressions, and the context in which they are depicted; the inclusion of crowds or national and European symbols; the purpose of the posted image (categorised as informative, electoral, thematic, or comparative); the presence of emotional appeals (negative, humorous, emotional, or rational); and the primary theme addressed by the image.

These labels enabled the extraction of image subsets corresponding to specific categories, which were then used to train the Residual Visual Transformer model (ReViT) (Diko et al., 2024) and CNN models for tasks such as image classification, emotion recognition, and optical character recognition (OCR). The trained model was subsequently tested on the corpus of political images from the 2024 European election campaign, which had been manually annotated with the same labels used in the 2019 analysis. The goal was to evaluate the effectiveness of deep learning models in detecting specific image characteristics by cross-referencing their outputs with manual cataloguing performed on the same images and for the same attributes.

In this phase, the labels that proved most effective for training were related to: a) the emotions expressed by politicians; b) the type of image, categorised as photographic or as (so-called) web cards (Novelli, 2018), which combine graphic and textual elements; and c) the presence of text within the images. These labels enabled the execution of various tasks, including identifying the predominant expression of the politician depicted in the image, determining which images in the sample contained textual elements, and extracting the text embedded within the images.

Results

For the first task, the BLIP-2 model was employed to perform sentiment analysis on images through a supervised classification task. A pre-trained model was utilised, to which images from the 2019 campaign, previously labelled to indicate emotions, were submitted. The labels used to train the model were (0 – happy, 1 – sad, 2 – angry, 3 – neutral). Images from the 2024 campaign were analysed using a specific prompt designed to guide the model in identifying and categorising the emotion conveyed in the image into one of the four predefined categories, yielding predictions aligned with the estimated category.

This approach leverages BLIP-2's capabilities to recognise visual patterns and generate predictions based on learned visual-linguistic representations, enhancing the classification of facial expressions within a political context. BLIP-2 is built upon a modular architecture that integrates pre-trained image encoders and frozen large language models (LLMs) with an intermediate component, the Querying Transformer (Q-Former). These features enabled the effective use of BLIP-2 for recognising visual patterns and processing predictions based on its learned visual-linguistic representations, thus optimising the classification of facial expressions in the political domain.

The preliminary results are highly promising, achieving an effective categorisation rate of approximately 90%. Some examples of classification outcomes are presented below (Figure 1).

Figure 1. Example of automatic emotion classification

The most significant challenges, as expected, lie in capturing the nuances of emotions, as illustrated in the second image presented in the example (Figure 1). The politician depicted displays a serious expression. However, it is certainly not a sad one. However, in this case, the model interpreted the expression as sad, likely because it aligned more closely with the patterns learned during training.

The second task focused on studying the type of image. For this purpose, the Shifted Window Transformer Tiny (Swin-T) model was employed. Initially pre-trained, the model was further fine-tuned on a 2019 dataset of web cards for 10 iterations, using a batch size of 32 and a learning rate of 0.0001. The classification goal was to distinguish between two classes: 0 for images and 1 for web cards. Swin-T, with its hierarchical architecture and innovative shifted windowing mechanism, proved particularly well suited to addressing the complexities of visual data.

The model's ability to create hierarchical feature maps and reduce computational complexity by calculating attention within non-overlapping local windows allowed for efficient and accurate processing of high-resolution images from the corpus. This approach achieved a high level of precision in classifying web cards, significantly facilitating the analysis of visual materials related to the electoral campaign. Although the model is still undergoing refinement, it currently achieves a classification accuracy of 93.4% for identifying emotions in the 2024 images. Some examples of misclassifications are presented below (Figure 2).

Figure 2. Example of misclassifications in image type classification

Predicted label: Photo
True label: Webcard

Predicted label: Photo
True label: Webcard

Predicted label: Webcard
True label: Photo

In the first and second cases, the issue likely stems from the predominance of graphical elements over textual ones, with the textual component being minimal, particularly in the first image. In the third type of misclassification, however, the error is less intuitive from this perspective. It indicates the need for further refinement during the model training phase to improve classification accuracy.

Finally, the last task employed an OCR model to extract text from images previously categorised as web cards. The results in this case were also satisfactory, achieving the following performance metrics: text recognition accuracy of 93.8%, text detection precision of 91.39%, and recall of 83.77%. An example of this process is illustrated in the graphic below (Figure 3).

Figure 3. Example of text extraction using the OCR technique

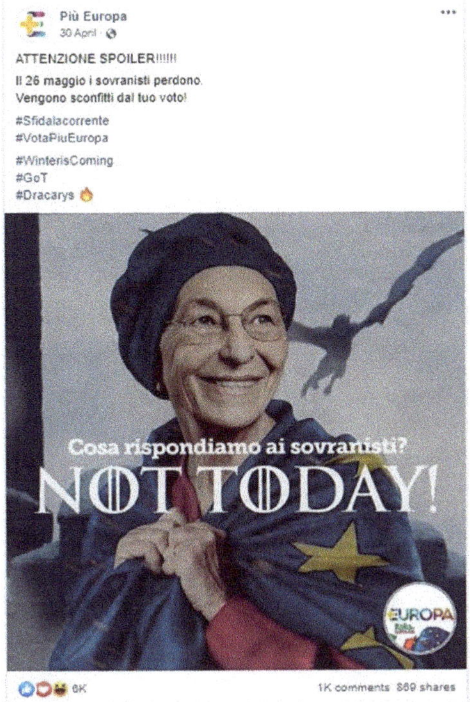

[[[49.0, 9.0], [111.0, 11.0], [111.0, 25.0], [49.0, 22.0]], ('Più Europa', 0.8610519170761108)]
[[[47.0, 23.0], [102.0, 26.0], [102.0, 39.0], [47.0, 37.0]], ('30April', 0.873740017414093)]
[[[12.0, 48.0], [153.0, 48.0], [153.0, 61.0], [12.0, 61.0]], ('ATTENZIONESPOILER!!!!', 0.9196048974990845)]
[[[10.0, 65.0], [176.0, 67.0], [176.0, 81.0], [10.0, 79.0]], ('26 maggio i sovranisti perdono.', 0.9475921988487244)]
[[[10.0, 81.0], [162.0, 82.0], [162.0, 96.0], [10.0, 95.0]], ('Vengono sconfitti dal tuo voto!', 0.9264500141143799)]
[[[10.0, 101.0], [94.0, 102.0], [94.0, 115.0], [10.0, 114.0]], ('#Sfidalacorrente', 0.987281322479248)]
[[[10.0, 117.0], [94.0, 117.0], [94.0, 130.0], [10.0, 130.0]], ('#VotaPiuEuropa', 0.9926881790161133)]
[[[9.0, 136.0], [98.0, 138.0], [98.0, 151.0], [9.0, 148.0]], ('#WinterisComing', 0.9914668202400208)]
[[[11.0, 151.0], [42.0, 151.0], [42.0, 165.0], [11.0, 165.0]], ('#GoT', 0.9801514744758606)]
[[[9.0, 166.0], [80.0, 164.0], [81.0, 181.0], [9.0, 183.0]], ('#Dracarys', 0.9898154735565186)]
[[[56.0, 385.0], [351.0, 385.0], [351.0, 404.0], [56.0, 404.0]], ('Cosa rispondiamo ai sovranisti?', 0.9654625654220581)]
[[[28.0, 416.0], [378.0, 413.0], [379.0, 449.0], [28.0, 452.0]], ('NOIATODDAY!', 0.7912795543670654)]
[[[334.0, 534.0], [390.0, 532.0], [391.0, 547.0], [334.0, 549.0]], ('EUROPA', 0.9927287697792053)]
[[[11.0, 596.0], [73.0, 596.0], [73.0, 613.0], [11.0, 613.0]], ('O6K', 0.7121278643608093)][[[272.0, 599.0], [393.0, 599.0], [393.0, 611.0], [272.0, 611.0]], ('1K comments 869 shares', 0.9674939513206482)]

DISCUSSION AND CONCLUSION

With the shift of political communication from traditional to digital media, images and videos have become even more essential in shaping politicians' public personas, surpassing text in both reach and impact. Digital platforms enable politicians to present themselves directly and personally, bypassing conventional media filters and establishing an immediate connection with their audience (Lilleker et al., 2019). This newfound accessibility has reshaped how citizens perceive their leaders, fostering intimacy and personality that drive empathetic dynamics of intimate politics (Stanyer, 2013), celebrity politics (Street, 2004) and lifestyle politics (Bennett, 1998).

The rapid proliferation of visual content as a symbolic vehicle for political messaging has catalysed the growth of Visual Politics research, thanks to ever-increasing data availability and accessibility (Farkas & Bene, 2021). Computer vision techniques have revolutionised this field by enabling researchers to analyse images and videos with unprecedented precision, automatically detecting visual features and nonverbal cues (Bucy, 2023). Integrating these tools enables deeper insights into how visuals influence political communication, collective identity, and power relations. Automation also makes it feasible to process vast visual archives far more efficiently than manual coding, supporting more nuanced and contextualised studies of political phenomena. However, this opportunity requires continuous methodological refinement and rigorous ethical oversight, ensuring that findings remain both scientifically robust and socially responsible.

Visual data are inherently complex: multimodal, rich in spatial and contextual information, and sensitive to lighting, angle, and resolution. This complexity complicates the extraction of reliable signals and often requires qualitative interpretation to capture deeper symbolic meanings and communicative intentions (Caiani, 2024). Going forward, we must both develop more precise, bias-aware algorithms and complement them with qualitative approaches capable of micro-level semantic analysis—subtleties that current models still struggle to grasp. This overview of the application of computer vision in the European electoral campaign, although still in the preliminary stages of the study, demonstrates the full potential of this approach for studying Italian politics as well. Building upon the datasets and these initial attempts at automated analysis, it will be possible to refine these tasks further and develop new ones. In this regard, the automatic analysis aimed at detecting and extracting text from images is particularly promising, as it can serve as a preliminary approach for conducting more in-depth analyses by combining these results with other tasks, such as sentiment analysis or the identification of symbolic elements related to Italy or Europe. In any case, the scope of these methodologies' applications to visual politics has only scratched the surface of their potential. Looking ahead, it is hoped that the call made by many scholars to promote an interdisciplinary perspective and increase investment in computational research in political communication will be heeded. This may lead to widespread and significant progress in understanding the role of visual politics on new digital platforms.

Methodologically, this work successfully demonstrated the effectiveness of applying a pre-trained Residual Vision Transformer (ReViT) to the task of classifying visual content from political campaigns as either webcards or images. By leveraging the strengths of the transformer architecture, particularly its ability to model global context and benefit from large-scale pre-training, and mitigating the feature collapse issue via ReViT's residual attention mechanism, we achieved high classification accuracy.

Future efforts could focus on developing more sophisticated systems that move beyond simple classification. A direction could be the development of interactive analysis systems, that is, creating tools that allow users to interact with the system, querying specific features or content within the images. This could involve integrating techniques like Visual Question Answering (VQA) (Kohavi, 1995) or object detection (Diko et al, 2024) tailored to elements relevant in a political context (e.g., identifying logos, specific individuals, analysing sentiment cues, etc.) – thus identifying possible paths for future interdisciplinary analysis of online visual political communication.

Despite this potential, applying computer vision to political imagery is not without risks, particularly regarding the accuracy of automated processing and the ethical implications for privacy and algorithmic bias. Future work must refine training regimes, mitigate discriminatory biases, and enhance model precision to develop shared best practices that ensure results are reliable, scalable, and, above all, fair and ethical.

REFERENCES

Altheide, D. L., & Snow, R. P. (1979). Media Logic. *Sage (Atlanta, Ga.)*.

Bene, M. (2017). Go Viral on the Facebook! Interactions between Candidates and Followers on Facebook during the Hungarian General Election Campaign of 2014. *Information Communication and Society*, *20*(4), 513–529.

Bennett, L. (1998). The UnCivic Culture: Communication, Identity, and the Rise of Lifestyle Politics. *PS, Political Science & Politics*, *31*(4), 740–761.

Bennett, W. L., & Segerberg, A. (2012). THE LOGIC OF CONNECTIVE ACTION: Digital media and the personalisation of contentious politics. *Information Communication and Society*, *15*(5), 739–768. DOI: 10.1080/1369118X.2012.670661

Blumler, J. G. (2016). The Fourth Age of Political Communication. *Political Communication*, *6*(1), 19–30. DOI: 10.3917/pdc.006.0019

Bonini, T., & Trerè, E. (2024). *Algorithms of Resistance. The Everyday Fight against Platform Power*. MIT Press.

Bossetta, M. (2018). The Digital Architectures of Social Media: Comparing Political Campaigning on Facebook, Twitter, Instagram, and Snapchat in the 2016 U.S. Election. *Journalism & Mass Communication Quarterly*, *95*(2), 471–496.

Bossetta, M., & Schmøkel, R. (2023). Cross-platform emotions and audience engagement in social media political campaigning: Comparing candidates' Facebook and Instagram images in the 2020 U.S. election. *Political Communication*, *40*(1), 48–68.

Bourdieu, P. (1996). *Sur la télévision*, Liber-Raisons d'agir.

Boussalis, C., & Coan, T. G. (2021). Facing the electorate: Computational approaches to the study of nonverbal communication and voter impression formation. *Political Communication*, *38*(1&2), 75–97.

Bucy, E. P. (2020). Invitations to Participation: How Immersive Presentations and Emotional Displays Promote Political Involvement. In Filimowicz, M., & Tzankova, V. (Eds.), *Reimagining Communication: Action* (Vol. 3, pp. 320–337). Routledge.

Bucy, E. P. (2023). Politics through machine eyes: What computer vision allows us to see. *Journal of Visual Political Communication*, *10*(1), 59–68.

Bucy, E. P., & Joo, J. (2021). Editors' Introduction: Visual Politics, Grand Collaborative Programs, and the Opportunity to Think Big. *The International Journal of Press/Politics*, *26*(1), 5–21.

Bull, P. E. (1987). *Posture and Gesture*. Pergamon Press.

Bull, P. E., & Mayer, K. (1993). How not to answer questions in political interviews. *Political Psychology*, *12*(1), 651–666.

Castells, M. (2009). *Communication Power*. Oxford University Press.

Chadwick, A. (2013). *The Hybrid Media System: Politics and Power*. Oxford University Press.

Chen, D., Park, K., & Joo, J. (2020). Understanding gender stereotypes and electoral success from visual self-presentations of politicians in social media. In *ATQAM/MAST'20: Joint Workshop on Aesthetic and Technical Quality Assessment of Multimedia and Media Analytics for Societal Trends*, (pp. 21–25) Association for Computing Machinery. https://dl.acm.org/doi/10.1145/3423268.3423583

Chollet, F. (2017). *Deep Learning with Python*. Manning Publications.

Dieterich, W., Mendoza, C., & Brennan, T. (2016). *Compas risk scales: Demonstrating accuracy, equity, and predictive parity*. Northpointe Inc.

Diko, A., Avola, D., Cascio, M., & Cinque, L. (2024). ReViT: Enhancing Vision Transformers Feature Diversity with Attention Residual Connections. *arXiv preprint arXiv:2402.11301*.

Dumitrescu, D. (2016). Nonverbal Communication in Politics: A Review of Research Developments, 2005–2015. *The American Behavioral Scientist*, *60*(14), 1656–1675.

Enli, G. S., & Skogerbø, E. (2013). Personalised Campaigns in Party-Centered Politics: Twitter and Facebook as Arenas for Political Communication. *Information Communication and Society*, *16*(5), 757–774.

Farkas, X. (2023). Visual Political Communication Research: A Literature Review from 2012 to 2022. *Journal of Visual Political Communication*, *10*(2), 95–126. https://doi.org/https://doi.org/10.1386/jvpc_00027_1

Farkas, X., & Bene, M. (2021). Images, Politicians, and Social Media: Patterns and Effects of Politicians' Image-Based Political Communication Strategies on Social Media. *The International Journal of Press/Politics*, *26*(1), 119–142.

Franklin, B. (1994). *Packaging Politics. Political Communication in Britain's Media Democracy*. Arnold.

Gerbaudo, P. (2024). TikTok and the algorithmic transformation of social media publics: From social networks to social interest clusters. *New Media & Society*, *0*(0). Advance online publication. DOI: 10.1177/14614448241304106

Glaeser, E. L., Kominers, S. D., Luca, M., & Naik, N. (2018). Big data and big cities: The promises and limitations of improved measures of urban life. *Economic Inquiry*, *56*(1), 114–137.

Grabe, M. E., & Bucy, E. P. (2009). *Image Bite Politics: News and the Visual Framing of Elections*. Oxford University Press.

Graber, D. A. (1988). *Processing the News: How People Tame The Information Tide*. Longman.

Haim, M., & Jungblut, M. (2021). Politicians' self-depiction and their news portrayal: Evidence from 28 countries using visual computational analysis. *Political Communication*, *38*(1&2), 55–74.

Hallin, D. C. (1992). Sound Bite News: Television Coverage of Elections, 1968–1988. *Journal of Communication*, *42*(2), 5–24.

Heiss, R., Schmuck, D., & Matthes, J. (2019). What Drives Interaction in Political Actors' Facebook Posts? Profile and Content Predictors of User Engagement and Political Actors' Reaction. *Information Communication and Society*, *22*(10), 1497–1513.

Hermans, L., & Vergeer, M. (2012). Personalisation in E-Campaigning: A Cross-National Comparison of Personalization Strategies Used on Candidate Websites of 17 Countries in EP Elections 2009. *New Media & Society*, *15*(1), 72–92.

Hoffmann, M., Santos, F. G., Neumayer, C., & Mercea, D. (2022). Lifting the veil on the use of big data news repositories: A documentation and critical discussion of a protest event analysis. *Communication Methods and Measures*, *16*(4), 283–302.

Hunziker, P., & Cederman, L. E. (2017). No extraction without representation: The ethno-regional oil curse and secessionist conflict. *Journal of Peace Research*, *54*(3), 365–381.

Joo, J., Li, W., Steen, F. F., & Zhu, S. C. (2014). Visual Persuasion: Inferring Communicative Intents of Images. In *Proceedings of the IEEE Conference on Computer Vision and Pattern Recognition* (pp. 216–223). OH: IEEE.

Joo, J., & Steinert-Threlkeld, Z. C. (2018). Image as data: Automated visual content analysis for political science. *arXiv preprint arXiv:1810.01544*. https://doi.org/DOI: 10.48550/ARXIV.1810.01544

Krause, A., & Bucy, E. P. (2018). Visual Framing of Fracking: How Standing Attitudes Shape Perceptions of Environmental Risk and Economic Benefit. *Environmental Communication*, *12*(3), 322–343.

Krizhevsky, A., Sutskever, I., & Hinton, G. E. (2012). ImageNet classification with deep convolutional neural networks. In *NIPS'12: Proceedings of the 25th International Conference on Neural Information Processing Systems,* (vol. 1, 3 December) pp. 1097–105. Curran Associates https://dl.acm.org/doi/10.5555/2999134.2999257

Laranjeira, C., Mota, V. F., & dos Santos, J. A. (2021). Machine learning Bias in Computer Vision: Why do I have to care? In Proceedings of the 34th SIBGRAPI Conference on Graphics, Patterns and Images (SIBGRAPI). Doi: DOI: 10.1109/SIBGRAPI54419.2021.00010

Lilleker, D. G. (2019). The Power of Visual Political Communication: Pictorial Politics Through the Lens of Communication Psychology. In A. Veneti, D. Jackson, & D. G. Lilleker (Eds.), *Visual Political Communication* (pp. 37-51). Palgrave.

Louw, E. (2010). The Media and Political Process. *Sage (Atlanta, Ga.).*

Manin, B. (1995). *Principes du Gouvernement Représentatif.* Calmann-Lévy.

Mazzoleni, G. (1998). *La Comunicazione Politica*. il Mulino.

McLuhan, M. (1964). *Understanding Media. The Extension of Man*. McGraw-Hill.

McQuail, D. (1992). Media Performance. Mass Communication and the Public Interest. *Sage (Atlanta, Ga.).*

Nimmo, D. D., & Combs, J. E. (1990). *Mediated Political Realities*. Longman.

Norris, P. (2000). *A Virtuous Circle. Political Communications in Postindustrial Societies*. Cambridge University Press.

Otterbacher, J., Bates, J., & Clough, P. (2017). "Competent men and warm women: Gender stereotypes and backlash in image search results," in *Proceedings of the 2017 chi conference on human factors in computing systems.* pp. 6620–6631.

Peng, Y. (2018). Same candidates, different faces: Uncovering media bias in visual portrayals of presidential candidates with computer vision. *Journal of Communication, 68*(5), 920–941.

Peng, Y. (2021). What makes politicians' Instagram posts popular? Analysing social media strategies of candidates and office holders with computer vision. *The International Journal of Press/Politics, 26*(1), 143–166.

Peng, Y., & Lu, Y. (2023). Computational visual analysis in political communication. In Lilleker, D., & Veneti, A. (Eds.), *Research Handbook on Visual Politics* (pp. 42–54). Edwin Elgar.

Plasser, F., & Plasser, G. (2002). *Global Political Campaigning: A Worldwide Analysis of Campaign Professionals and Their Practices*. Praeger.

Postman, N. (1985). *Amusing Ourselves to Death: Public Discourse in the Age of Show Business*. Penguin.

Prior, M. (2007). *Post-Broadcast Democracy: How Media Choice Increases Inequality in Political Involvement and Polarised Elections*. Cambridge University Press.

Risser, L., Picard, A., Hervier, L., & Loubes, J. (2022). A survey of Identification and mitigation of Machine Learning algorithmic biases in Image Analysis. *10.48550/arXiv.2210.04491.*

Ritter, E. H., & Conrad, C. R. (2016). Preventing and responding to dissent: The observational challenges of explaining strategic repression. *The American Political Science Review*, *110*(1), 85–99.

Rojas, M., David, M., Todorov, A., & Vitria, J. (2011). Automatic prediction of facial trait judgments: Appearance vs. structural models. *PLoS One*, *6*(8). PMID: 21858069

Sartori, G. (1997). *Homo Videns. Televisione e Post-Pensiero*. Laterza.

Schill, D. (2012). The Visual Image and the Political Image: A Review of Visual Communication Research in the Field of Political Communication. *The Review of Communication*, *12*(2), 118–142. DOI: 10.1080/15358593.2011.653504

Sobolev, A., Chen, M. K., Joo, J., & Steinert-Threlkeld, Z. C. (2020). News and geolocated social media accurately measure protest size variation. *The American Political Science Review*, *114*(4), 1343–1351.

Somit, A., & Peterson, S. A. (1998). Biopolitics After Three Decades: A Balance Sheet. *Journal of Political Science*, *25*, 59–571.

Stanyer, J. (2013). *Intimate Politics: Publicity, Privacy and the Personal Lives of Politicians in Media Saturated Democracies*. John Wiley & Sons.

Statera, G. (1986). *La Politica Spettacolo. Politici e Mass Media nell'era dell'immagine*. Mondadori.

Steinert-Threlkeld, Z. C. (2017). Spontaneous collective action: Peripheral mobilisation during the Arab Spring. *The American Political Science Review*, *111*(2), 379–403.

Street, J. (2004). Celebrity Politicians: Popular Culture and Political Representation. *British Journal of Politics and International Relations*, *6*(4), 435–452.

Strömbäck, J. (2008). Four Phases of Mediatization: An Analysis of the Mediatization of Politics. *The International Journal of Press/Politics*, *13*(3), 228–246.

Tapiador, F. J., Avelar, S., Tavares-Corrêa, C., & Zah, R. (2011). Deriving fine-scale socioeconomic information of urban areas using very high-resolution satellite imagery. *International Journal of Remote Sensing*, *32*(21), 6437–6456.

Thompson, J. B. (1995). *The Media and Modernity: A Social Theory of the Media*. Stanford University Press.

Towner, T. L., & Dulio, D. A. (2012). New Media and Political Marketing in the United States: 2012 and Beyond. *Journal of Political Marketing*, *1*(1-2), 95–119.

Van Aelst, P., Sheafer, T., & Stanyer, T. (2012). The Personalization of Mediated Political Communication: A Review of Concepts, Operationalisations and Key Findings. *Journalism*, *13*(2), 203–220.

Van Dijck, J., Poell, T., & De Waal, M. (2018). *The Platform Society: Public Values in a Connective World*. Oxford University Press.

Van Zoonen, L. (1998). "Finally, I have my mother back": Politicians and Their Families in Popular Culture. *The Harvard International Journal of Press/Politics*, *3*(1), 48–64.

Vernon, R. J., Sutherland, C. A., Young, A. W., & Hartley, T. (2014). Modelling first impressions from highly variable facial images. *Proceedings of the National Academy of Sciences of the United States of America*, *111*(32), E3353–E3361. PMID: 25071197

Weidmann, N. B., & Schutte, S. (2017). Using night light emissions for the prediction of local wealth. *Journal of Peace Research*, *54*(2), 125–140.

Williams, R. (1974). *Television: Technology and Cultural Form*. HarperCollins.

Xi, N., Ma, D., Liou, M., Steinert-Threlkeld, Z. C., Anastasopoulos, J., & Joo, J. (2020). Understanding the political ideology of legislators from social media images. In *Proceedings of the International AAAI Conference on Web and Social Media*, (14(1), pp. 726–37), https://doi.org/DOI: 10.1609/icwsm.v14i1.7338

Yoon, J., Joo, J., Park, E., & Han, J. (2020). *Cross-domain classification of facial appearance of leaders. Social Informatics, SocInfo 2020* (Vol. 12467). Lecture Notes in Computer Science. Springer., DOI: 10.1007/978-3-030-60975-7_32

Compilation of References

Aboshady, D., Ghannam, N., Elsayed, E., & Diab, L. (2022). The Malware Detection Approach in the Design of Mobile Applications. *Symmetry 2022, Vol. 14, Page 839, 14*(5), 839. https://doi.org/DOI: 10.3390/SYM14050839

Adair, B., Li, C., Yang, J., & Yu, C. (2017). Progress Toward "the Holy Grail": The Continued Quest to Automate Fact-Checking. *Northwestern University*. https://cj2017.northwestern.edu/documents/progress-cj2017-paper-18.pdf

Adhikary, I. (2024). Dress to Impress Confirms 'Brat' Collab with Charli XCX. *Bee Bom*. Available at: https://beebom.com/dress-to-impress-charli-xcx-brat-collab-teased/

Adner, R. (2002). When are technologies disruptive? A demand-based view of the emergence of competition. *Strategic Management Journal*, *23*, 667–688.

Adrian, A. (2012). An old narrative for a new culture: What cyber-culture can learn from indigenous culture. *International Journal of Liability and Scientific Enquiry*, *5*(3/4), 242. DOI: 10.1504/IJLSE.2012.051959

Agüero Santangelo, F., Nebot Adell, M., Pérez Giménez, A., López Medina, M. J., & García Continente, X. (2011). Actitudes y comportamientos preventivos durante la pandemia de gripe (H1N1) 2009 en España. *Revista Española de Salud Pública*, *85*(1), 73–80. https://www.mscbs.gob.es/biblioPublic/publicaciones/recursos_propios/resp/revista_cdrom/vol85/vol85_1/RS851C_portada.pdf PMID: 21750845

Airan, A., Patil, V., Narayan, V., & Kappor, A. (2024, July 2). Examining the Impact of Generative Artificial Intelligence on the Video Creation Industry. *aapti institute*. https://aapti.in/blog/examining-the-impact-of-generative-artificial-intelligence-on-the-video-creation-industry/, Retrieved 25.07.2024.

Akademie, D. W. (2022). "In Armenia, factchecking disinformation on politics and war", May 12. Retrieved from https://akademie.dw.com/en/in-armenia-factchecking-disinformation-on-politics-and-war/a-61757306 (Accessed 25/8/2025).

Akdemir, N. (2018). Visible Expression of Social Identity: The Clothing and Fashion. *Gaziantep University Journal of Social Sciences*, *17*(4), 1389–1397. DOI: 10.21547/jss.411181

Albeanu, C. (2016a, Feb., 24). Around two thirds of Independent's UK readers access it only from mobile. https://www.journalism.co.uk/news/around-two-thirds-of-independent-readers-in-the-uk-access-the-title-only-from-their-mobile-devices/s2/a613542/ Retrieved 10.08.2024.

Albeanu, C. (2016b, Aug., 17). The Washington Post rolls out automation for the Olympic Games, gearing up for US election coverage. https://www.journalism.co.uk/news/the-washington-post-rolls-out-automation-for-the-rio-olympic-games-gearing-up-for-us-election-coverage/s2/a665769/ Retrieved 10.08.2024.

Ali, W., & Hassoun, M. (2019). Artificial Intelligence and Automated Journalism: Contemporary Challenges and New Opportunities. *International Journal of Media* [IJMJMC]. *Journalism and Mass Communications*, *5*(1), 40–49. DOI: 10.20431/2454-9479.0501004

Altheide, D. L., & Snow, R. P. (1979). Media Logic. *Sage (Atlanta, Ga.).*

Amazeen, M. A. (2020). Journalistic interventions: The structural factors affecting the global emergence of fact-checking. *Journalism*, *21*(1), 95–111. DOI: 10.1177/1464884917730217

Anderson, C. W. (2016). News Ecosystems. In Witschge, T., Anderson, C. W., Domingo, D., & Hermida, A. (Eds.), *The Sage Handbook of Digital Journalism*. SAGE.

Anderson, M. (2020). Digital Fashion and the Ethics of Cultural Appropriation in Virtual Worlds. *Journal of Visual Culture*, *3*(1), 25–38.

Andersson, T. (2023). Nordic media systems. In Papathanassopoulos, S., & Miconi, A. (Eds.), *The media systems in Europe: Continuities and Discontinuities* (pp. 99–131). Springer., DOI: 10.1007/978-3-031-32216-7_5

Anzu blog (2023). From Runway To Gameplay: The Rise of Fashion in Gaming. 17/01/2023. Available at: https://www.anzu.io/blog/fashion-brands-in-game-advertising

Appio, F. P., Frattini, F., Petruzzelli, A. M., & Neirotti, P. (2021). Digital transformation and innovation management: A synthesis of existing research and an agenda for future studies. *. *Journal of Product Innovation Management*, *38*(1), 4–20. DOI: 10.1111/jpim.12526

Ardichvili, A., Maurer, M., Li, W., Wentling, T., & Stuedemann, R. (2006). Cultural influences on knowledge sharing through online communities of practice. *Journal of Knowledge Management*, *10*(1), 94–107. DOI: 10.1108/13673270610650139

Ardielli, E. (2017). Usage of ICT in EU Culture Sector as a Tool of Cultural Participation. *International Journal of Entrepreneurial Knowledge*, *5*(2), 33–46. DOI: 10.1515/ijek-2017-0009

Arguedas, A. R., & Simon, F. M. (2023). *Automating Democracy: Generative AI, Journalism, and the Future of Democracy* (p. 21). Balliol Interdisciplinary Institute, University of Oxford. Associated Press (n.d). ARTIFICIAL INTELLIGENCE: Leveraging AI to advance the power of facts. *Associated Press*. https://www.ap.org/solutions/artificial-intelligence/ Retrieved 10.08.2024.

Arora, T., & Grey, I. (2020). Health behaviour changes during COVID-19 and the potential consequences: A mini-review. *Journal of Health Psychology*, *25*(9). Advance online publication. DOI: 10.1177/1359105320937053 PMID: 32551944

Aswal, P. (2023). The Impact Of Pop Culture On Fashion & Aesthetics. *Indian Institute of Art and Design*. Available at:https://www.iiad.edu.in/the-circle/the-impact-of-pop-culture-on-fashion-and-aesthetics/

Ataguba, O., & Ataguba, J. E. (2020). Social determinants of health: The role of effective communication in the COVID-19 pandemic in developing countries. *Global Health Action*, *13*(1), 1788263. DOI: 10.1080/16549716.2020.1788263 PMID: 32657669

Automated Insights (n.d) Wordsmith. *Automated Insights*.https://automatedinsights.com/wordsmith/, Retrieved 07.07.2024.

Azam, M. S., & Quaddus, M. (2012). Adoption of e-commerce by the SMEs in Bangladesh: The effects of innovation characteristics and perceived risk. **Proceedings of the 15th International Business Research Conference**, 1–15.

Babarykina, N., Demianenko, O., & Mahda, Y. (2022). Decentralization as a global trend of democracy development. *Cuestiones Políticas*, *40*(73), 454–467. DOI: 10.46398/cuestpol.4073.24

Bagui, S., & Brock, H. (2023). Machine Learning for Android Scareware Detection. Https://Services.Igi-Global.Com/Resolvedoi/Resolve.Aspx?Doi=10.4018/JITR.298326, *15*(1), 1–15. https://doi.org/DOI: 10.4018/JITR.298326

Bahroun, Z., Anane, C., Ahmed, V., & Zacca, A. (2023). Transforming Education: A Comprehensive Review of Generative Artificial Intelligence in Educational Settings through Bibliometric and Content Analysis. [Switzerland]. *Sustainability*, *15*, 12983. Advance online publication. DOI: 10.3390/su151712983

Balch, K. (2022). Gaming and Fashion: Virtual Realities and the Democratization of Style. *Game Culture Review*, *8*(2), 112–129.

Bandura, A. (1985). *Social Foundations of Thought and Action: A Social Cognitive Theory*. Prentice Hall.

Baraniuk, C. (2018, Nov.8). *China's Xinhua agency unveils AI news presenter.* https://www.bbc.com/news/technology-46136504 Retrieved 15.10.2024

Barthes, R. (1967). *Système de la mode*. Seuil.

Bayazit, E. C., Sahingoz, O. K., & Dogan, B. (2022). A Deep Learning Based Android Malware Detection System with Static Analysis. *HORA 2022 - 4th International Congress on Human-Computer Interaction, Optimization and Robotic Applications, Proceedings*. https://doi.org/DOI: 10.1109/HORA55278.2022.9800057

Beattie, A., Edwards, A. P., & Edwards, C. (2020). A bot and a smile: Interpersonal impressions of chatbots and humans using emoji in computer-mediated communication. *Communication Studies*, *71*(3), 409–427. DOI: 10.1080/10510974.2020.1725082

Becker, S. J., Nemat, A. T., Lucas, S., Heinitz, R. M., Klevesath, M., & Charton, J. E. (2023). A Code of Digital Ethics: Laying the foundation for digital ethics in a science and technology company. *AI & Society*, *38*(6), 2629–2639. DOI: 10.1007/s00146-021-01376-w

Bene, M. (2017). Go Viral on the Facebook! Interactions between Candidates and Followers on Facebook during the Hungarian General Election Campaign of 2014. *Information Communication and Society*, *20*(4), 513–529.

Bennett, L. (1998). The UnCivic Culture: Communication, Identity, and the Rise of Lifestyle Politics. *PS, Political Science & Politics*, *31*(4), 740–761.

Bennett, W. L., & Segerberg, A. (2012). THE LOGIC OF CONNECTIVE ACTION: Digital media and the personalisation of contentious politics. *Information Communication and Society*, *15*(5), 739–768. DOI: 10.1080/1369118X.2012.670661

Bensaoud, A., & Kalita, J. (2022). Deep multi-task learning for malware image classification. *Journal of Information Security and Applications*, *64*, 103057. DOI: 10.1016/J.JISA.2021.103057

Bertola, P., & Teunissen, J. (2018). Fashion 4.0. Innovating Fashion Industry through Digital Transformation. *Research Journal of Textile and Apparel*, *22*, 352–369. DOI: 10.1108/RJTA-03-2018-0023

Bharati, P., & Chaudhury, A. (2019). SMEs and competitiveness: The role of information systems. *. *International Journal of Management and Enterprise Development*, *1*(1), 1–19.

Bigman, Y. E., Wilson, D., Arnestad, M. N., Waytz, A., & Gray, K. (2023). Algorithmic discrimination causes less moral outrage than human discrimination. *Journal of Experimental Psychology. General*, *152*(1), 4–27. DOI: 10.1037/xge0001250 PMID: 35758989

Bird, S., Barocas, S., Crawford, K., Diaz, F., & Wallach, H. (2016). Exploring or exploiting? Social *and ethical implications of autonomous experimentation in AI*. Workshop on Fairness, Accountability, and Transparency in Machine Learning. https://papers.ssrn.com/sol3/papers.cfm?abstract_id=2846909

Bish, A., & Michie, S. (2010). Demographic and attitudinal determinants of protective behaviours during a pandemic: A review. *British Journal of Health Psychology*, *15*, 797–824. DOI: 10.1348/135910710X485826 PMID: 20109274

Bitzer, T., Wiener, M., & Cram, W. A. (2023). Algorithmic transparency: Concepts, antecedents, and consequences. A review and research framework. *Communications of the Association for Information Systems*, *52*, 293–331.

Blumler, J. G. (2016). The Fourth Age of Political Communication. *Political Communication*, *6*(1), 19–30. DOI: 10.3917/pdc.006.0019

Bolander, E. C. (2024). Online Learning: Engaging Families Using Culturally Relevant Pedagogy. *Intervention in School and Clinic*, *59*(3), 199–206. DOI: 10.1177/10534512231156889

Bongomin, O., Gilibrays Ocen, G., Oyondi Nganyi, E., Musinguzi, A., & Omara, T. (2020). Exponential Disruptive Technologies and the Required Skills of Industry 4.0. *Journal of Engineering*, *4280156*, 1–17. DOI: 10.1155/2020/4280156

Bonini, T., & Trerè, E. (2024). *Algorithms of Resistance. The Everyday Fight against Platform Power*. MIT Press.

Boslaugh, S. E., Kreuter, M. W., Nicholson, R. A., & Naleid, K. (2005). Comparing demographic, health status and psychosocial strategies of audience segmentation to promote physical activity. *Health Education Research*, *20*(4), 430–438. DOI: 10.1093/her/cyg138 PMID: 15572439

Bossetta, M. (2018). The Digital Architectures of Social Media: Comparing Political Campaigning on Facebook, Twitter, Instagram, and Snapchat in the 2016 U.S. Election. *Journalism & Mass Communication Quarterly*, *95*(2), 471–496.

Bossetta, M., & Schmøkel, R. (2023). Cross-platform emotions and audience engagement in social media political campaigning: Comparing candidates' Facebook and Instagram images in the 2020 U.S. election. *Political Communication*, *40*(1), 48–68.

Bourdieu, P. (1996). *Sur la télévision*, Liber-Raisons d'agir.

Boussalis, C., & Coan, T. G. (2021). Facing the electorate: Computational approaches to the study of nonverbal communication and voter impression formation. *Political Communication*, *38*(1&2), 75–97.

Bower, L. J., & Christensen, M. C. (1995). Disruptive technologies: Catching the wave. *Harvard Business Review*, *73*, 43–53. https://hbr.org/1995/01/disruptive-technologies-catching-the-wave

Braojos, J., Benitez, J., & Llorens, J. (2019). How do social commerce-IT capabilities influence firm performance? Theory and empirical evidence. *. *Information & Management*, *56*(2), 155–171. DOI: 10.1016/j.im.2018.04.006

Braun, V., & Clarke, V. (2006). Using thematic analysis in psychology. *Qualitative Research in Psychology*, *3*(2), 77–101. DOI: 10.1191/1478088706qp063oa

Bridgman, A., Merkley, E., Loewen, P. J., Owen, T., Ruths, D., Teichmann, L., & Zihilin, O. (2020). The causes and consequences of COVID-19 misperceptions: Understanding the role of news and social media. *Harvard Kennedy School Misinformation Review*, *1*. Advance online publication. DOI: 10.37016/mr-2020-028

Broeder, P. (2021). Informed Communication in High Context and Low Context Cultures. *Journal of Education, Innovation, and Communication*, *3*(1). Advance online publication. DOI: 10.34097/jeicom-3-1-june21-1

Broussard, M., Diakopoulos, N., Guzman, A. L., Abebe, R., Dupagne, M., & Chuan, C. H. (2019). Artificial Intelligence and Journalism. *Journalism & Mass Communication Quarterly*, *96*(3), 673–695. DOI: 10.1177/1077699019859901/ASSET/1077699019859901.FP.PNG_V03

Brown-Jeffy, S., & Cooper, J. E. (2011). Toward a Conceptual Framework of Culturally Relevant Pedagogy. *Teacher Education Quarterly*. https://files.eric.ed.gov/fulltext/EJ914924.pdf

Brug, J., Aro, A. R., Oenema, A., Zwart, O., Richardus, J. H., & Bishop, G. D. (2004). SARS Risk Perception, Knowledge, Precautions, and Information Sources, the Netherlands. *Emerging Infectious Diseases*, *10*(8), 1486–1489. 10.3201%2Feid1008.040283 PMID: 15496256

Brunskill, D. (2013). Introducing the Social Avatar. *The British Journal of Psychiatry*, *203*, 386.

Bucy, E. P. (2020). Invitations to Participation: How Immersive Presentations and Emotional Displays Promote Political Involvement. In Filimowicz, M., & Tzankova, V. (Eds.), *Reimagining Communication: Action* (Vol. 3, pp. 320–337). Routledge.

Bucy, E. P. (2023). Politics through machine eyes: What computer vision allows us to see. *Journal of Visual Political Communication*, *10*(1), 59–68.

Bucy, E. P., & Joo, J. (2021). Editors' Introduction: Visual Politics, Grand Collaborative Programs, and the Opportunity to Think Big. *The International Journal of Press/Politics*, *26*(1), 5–21.

Bugg, J. (2010). The shifting focus: culture, fashion and identity. *24th International Costume Congress: Fashion, Culture and Identity*, (pp. 23-30). Yonsei University, Seoul, Korea.

Bull, P. E. (1987). *Posture and Gesture*. Pergamon Press.

Bull, P. E., & Mayer, K. (1993). How not to answer questions in political interviews. *Political Psychology*, *12*(1), 651–666.

Busch, T., & Götz, D. (2020). Virtual Economies and Digital Fashion. *Journal of Digital Culture Studies*, *5*(3), 210–225.

Cambosa, T. (2020). Avatars' wardrobe on Bitmoji, Snapchat gets upgrade with new Levi's tie-up. Available at: https://marketech-apac.com/avatars-wardrobe-on-bitmoji-snapchat-gets-upgrade-with-new-levis-tie-up/

Cao, Y., Chen, M., & Hershcovich, D. (2024). *Bridging Cultural Nuances in Dialogue Agents through Cultural Value Surveys* (Version 2). arXiv. DOI: 10.48550/ARXIV.2401.10352

Cao, Q., & Zhang, F. (2011). The impact of digital transformation on business performance: Evidence from financial firms. *. *Management Information Systems Quarterly*, *35*(4), 809–822. DOI: 10.2307/23043418

Cappellazzo, N. (2022, March 23). Algoritmi, automazione e meccanismi di intelligenza artificiale: la classificazione proposta dal Consiglio di Stato. Retrieved from https://www.federalismi.it/nv14/articolo-documento.cfm?Artid=46943

Caravaggion, G., & Orofino, M. (2023). Lingua e costituzione: L'irrompere dei linguaggi algoritmici. *Rivista AIC*, *4*, 155–184.

Carnahan, D., & Bergan, D. E. (2021). Correcting the Misinformed: The Effectiveness of Fact-checking Messages in Changing False Beliefs. *Political Communication*, *39*(2), 166–183. DOI: 10.1080/10584609.2021.1963358

Casonato, C., & Marchetti, B. (2021). Prime osservazioni sulla proposta di regolamento dell'Unione Europea in materia di intelligenza artificiale. *BioLaw Journal*, *3*, 415–437.

Castells, M. (2011b). Introduction in International Seminar on Network Theory: Network Multidimensionality in the Digital Age. The Annenberg Networks Network (ANN), University of Southern California. YouTube video, https://www.youtube.com/watch?v=vDh7eGJwhwg

Castells, M. (2000). *The Rise of the Network Society*. Wiley Blackwell.

Castells, M. (2009). *Communication Power*. Oxford University Press.

Castells, M. (2011a). Introduction to the Workshop: The Promise of Network theory. *International Journal of Communication*, *2011*(5), 794–795.

Çayır, A., Ünal, U., & Dağ, H. (2021). Random CapsNet forest model for imbalanced malware type classification task. *Computers & Security*, *102*, 102133. DOI: 10.1016/J.COSE.2020.102133

Chadwick, A. (2013). *The Hybrid Media System: Politics and Power*. Oxford University Press.

Chaki, J., & Dey, N. (2021). Data Tagging in Medical Images: A Survey of the State-of-Art. *Current Medical Imaging Formerly Current Medical Imaging Reviews*, *16*(10), 1214–1228. DOI: 10.2174/1573405616666200218130043 PMID: 32108002

Chameroy, F., Salgado, S., De Barnier, V., & Chaney, D. (2024). In platform we trust: How interchangeability affects trust decisions in collaborative consumption. *Technological Forecasting and Social Change*, *198*, 122997. DOI: 10.1016/j.techfore.2023.122997

Chemero, A., & Turvey, M. T. (2007). Complexity, hypersets, and the ecological perspective on perception-action. *Biological Theory*, *2*, 23–36.

Chen, D., Park, K., & Joo, J. (2020). Understanding gender stereotypes and electoral success from visual self-presentations of politicians in social media. In *ATQAM/MAST'20: Joint Workshop on Aesthetic and Technical Quality Assessment of Multimedia and Media Analytics for Societal Trends*, (pp. 21–25) Association for Computing Machinery. https://dl.acm.org/doi/10.1145/3423268.3423583

Cheong, P. H. (2020). Religion, Robots and Rectitude: Communicative Affordances for Spiritual Knowledge and Community. *Applied Artificial Intelligence, 34*(5), 412–431. DOI: 10.1080/08839514.2020.1723869

Cherubini, F., & Sharma, R. (2023, December 5). Changing Newsrooms 2023: Media leaders struggle to embrace diversity in full and remain cautious on AI disruption. *Reuters Institute* https://reutersinstitute.politics.ox.ac.uk/changing-newsrooms-2023-media-leaders-struggle-embrace-diversity-full-and-remain-cautious-ai#header--2, Retrieved 10/08/2024.

Chollet, F. (2017). *Deep Learning with Python*. Manning Publications.

Chomanski, B. (2022). Legitimacy and automated decisions: The moral limits of algocracy. *Ethics and Information Technology, 3*, 1–9.

Chong, I., & Proctor, R. W. (2020). On the Evolution of a Radical Concept: Affordances According to Gibson and Their Subsequent Use and Development. *Perspectives on Psychological Science, 15*(1), 117–132. DOI: 10.1177/1745691619868207

Christensen, C. M. (2014, January 1). *Disruptive Innovation*. Interaction Design Foundation - IxDF. https://www.interaction-design.org/literature/book/the-encyclopedia-of-human-computer-interaction-2nd-ed/disruptive-innovation

Christensen, C. (1997). *The Innovator's Dilemma*. Harvard Business School Press.

Christensen, C. M. (1997). *The Innovator's Dilemma: The Revolutionary Book that Will Change the Way You Do Business*. Harvard Business School Press.

Christopoulos, A., Styliou, M., Ntalas, N., & Stylios, C. (2024). The Impact of Immersive Virtual Reality on Knowledge Acquisition and Adolescent Perceptions in Cultural Education. *Information (Basel), 15*(5), 261. DOI: 10.3390/info15050261

Ciampaglia, G., Shiralkar, P., Rocha, L. M., Bollen, J., Menczer, F., & Flammini, A. (2015). Computational Fact Checking from Knowledge Networks. *PLoS One, 10*(6), e0128193. DOI: 10.1371/journal.pone.0128193

Cirksena, M. K., & Flora, J. A. (1995). Audience segmentation in worksite health promotion: A procedure using social marketing concepts. *Health Education Research, 10*(2), 211–224. DOI: 10.1093/her/10.2.211

Cohen, S., Kamarck, T., & Mermelstein, R. (1983). A Global Measure of Perceived Stress. *Journal of Health and Social Behavior*, *24*(4), 385–396. DOI: 10.2307/2136404 PMID: 6668417

Colapietro, C. (2023). Gli algoritmi tra trasparenza e protezione dei dati personali. *Federalismi.it, 5*, 151-174.

Cooper, C. (2024). Understanding Web3:The Decentralised Revolution of the Internet. *Medium.* Available at: https://medium.com/@colin-cooper/understanding-web3-the-decentralised-revolution-of-the-internet-2c867be44415

COPARMEX. (2020, junio 22). *Confederación Patronal de la República Mexicana.* Retrieved febrero 23, 2021, from Gobierno de México mala estrategia y errática comunicación frente al Covid-19: https://coparmex.org.mx/54308-2/

Corrado, A. (2023). La trasparenza necessaria per infondere fiducia in una amministrazione algoritmica e antropocentrica. *Federalismi.it, 5*, 175-215.

Council of Europe. Parliamentary Assembly. (2020, October 22). The brain-computer interface: new rights or new threats to fundamental freedoms? Resolution No. 2344. Retrieved from https://pace.coe.int/en/files/28810

COVIDiSTRESS Global Survey Network. (2020, mayo 10). *COVIDiSTRESS Global Survey*. https://doi.org/DOI: 10.17605/OSF.IO/Z39US

Cowling, B. J., Ng, D. M., Ip, D. K., Liao, Q., Lam, W. W., Wu, J. T., & Fielding, R. (2010). Community Psychological and Behavioral Responses through the First Wave of the 2009 Influenza A(H1N1) Pandemic in Hong Kong. *The Journal of Infectious Diseases*, *202*(6), 867–876. DOI: 10.1086/655811 PMID: 20677945

Cristianini, N., & Scartamburlo, T. (2020). On social machines for algorithmic regulation. *AI & Society*, *35*, 645–662.

Crosier, A., McVey, D., & French, J. (2014). 'By failing to prepare you are preparing to fail': Lessons from the 2009 H1N1 'swine flu' pandemic. *European Journal of Public Health*, *25*(1), 135–139. DOI: 10.1093/eurpub/cku131 PMID: 25125575

Dahlman, C. (2018). *Technology, globalization, and international competitiveness: Challenges for developing countries*. United Nations Publications.

Danaher, J. (2016). The threat of algocracy: Reality, resistance and accommodation. *Philosophy & Technology*, *3*, 245–268.

Datta, A., & Roumani, Y. (2015). Knowledge sharing in virtual communities: A study of citizenship behavior and social learning. *. *The Journal of Strategic Information Systems*, *24*(1), 57–74.

De Minico, G. (2021). Fundamental rights, European digital regulation and algorithmic challenge. *Medialaws*, *1*, 9–38.

De Vivo, I. (2024). Il sé allo specchio dell'algoritmo. Libertà epistemica e identità individuale. In Sterpa, A. (Ed.), *L'ordine giuridico dell'algoritmo* (pp. 51–69). Editoriale Scientifica.

Deck, A. (2024, May, 9). For the first time, two Pulitzer winners disclosed using AI in their reporting. https://www.niemanlab.org/2024/05/for-the-first-time-two-pulitzer-winners-disclosed-using-ai-in-their-reporting/ Retrieved 11.10.2024.

de-Lima-Santos, M. F., Yeung, W. N., & Dodds, T. (2024). *Guiding the Way: A Comprehensive Examination of AI Guidelines in Global Media.* arXiv:2405.04706. https://doi.org//arXiv.2405.04706DOI: 10.48550

Deuze, M., & Witschge, T. (2018). Beyond journalism: Theorizing the transformation of journalism. *Journalism*, *19*(2), 165–181. DOI: 10.1177/1464884916688550 PMID: 29417952

Dhiman, D. Bharat, Does Artificial Intelligence Help Journalists: A Boon or Bane? (March 24, 2023). Available at *SSRN*: https://ssrn.com/abstract=4401194

Diakopoulos, N. (2019). *Automating the news: how algorithms are rewriting the media.* Harvard University Press.

Diakopoulos, N., & Koliska, M. (2017). Algorithmic transparency in the news media. *Digital Journalism (Abingdon, England)*, *5*(7), 809–828.

Dias, N., & Sippitt, A. (2020). Researching Fact Checking: Present Limitations and Future Opportunities. *The Political Communication Quarterly*, *91*(3), 605–613. DOI: 10.1111/1467-923X.12892

Dierickx, L., Opdahl, A. L., & Khan, S. A.. (2024). A data-centric approach for ethical and trustworthy AI in journalism. *Ethics and Information Technology*, *26*, 64. DOI: 10.1007/s10676-024-09801-6

Dierickx, L., Sirén-Heikel, S., Lindén, C.-G., & Guerrero Rojas, D. C. (2024). Outsourcing, Augmenting, or Complicating: The Dynamics of AI in Fact-Checking Practices in the Nordics. *Emerging Media*, *2*(3), 449–473. DOI: 10.1177/27523543241288846

Dieterich, W., Mendoza, C., & Brennan, T. (2016). *Compas risk scales: Demonstrating accuracy, equity, and predictive parity*. Northpointe Inc.

Dieudonné, D. (2017). "Fact-checking the French election: lessons from CrossCheck, a collaborative effort to combat misinformation". Retrieved from https://blog.google/outreach-initiatives/google-news-initiative/fact-checking-french-election-lessons-crosscheck-collaborative-effort-combat-misinformation/ (Accessed 25/8/2025).

Diko, A., Avola, D., Cascio, M., & Cinque, L. (2024). ReViT: Enhancing Vision Transformers Feature Diversity with Attention Residual Connections. *arXiv preprint arXiv:2402.11301.*

Dimitriadis, D. (2023). *Disinformation Landscape in Greece*, EU DisinfoLab, June. (Project funded by the Friedrich Naumann Foundation for Freedom). Retrieved from www.disinfo.eu/wp-content/uploads/2023/06/20230623_GreeceDisinfoFS.pdf (Accessed 25/8/2025).

Dixon, C. (2024). *Read write own: building the next era of the Internet*. Random House.

Dul, J., & Hak, T. (2007). *Case study methodology in business research*. Elsevier.

Dumitrescu, D. (2016). Nonverbal Communication in Politics: A Review of Research Developments, 2005–2015. *The American Behavioral Scientist*, *60*(14), 1656–1675.

Ebrahimi, N., Singh, S. K. G., & Tabrizi, R. S. (2010). *Cultural Effect On Using New Technologies*. DOI: 10.5281/ZENODO.1056192

Efremov, J., Kertakova, M., & Kuzmanovska, V. D. (2021). Expression of Personality Through Dressing. *Tekstilna Industrija*, *69*(1), 28–35.

Ehrlinger, L., Haunschmid, V., Palazzini, D., & Lettner, C. (2019). A DaQL to monitor data quality in machine learning applications. In *Lecture Notes in Computer Science* (pp. 227–237). Springer.

Eisenstein, E. (1979). *The Printing Press as an Agent of Change* (Vol. I and II). Cambridge University Press.

Elguera Echegaray, E. (2020). *C&E*. Retrieved febrero 23, 2021, from Comunicación y estrategia política en tiempos de COVID-19: https://www.ceonline.com.mx/post/comunicacion-y-estrategia-politica-en-tiempos-de-covid-19

Enli, G. S., & Skogerbø, E. (2013). Personalised Campaigns in Party-Centered Politics: Twitter and Facebook as Arenas for Political Communication. *Information Communication and Society*, *16*(5), 757–774.

Esposito, J. (2010). Publishing After the Apocalypse. *Logos*, *21*(3-4), 13–19.

Eule, A. (2013, Aug.31). Big Data and Yahoo's Quest for Mass Personalization. *BARRON'S*. https://www.barrons.com/articles/SB500014240527487047192045790287524666630862. Retrieved 07.08.2024.

European Media and Information Fund. (2025). "Fact-Checking and Countering Disinformation: Poland's 2025 Presidential Election", March 25. Retrieved from https://gulbenkian.pt/emifund/projects/fact-checking-and-countering-disinformation-polands-2025-presidential-election/#:~:text=The%20project%20also%20emphasizes%20disinformation,Updated%20on%2025%20march%202025 (Accessed 25/8/2025).

European Parliament Research Service (EPRS). (2023, March). General-purpose artificial intelligence. PE 745.708. Retrieved from https://www.europarl.europa.eu/RegData/etudes/ATAG/2023/745708/EPRS_ATA(2023)745708_EN.pdf

European Parliament. (2020). Framework of ethical aspects of artificial intelligence, robotics and related technologies. Resolution P9_TA 0275. Retrieved from https://www.europarl.europa.eu/doceo/document/TA-9-2020-0275_EN.html

European Parliament. (2022). Artificial intelligence in a digital age. Resolution P9_TA 0140. Retrieved from https://www.europarl.europa.eu/doceo/document/TA-9-2022-0140_EN.html

Evans, W. D. (2006). How social marketing works in health care. *BMJ (Clinical Research Ed.)*, *332*(7551), 1207–1210. 10.1136%2Fbmj.332.7551.1207-a PMID: 16710002

Fan, S., Barlas, P., Christoforou, E., Otterbacher, J., Sadiq, S., & Demartini, G. (2022). Socio-Economic Diversity in Human Annotations. *14th ACM Web Science Conference 2022*, 98–109. DOI: 10.1145/3501247.3531588

Farhi, P. (2023, Jun, 17). A news site used AI to write articles. It was a journalistic disaster. *The Washington Post*. https://www.washingtonpost.com/media/2023/01/17/cnet-ai-articles-journalism-corrections/ Retrieved 12.10.2024.

Farkas, X. (2023). Visual Political Communication Research: A Literature Review from 2012 to 2022. *Journal of Visual Political Communication*, *10*(2), 95–126. https://doi.org/https://doi.org/10.1386/jvpc_00027_1

Farkas, X., & Bene, M. (2021). Images, Politicians, and Social Media: Patterns and Effects of Politicians' Image-Based Political Communication Strategies on Social Media. *The International Journal of Press/Politics*, *26*(1), 119–142.

Fasan, M. (2022). I principi costituzionali nella disciplina dell'Intelligenza Artificiale. Nuove prospettive interpretative. *DPCE Online*, *1*, 181–199.

Fingas, J. (2019). A digital 'dress' sold for $9,500. *Engagement*. Available at: https://www.engadget.com/2019-05-27-fabricant-blockchain-digital-dress.html?guccounter=1&guce_referrer=aHR0cHM6Ly93d3cuZ29vZ2xlLmNvbS8&guce_referrer_sig=AQAAAMfAOtfKtFW0tRw8QZFHD2n-uCj_FfP9d3sXDxSpPpQF80HY21MQDp_vABlFcrxw2KeJw0EToxjx7V5Db8YT_wyvDCN9yhha3SyFGuqtrod2CSWSdF72JQoz_SnMnXxP1QvGAYnt6VD9wb2xxMklNMBta6hax_AWaJwsM_kPLZrb

Fishbein, M., & Ajzen, I. (2015). *Predicting and Changing Behavior: The Reasoned Action Approach*. Routledge.

Fletcher, R., & Nielsen, R. K. (2024). *AI and the Future of News: What does the public in six countries think of generative AI in news?* Reuters Institute for the Study of Journalism.

Flores Villanueva, C. A., Garza Villegas, J. B., & Olague, J. T. (2017). Identifying strategic groups: An assessment in Mexican franchises. *Revista Ibero-Americana de Estratégia*, *16*(2), 20–34. DOI: 10.5585/riae.v16i2.2352

Floridi, L., Cowls, J., Beltrametti, M., Chatila, R., Chazerand, P., Dignum, V., Luetge, C., Madelin, R., Pagallo, U., Rossi, F., Schafer, B., Valcke, P., & Vayena, E. (2018). AI4People – An Ethical Framework for a Good AI Society: Opportunities, Risks, Principles, and Recommendations. *Minds and Machines*, *28*(1), 689–707.

Fofana, N. K., Latif, F., & Sarfraz, S., Bilal, Bashir, M. F., & Komal, B. (2020). Fear and agony of the pandemic leading to stress and mental illness: An emerging crisis in the novel coronavirus (COVID-19) outbreak. *Psychiatry Research*, *291*, 113230. DOI: 10.1016/j.psychres.2020.113230 PMID: 32593067

Forthofer, M. S., & Bryant, C. A. (2000). Using Audience-Segmentation Techniques to Tailor Health Behavior Change Strategies. *American Journal of Health Behavior*, 36 - 43. https://doi.org/DOI: 10.5993/AJHB.24.1.6

Franklin, B. (1994). *Packaging Politics. Political Communication in Britain's Media Democracy*. Arnold.

Freimuth, V., Linnan, H. W., & Potter, P. (2000). Communicating the Threat of Emerging Infections to the Public. *Perspectives*, 337 - 347. https://doi.org/DOI: 10.3201/eid0604.000403

Fridkin, K., Kenney, P. J., & Wintersieck, A. (2015). Liar, Liar, Pants on Fire: How Fact-Checking Influences Citizens' Reactions to Negative Advertising. *Political Communication*, *32*(1), 127–151. DOI: 10.1080/10584609.2014.914613

Gajendran, T., & Brewer, G. (2012). Cultural consciousness and the effective implementation of information and communication technology. *Construction Innovation, 12*(2), 179–197. DOI: 10.1108/14714171211215930

Gallofré Ocaña, M., & Opdahl, A. L. (2020). "Challenges and Opportunities for Journalistic Knowledge Platforms". *Proceedings of the CIKM 2020 Workshops*, October 19-20, Galway, Ireland.

Garber, D. (2004). Growing Virtual Communities. *International Review of Research in Open and Distance Learning*, *5*(2). Advance online publication. DOI: 10.19173/irrodl.v5i2.177

García-Gordillo, M., Rivas-de-Roca, R., & de-Lima-Santos, M. (2025). The Fact-Checking Initiatives in the EU: A Diverse Ecosystem Against Disinformation. *Media and Communication*, *13*, 9421. Advance online publication. DOI: 10.17645/mac.9421

García-Marín, D., Elías, C., & Soengas-Pérez, X. (2022). "Big Data and Disinformation: Algorithm Mapping for Fact Checking and Artificial Intelligence". In J. Vázquez-Herrero, A. Silva-Rodríguez, M. C. Negreira-Rey, C. Toural-Bran, & X. López-García (Eds.), *Total Journalism*. Studies in Big Data, vol 97 (pp. 123-135). Springer, Cham. https://doi.org/DOI: 10.1007/978-3-030-88028-6_10

Garrett, R. K., & Weeks, B. E. (2013). "The promise and peril of real-time corrections to political misperceptions". In Conference Proceedings of *the 2013 Conference on Computer-supported Cooperative Work*. New York: Association for Computing Machinery (pp. 1047-1058). https://doi.org/DOI: 10.1145/2441776.244189

Gaver, W. W. (1991). Technology affordances. In Proceedings of the *SIGCHI Conference on Human Factors in Computing Systems: Reaching through Technology*. New York: Association for Computing Machinery (pp. 79–84).

Gerbaudo, P. (2024). TikTok and the algorithmic transformation of social media publics: From social networks to social interest clusters. *New Media & Society*, *0*(0). Advance online publication. DOI: 10.1177/14614448241304106

Gholampour Rad, M. (2017). Disruptive innovation in media industry ecosystem and need for improving managerial cognitive capabilities in polymediation era. *Cogent Business & Management*, *4*(1). Advance online publication. DOI: 10.1080/23311975.2017.1352183

Gianni, R., Lehtinen, S., & Nieminen, M. (2022). Governance of responsible AI: From ethical guidelines to cooperative policies. *Frontiers of Computer Science*, *4*, 873437. DOI: 10.3389/fcomp

Gibbs, J. L., Gibson, C. B., Grushina, S. V., & Dunlop, P. D. (2021). Understanding orientations to participation: Overcoming status differences to foster engagement in global teams. *European Journal of Work and Organizational Psychology*, *30*(5), 653–671. DOI: 10.1080/1359432X.2020.1844796

Gibson, J. J. (1979). *The ecological approach to visual perception*. Houghton Mifflin.

Gildert, N., Pomfret, A., & Timmis, J. (2019). Comparing The Performance of Explicit and Implicit Communication in Simple Object Manipulation Tasks. *2019 IEEE 15th International Conference on Automation Science and Engineering (CASE)*, 1456–1462. DOI: 10.1109/COASE.2019.8843011

Girasa, R. (2020). *Artificial Intelligence as a Disruptive Technology: Economic Transformation and Government Regulation*. Palgrave Macmillan., DOI: 10.1007/978-3-030-35975-1

Glaeser, E. L., Kominers, S. D., Luca, M., & Naik, N. (2018). Big data and big cities: The promises and limitations of improved measures of urban life. *Economic Inquiry*, *56*(1), 114–137.

Glik, D. C. (2007). Risk Communication for Public Health Emergencies. *Annual Review of Public Health*, *28*(1), 33–54. DOI: 10.1146/annurev.publhealth.28.021406.144123 PMID: 17222081

Gobierno de México. (2020a). *Coronavirus*. Retrieved octubre 1, 2020, from Todo sobre el coronavirus: https://coronavirus.gob.mx/

Gobierno de México. (2020b, marzo 23). *Gobierno de México*. Retrieved mayo 1, 2020, from Jornada Nacional de Sana Distancia: https://www.gob.mx/cms/uploads/attachment/file/541687/Jornada_Nacional_de_Sana_Distancia.pdf

Goerzig, D., & Bauernhansl, T. (2018). Industry 4.0: How it changes the society and the economy. In *Digital Economy* (pp. 89–102). Springer.

Goffman, E. (1959). *The Presentation of Self in Everyday Life*. Anchor Books.

Gollmitzer, M. (2023). Journalism Ethics with Foucault: Casually Employed Journalists' Constructions of Professional Integrity. *Journalism*, *24*, 1015–1033. DOI: 10.1177/14648849211036301

Gómez, A., Loar, R., Kramer, A. E., & Garnett, G. P. (2019). Reaching and targeting more effectively: The application of market segmentation to improve HIV prevention programmes. *Journal of the International AIDS Society*, *22*(S4), e25318. DOI: 10.1002/jia2.25318 PMID: 31328397

Gonzales, A. L., & Hancock, J. T. (2011). Mirror, Mirror on My Facebook Wall: Effects of Exposure to Facebook on Self-Esteem. *Cyberpsychology, Behavior, and Social Networking*, *14*, 79–83. DOI: 10.1089/cyber.2009.0411

González Díaz, M. (2020, marzo 17). *BBC News*. Retrieved febrero 23, 2021, from Mundo: https://www.bbc.com/mundo/noticias-america-latina-51921323

Grabe, M. E., & Bucy, E. P. (2009). *Image Bite Politics: News and the Visual Framing of Elections*. Oxford University Press.

Graber, D. A. (1988). *Processing the News: How People Tame The Information Tide*. Longman.

Granzin, K. L., Olsen, J. E., & Painter, P. J. (1998). Marketing to consumer segments using health-promoting lifestyles. *Journal of Retailing and Consumer Services*, *5*(3), 131–141. DOI: 10.1016/S0969-6989(96)00059-8

Graves, L., & Cherubini, F. (2016). *The Rise of Fact-Checking Sites in Europe – Digital News Project 2016*. Reuters Institute for the Study of Journalism, with the support of Google Digital News Initiative. DOI: . Retrieved from https://reutersinstitute.politics.ox.ac.uk/sites/default/files/research/files/The%2520Rise%2520of%2520Fact-Checking%2520Sites%2520in%2520Europe.pdfDOI: 10.60625/risj-tdn4-p140

Graves, L. (2007). The affordances of blogging: A case study in culture and technological effects. *The Journal of Communication Inquiry*, *31*(4), 331–346.

Graves, L. (2016). *Deciding what's true: The rise of political fact-checking in American journalism*. Columbia University Press.

Graves, L., & Amazeen, M. A. (2019). Fact-checking as idea and practice in journalism. In Nussbaum, J. (Ed.), *Oxford Research Encyclopedia of Communication*. Oxford University Press., DOI: 10.1093/acrefore/9780190228613.013.808

Graves, L., & Anderson, C. (2020). Discipline and promote: Building infrastructure and managing algorithms in a "structured journalism" project by professional fact-checking groups. *New Media & Society*, *22*(2), 342–360. DOI: 10.1177/1461444819856916

Gravili, G. (2016). Sharing Knowledge through Social Media: The Influence of National Cultures. *Proceedings of the 2016 International Conference on Artificial Intelligence: Technologies and Applications*. 2016 International Conference on Artificial Intelligence: Technologies and Applications, Bangkok, Thailand. DOI: 10.2991/icaita-16.2016.34

Grgić-Hlača, N., Zafar, M. B., Gummadi, K. P., & Weller, A. (2017). *On Fairness, Diversity and Randomness in Algorithmic Decision Making* (No. arXiv:1706.10208). arXiv. https://arxiv.org/abs/1706.10208

Grier, S., & Bryant, C. A. (2005). Social marketing in public health. *Annual Review of Public Health, 26*, 319–339. DOI: 10.1146/annurev.publhealth.26.021304.144610 PMID: 15760292

Grimaldi, L. (2022). Costituzionalismo "post-umano" alla prova della decisione pubblica algoritmica. *Federalismi.it, 34*, 75-95.

Guidotti, R., Monreale, A., Ruggieri, S., Turini, F., Giannotti, F., & Pedreschi, D. (2019). A survey of methods for explaining black box models. *ACM Computing Surveys, 5*, 1–42.

Gupta, R., & Pal, S. K. (2020). Learning and Developing With Each Other: Case of Collaborative Innovation Through ICT in India and Canada. In Vij Mali, N. (Ed.), (pp. 132–146). Advances in Electronic Government, Digital Divide, and Regional Development. IGI Global., DOI: 10.4018/978-1-5225-5412-7.ch006

Guţă, A. J. (2021). The role of social media platforms in increasing the efficiency of online communication activity. *MATEC Web of Conferences, 342*, 08011. DOI: 10.1051/matecconf/202134208011

Gutiérrez-Caneda, B., & Vázquez-Herrero, J. (2024). Redrawing the Lines Against Disinformation: How AI Is Shaping the Present and Future of Fact-checking. *Tripodos*, (55), 55–74. DOI: 10.51698/tripodos.2024.55.04

Gutiérrez-Caneda, B., Vázquez-Herrero, J., & López-García, X. (2023). AI application in journalism: ChatGPT and the uses and risks of an emergent technology. *El Profesional de la Información, 32*(5), e320514. DOI: 10.3145/epi.2023.sep.14

Hagger, M. S., Smith, S. R., Keech, J. J., Moyers, S. A., & Hamilton, K. (2020). Predicting Social Distancing Intention and Behavior During the COVID-19 Pandemic: An Integrated Social Cognition Model. *Annals of Behavioral Medicine, 54*, 713–727. DOI: 10.1093/abm/kaaa073 PMID: 32914831

Haim, M., & Jungblut, M. (2021). Politicians' self-depiction and their news portrayal: Evidence from 28 countries using visual computational analysis. *Political Communication, 38*(1&2), 55–74.

Hair, J. F., Hult, G. T. M., Ringle, C., & Sarstedt, M. (2017). *A Primer on Partial Least Squares Structural Equation Modeling (PLS-SEM)* (2nd ed.). Sage Publications.

Hair, J. F.Jr, Black, W. C., Babin, B. J., & Anderson, R. E. (1999). *Análisis Multivariante* (Prentice, E., & Cano, D., Trans.). Prentice Hall.

Häkkilä, J., Wiberg, M., Eira, N. J., Seppänen, T., Juuso, I., Mäkikalli, M., & Wolf, K. (2020). Design Sensibilities—Designing for Cultural Sensitivity. *Proceedings of the 11th Nordic Conference on Human-Computer Interaction: Shaping Experiences, Shaping Society*, 1–3. DOI: 10.1145/3419249.3420100

Hallin, D. C. (1992). Sound Bite News: Television Coverage of Elections, 1968–1988. *Journal of Communication*, *42*(2), 5–24.

Hannaford, E. D., Schlegel, V., Lewis, R., Ramsden, S., Bunn, J., Moore, J., Alexander, M., Barker, H., Batista-Navarro, R., Hughes, L., & Nenadic, G. (2024). Our Heritage, Our Stories: Developing AI tools to link and support community-generated digital cultural heritage. *The Journal of Documentation*. Advance online publication. DOI: 10.1108/JD-03-2024-0057

Harper, C. A., Satchell, L. P., Fido, D., & Latzman, R. D. (2020). Functional Fear Predicts Public Health Compliance. *International Journal of Mental Health and Addiction*. Advance online publication. DOI: 10.1007/s11469-020-00281-5 PMID: 32346359

Hassim, M. N., & Mohamad Nasir, N. N. A. (2025). The Rise of AI in Journalism: Ethical and Practical Implications. *E- Journal of Media and Society, 8*(2), 46–58. https://doi.org/DOI: 10.24191/ejoms.v8i2.5841

Heiss, R., Schmuck, D., & Matthes, J. (2019). What Drives Interaction in Political Actors' Facebook Posts? Profile and Content Predictors of User Engagement and Political Actors' Reaction. *Information Communication and Society*, *22*(10), 1497–1513.

Helberger, N., & Diakopoulos, N. (2022). The European AI Act and how it matters for research into AI in media and journalism. *Digital Journalism, online first*. https://doi.org/DOI: 10.1080/21670811.2022.2 082505

Hellmueller, L., Vos, T. P., & Poepsel, M. A. (2013). Shifting Journalistic Capital?: Transparency and Objectivity in the Twenty-First Century. *Journalism Studies*, *14*(3), 287–304. DOI: 10.1080/1461670X.2012.697686

Hermans, L., & Vergeer, M. (2012). Personalisation in E-Campaigning: A Cross-National Comparison of Personalization Strategies Used on Candidate Websites of 17 Countries in EP Elections 2009. *New Media & Society*, *15*(1), 72–92.

Hess, T., Matt, C., Benlian, A., & Wiesböck, F. (2016). Options for formulating a digital transformation strategy. *. *MIS Quarterly Executive*, *15*(2), 123–139.

Hines, Kr. (2023, June 4). History Of ChatGPT: A Timeline OfThe Meteoric Rise Of Generative AI Chatbots. *Search Engine Journal*. https://www.searchenginejournal.com/history-of-chatgpt-timeline/488370/, Retrieved 07.07.2024.

Hoffmann, M., Santos, F. G., Neumayer, C., & Mercea, D. (2022). Lifting the veil on the use of big data news repositories: A documentation and critical discussion of a protest event analysis. *Communication Methods and Measures*, *16*(4), 283–302.

Hopster, J. (2021). What are socially disruptive technologies? *Technology in Society*, *67*(101750), 1–8. DOI: 10.1016/j.techsoc.2021.101750

Horska, K. (2020). A New Test of Artificial Intelligence: Should the Media Industry Be Afraid? *Science and Education a New Dimension*, *VIII*(39), 26–29. DOI: 10.31174/send-hs2020-231viii39-06

Hrckova, A., Moro, R., Srba, I., Simko, J., & Bielikova, M. (2024). Autonomation, not Automation: Activities and Needs of Fact-checkers as a Basis for Designing Human-Centered AI Systems. *arXiv:2211.12143v2 [cs.CY]*. https://arxiv.org/html/2211.12143v2

Huang, Y., Lv, S., Tseng, K., Tseng, P., Xie, X., & Lin, R. F. (2023). Recent advances in artificial intelligence for video production system. *Enterprise Information Systems*, *17*(2). Advance online publication. DOI: 10.1080/17517575.2023.2246188

Humprecht, E. (2019). How Do They Debunk "Fake News"? A Cross-National Comparison of Transparency in Fact Checks. *Digital Journalism (Abingdon, England)*, *8*(3), 310–327. DOI: 10.1080/21670811.2019.1691031

Hunziker, P., & Cederman, L. E. (2017). No extraction without representation: The ethno-regional oil curse and secessionist conflict. *Journal of Peace Research*, *54*(3), 365–381.

Hutchby, I. (2014). Communicative affordances and participation frameworks in mediated interaction. *Journal of Pragmatics*, *72*, 86–89.

Hwang, G.-J., & Chien, S.-Y. (2022). Definition, roles, and potential research issues of the metaverse in education: An artificial intelligence perspective. *Comput. Educ. Artif. Intell.*, *3*, 100082. DOI: 10.1016/j.caeai.2022.100082

Hyun, K. K., & Yoan, R. S. (2018). Variations in the Asian Collectivistic Working Culture in Intercultural Collaboration: A Case of a South Korean Company in Malaysia. *Social Sciences & Humanities*, *26*(3), 1633–1650.

Hyvärinen, J., & Vos, M. (2016). Communication Concerning Disasters and Pandemics: Co-producing community resilience and crisis response. In A. Schwartz, M. Seeger, & C. Auer (Eds.), *The Handbook of International Crisis Communication Research* (p. 552). Hoboken, EE.UU.: Wiley - Blackwell.

Idoiaga, N., Berasategi, N., Eiguren, A., & Picaza, M. (2020). Exploring Children's Social and Emotional Representations of the COVID-19 Pandemic. *Frontiers in Psychology*, *11*, 1952. DOI: 10.3389/fpsyg.2020.01952 PMID: 32922334

IFCN. (2025). The commitments of the Code of Principles: Eligibility to be a signatory. *International Fact-Checking Network – Poynter*. https://ifcncodeofprinciples.poynter.org/the-commitments (last access 10/1/2025).

Ihm, J., & Lee, C.-J. (2021). Toward More Effective Public Health Interventions during the COVID-19 Pandemic: Suggesting Audience Segmentation Based on Social and Media Resources. *Health Communication*, *36*(1), 98–108. DOI: 10.1080/10410236.2020.1847450 PMID: 33198528

Immonen, M., & Sintonen, S. (2015). Customer experience and digital transformation: A framework for understanding the role of technology in customer interaction. *. *International Journal of Information Management*, *35*(2), 215–224.

Ishikawa, Y., Nishiuchi, H., Hayashi, H., & Viswanath, K. (2012). Socioeconomic status and health communication inequalities in Japan: A nationwide cross-sectional survey. *PLoS One*, *7*(7), e40664. DOI: 10.1371/journal.pone.0040664 PMID: 22808229

Izotova, N. N. (2012). Ethno-cultural characteristics of Japanese communication style. *MGIMO Review of International Relations*, *6*(27), 179–182. DOI: 10.24833/2071-8160-2012-6-27-179-182

Jamieson, K., & Albarracin, D. (2020). The Relation between Media Consumption and Misinformation at the Outset of the SARS-CoV-2 Pandemic in the US. *Harvard Kennedy School Misinformation Review*, *1*. Advance online publication. DOI: 10.37016/mr-2020-012

Jamil, S. (2023). Evolving newsrooms and the second level of digital divide: Implications for journalistic practice in Pakistan. *Journalism Practice*, *17*(9), 1864–1881. DOI: 10.1080/17512786.2022.2026244

Jekov, B., Petkova, P., Parusheva, Y., & Shoikova, E. (2022). Disruptive Technologies—Artificial Intelligence and Blockchain in Education. In Proceedings of the *11th Annual International Conference of Education, Research and Innovation (ICERI)*. IATED Academy (pp. 6784–6793).

Johns Hopkins University. (2021). *Center for Systems Science and Engineering (CSSE).* Retrieved octubre 14, 2020, from COVID-19 Dashboard: https://gisanddata.maps.arcgis.com/apps/opsdashboard/index.html#/bda7594740fd40299423467b48e9ecf6

Joo, J., & Steinert-Threlkeld, Z. C. (2018). Image as data: Automated visual content analysis for political science. *arXiv preprint arXiv:1810.01544*. https://doi.org/DOI: 10.48550/ARXIV.1810.01544

Joo, J., Li, W., Steen, F. F., & Zhu, S. C. (2014). Visual Persuasion: Inferring Communicative Intents of Images. In *Proceedings of the IEEE Conference on Computer Vision and Pattern Recognition* (pp. 216–223). OH: IEEE.

Joshua, R. (2011). *How Competition over Limited Natural Resources Impacts the Formation of Culture* [Undergraduate Capstone Project / Master of Liberal Studies, University College, University of Denver]. https://digitalcommons.du.edu/ucol_gs/8/

Juneja, P., & Mitra, T. (2022). Human and Technological Infrastructures of Fact-checking. *Proceedings of the ACM on Human-Computer Interaction*, 6 (CSCW2), Article 418. https://doi.org/DOI: 10.1145/3555143

Kaiser, S. B., & Green, D. N. (2022). *Fashion and cultural studies*. Bloomsbury.

Kalogeropoulos, A. (17 June 2024). Digital Report 2024, Greece. Reuters Institute for the Study of Journalism. https://reutersinstitute.politics.ox.ac.uk/digital-news-report/2024/greece retrieved 15/11/2024.

Kamate, S. K., Agrawal, A., Chaudhary, H., Singh, K., Mishra, P., & Asawa, K. (2010). Public knowledge, attitude and behavioural changes in an Indian population during the Influenza A (H1N1) outbreak. *Journal of Infection in Developing Countries*, *4*(1), 7–14. DOI: 10.3855/jidc.501 PMID: 20130372

Kammer, A. (2020). Researching Affordances. In Hunsinger, J., Allen, M., & Klastrup, L. (Eds.), *Second International Handbook of Internet Research*. Springer., DOI: 10.1007/978-94-024-1555-1_7

Kampermann, A., Opdenakker, R., Heijden, B. V. D., & Bücker, J. (2021). Intercultural Competencies for Fostering Technology-Mediated Collaboration in Developing Countries. *Sustainability*, *13*(14), 7790. DOI: 10.3390/su13147790

Kaperonis, S. (2024). How artificial intelligence (AI) is transforming the user experience in digital marketing. In *The use of artificial intelligence in digital marketing: Competitive strategies and tactics* (pp. 117–141). IGI Global.

Kaplan, J. (2025, January 7). "More Speech and Fewer Mistakes", about.meta.com. https://about.fb.com/news/2025/01/meta-more-speech-fewer-mistakes/

Kapos, P. (2023). *The Ecosystem of the book publishing industry in the 21st century: Digital transition and the Greek case (2009-2019)*. I. Sideris Publications. (in Greek)

Karagöz, E. (2024). Fashion in the Gaming Industry. *iion.io*. Available at: https://www.iion.io/blog/fashion-in-the-gaming-industry

Karimi, J., & Walter, Z. (2015). The Role of Dynamic Capabilities in Responding to Digital Disruption: A Factor-Based Study of the Newspaper Industry. *Journal of Management Information Systems*, *32*(1), 39–81. DOI: 10.1080/07421222.2015.1029380

Kasl, S. V., & Cobb, S. (1966). Health Behavior, Illness Behavior, and Sick-Role Behavior. *Archives of Environmental Health: An International Journal*, *12*(2), 246–266. DOI: 10.1080/00039896.1966.10664421 PMID: 5322534

Kawamura, Y. (2005). *Fashion-ology: An Introduction to Fashion Studies*. Berg Publishers.

Kent, E. (2024). Viral sensation Dress to Impress has a secret ingredient: fabulous video game jank. *Eurogamer*. Available at: https://www.eurogamer.net/viral-sensation-dress-to-impress-has-a-secret-ingredient-fabulous-video-game-jank

Keshavarz, H. (2021). Evaluating credibility of social media information: Current challenges, research directions and practical criteria. *Information Discovery and Delivery*, *49*(4), 269–279. DOI: 10.1108/IDD-03-2020-0033

Khaliq, L. N., & Manda, V. K. (2023). Customer Experience in the Web 3.0 Era: The Meeting of Blockchain and the Metaverse. In Majeed, M., Ofori, K. S., Amoako, G. K., Alolo, A.-R., & Awini, G. (Eds.), (pp. 258–274). Advances in Marketing, Customer Relationship Management, and E-Services. IGI Global., DOI: 10.4018/978-1-6684-7649-9.ch015

Khandabattu, H. (July 8 2025). The 2025 Hype Cycle for Artificial Intelligence Goes Beyond GenAI. *Gartner*. https://www.gartner.com/en/articles/hype-cycle-for-artificial-intelligence Retrieved 10.07.2025.

Khan, F., Ncube, C., Ramasamy, L. K., Kadry, S., & Nam, Y. (2020). A Digital DNA Sequencing Engine for Ransomware Detection Using Machine Learning. *IEEE Access : Practical Innovations, Open Solutions*, *8*, 119710–119719. DOI: 10.1109/ACCESS.2020.3003785

Kharpal, A. (2022). What is 'Web3'? Here's the vision for the future of the internet from the man who coined the phrase. *Cnbc*. Available at: https://www.cnbc.com/2022/04/20/what-is-web3-gavin-wood-who-invented-the-word-gives-his-vision.html

Khosrow-Pour, D. B. A. M. (Ed.). (2018). *Enhancing Art, Culture, and Design With Technological Integration:* IGI Global. DOI: 10.4018/978-1-5225-5023-5

Kim, S., & Eastin, M. S. (2011). Hedonic Tendencies and the Online Consumer: An Investigation of the Online Shopping Process. *Journal of Internet Commerce*, *10*(1).

Klenk, M. (2023). Algorithmic transparency and manipulation. *Philosophy & Technology*, *4*, 1–20.

Koffka, K. (1935). *Principles of Gestalt psychology*. Harcourt, Brace.

Koma, K. (2023). Fashion and Identity in Virtual Spaces. In Gardetti, M. Á., & Larios-Francia, R. P. (Eds.), *Sustainability Challenges in the Fashion Industry. Sustainable Textiles: Production, Processing, Manufacturing & Chemistry*. Springer., DOI: 10.1007/978-981-99-0349-8_12

Korda, H., & Itani, Z. (2013). Harnessing social media for health promotion and behavior *change. Health Promotion Practice*, *14*(1), 15–23. DOI: 10.1177/1524839911405850 PMID: 21558472

Korkut, P. K., Dolmaci, M., & Karaca, B. (2018). A Study on Communication Breakdowns: Sources of Misunderstanding in a Cross-Cultural Setting. *Eurasian Journal of Educational Research*, *18*(78), 1–20. DOI: 10.14689/ejer.2018.78.7

Kossow, N., Windwehr, S., & Jenkins, M. (2021). Algorithmic transparency and accountability. Retrieved from https://knowledgehub.transparency.org/assets/uploads/kproducts/Algorithmic-Transparency_2021.pdf

Kothari, A., & Cruikshank, S. A. (2022). Artificial intelligence and journalism: An agenda for journalism research in Africa. *African Journalism Studies*, *43*(1), 17–33. DOI: 10.1080/23743670.2021.1999840

Kotler, P., & Keller, K. L. (2016). *Marketing Management* (15th ed.). Pearson Education Limited.

Kotler, P., & Lee, N. R. (2008). *Social marketing: Influencing behaviors for good*. Sage Publications.

Krause, A., & Bucy, E. P. (2018). Visual Framing of Fracking: How Standing Attitudes Shape Perceptions of Environmental Risk and Economic Benefit. *Environmental Communication*, *12*(3), 322–343.

Krizhevsky, A., Sutskever, I., & Hinton, G. E. (2012). ImageNet classification with deep convolutional neural networks. In *NIPS'12: Proceedings of the 25th International Conference on Neural Information Processing Systems,* (vol. 1, 3 December) pp. 1097–105. Curran Associates https://dl.acm.org/doi/10.5555/2999134.2999257

Kucukemiroglu, O. (1999). Market segmentation by using consumer lifestyle dimensions and ethnocentrism. An empirical study. *European Journal of Marketing*, *33*(5/6), 470–487. DOI: 10.1108/03090569910262053

Kumari, P., & Nirban, V. S. (2018). Intercultural communication competencies in the Indian Information Technology industry. *International Journal of Cross Cultural Management*, *18*(3), 327–347. DOI: 10.1177/1470595818812996

Kwik, H., & Gondoputranto, O. (2022). Constructing Self Identity through the Lens of Fashion. Folio, 3(2), Retrieved from: https://journal.uc.ac.id/index.php/FOLIO/article/view/3488

Ladislav, Š. (2009). ICT influence on the formation of cultural knowledge. *Vysoká Škola Ekonomická v Praze*. https://core.ac.uk/outputs/30108111/?source=oai

Lahti, T. (2023). The Digital Runway: Virtual Fashion, Luxury, and the New Consumer. *Fashion Futures Journal*, *5*(3), 89–104.

Laranjeira, C., Mota, V. F., & dos Santos, J. A. (2021). Machine learning Bias in Computer Vision: Why do I have to care? In Proceedings of the 34th SIBGRAPI Conference on Graphics, Patterns and Images (SIBGRAPI). Doi: DOI: 10.1109/SIBGRAPI54419.2021.00010

Larsson-Book, B. (2022). Layers of Genius Behind Layered Clothing. *corp.roblox*. Available at: https://corp.roblox.com/newsroom/2022/04/layers-genius-behind-layered-clothing

Lau, J. T., Griffiths, S., Choi, K. C., & Tsui, H. Y. (2010). Avoidance behaviors and negative psychological responses in the general population in the initial stage of the H1N1 pandemic in Hong Kong. *BMC Infectious Diseases*, *10*, 139. DOI: 10.1186/1471-2334-10-139 PMID: 20509887

Laukyte, M. (2020). Disruptive Technologies and the Sport Ecosystem: A Few Ethical Questions. *Philosophies*, *5*(4), 24. DOI: 10.3390/philosophies5040024

Law, A. W. (2021). The Rise of Decentralized Autonomous Organizations: Opportunities and Challenges. *Stanford Journal of Blockchain Law & Policy*. Retrieved from: https://stanford-jblp.pubpub.org/pub/rise-of-daos https://corp.roblox.com/newsroom/2022/04/layers-genius-behind-layered-clothing

Lazar, M., Miron-Spektor, E., Agarwal, R., Erez, M., Goldfarb, B., & Chen, G. (2020). Entrepreneurial Team Formation. *The Academy of Management Annals*, *14*(1), 29–59. DOI: 10.5465/annals.2017.0131

Lazarski, E., Al-Khassaweneh, M., & Howard, C. (2021). Using NLP for Fact Checking: A Survey. *Designs*, *5*(3), 42. DOI: 10.3390/designs5030042

Lazcano Ponce, E., & Alpuche-Aranda, C. (2020). Alfabetización en salud pública ante la emergencia de la pandemia por Covid-19. *Salud Pública de México*, *62*(3), 331–340. DOI: 10.21149/11408 PMID: 32304371

Leandros, N. (2011). The International Media System in Transition. A Human Development Perspective. Retrieved from: https://www.newmedia21.eu/analizi/the-international-media-system-in-transition-a-human-development-perspective/

Leandros, N. (2010). Media Concentration and Systemic Failures in Greece. *International Journal of Communication*, *4*, 886–905.

Lee, J. J., & Park, S. M. (2011). *Fashion, Reflection of Culture*. Yekyong.

Leerssen, P. (2020). The soap box as a black box: Regulating transparency in social media recommender systems. *European Journal of Law and Technology*, *2*, 1–51.

Lemon, K. N., & Verhoef, P. C. (2016). Understanding customer experience throughout the customer journey. *. *Journal of Marketing*, *80*(6), 69–96. DOI: 10.1509/jm.15.0420

Liere-Netheler, K., Vogelsang, K., Packmohr, S., & Hoppe, U. (2018). A framework for customer-centric digital transformation. *. *Journal of Business Research*, *85*, 164–173. DOI: 10.1016/j.jbusres.2017.12.038

Lilleker, D. G. (2019). The Power of Visual Political Communication: Pictorial Politics Through the Lens of Communication Psychology. In A. Veneti, D. Jackson, & D. G. Lilleker (Eds.), *Visual Political Communication* (pp. 37-51). Palgrave.

Lin, C. A. (2019). The Challenge of Information and Communication Divides in the Age of Disruptive Technology. *Journal of Broadcasting & Electronic Media*, *63*(4), 587–594. DOI: 10.1080/08838151.2019.1699677

Lin, Y., Juang, L., Nie, S., Liu, Z., Yu, H., Yan, W., & Xu, Y. (2011). Knowledge, Attitudes and Practices (KAP) related to the Pandemic (H1N1) 2009 among Chinese General Population: A Telephone Survey. *BMC Infectious Diseases*, *11*, 128. DOI: 10.1186/1471-2334-11-128 PMID: 21575222

Litts, B. K., Searle, K. A., Brayboy, B. M. J., & Kafai, Y. B. (2021). Computing for all?: Examining critical biases in computational tools for learning. *British Journal of Educational Technology*, *52*(2), 842–857. DOI: 10.1111/bjet.13059

Liu, J. (2010). Culture and Knowledge Transfe: Theoretical Considerations. *Journal of Service Science and Management*, *03*(01), 159–164. DOI: 10.4236/jssm.2010.31020

Liu, X.-J., & Mesch, G. S. (2020). The Adoption of Preventive Behaviors during the COVID-19 Pandemic in China and Israel. *International Journal of Environmental Research and Public Health, 17*, 7170. DOI: 10.3390/ijerph17197170 PMID: 33007985

Li, Y., Rho, E. H. R., & Kobsa, A. (2022). Cultural differences in the effects of contextual factors and privacy concerns on users' privacy decision on social networking sites. *Behaviour & Information Technology, 41*(3), 655–677. DOI: 10.1080/0144929X.2020.1831608

Llano-Guibarra, N. I., & Águila-Sánchez, J. C. (2020). Conferencias de prensa y COVID-19: explorando la respuesta gubernamental mexicana desde la comunicación en salud. *Revista Española de Comunicación en Salud*(1), 128-141. https://doi.org/doi.org/10.20318/recs.2020.5451

Lo Sapio, G. (2021). La trasparenza sul banco di prova dei modelli algoritmici. *Federalismi.it, 11*, 239-252.

Lombardi, E. M. (2020). Norma e algoritmo: alcune considerazioni sul nuovo ordine tecnologico. *Giustizia Civile.com, 7*, 1-16.

Louw, E. (2010). The Media and Political Process. *Sage (Atlanta, Ga.).*

Lu, J., Dokyun, L., Tae Wan, K., & Danks, D. (2019). Good explanation for algorithmic transparency. Retrieved from https://papers.ssrn.com/sol3/papers.cfm?abstract_id=3503603

Mahaffey, C. J., Heim, E., & Wu, C. (2011). *Asian American Social Interaction at the University of Illinois at Urbana Champaign*. University of Illinois. http://hdl.handle.net/2142/34683

Maibach, E. W., Maxfield, A., Ladin, K., & Slater, M. (1996). Translating Health Psychology into Effective Health Communication: The American Healthstyles Audience Segmentation Project. *Journal of Health Psychology, 1*(3), 261–277. DOI: 10.1177/135910539600100302 PMID: 22011991

Majchrzak, M., Faraj, S., Kane, G. C., & Azad, B. (2013). The Contradictory Influence of Social Media Affordances on Online Communal Knowledge Sharing. *Journal of Computer-Mediated Communication, 19*(1), 1, 38–55. DOI: 10.1111/jcc4.12030

Majumdar, D., Banerji, P. K., & Chakrabarti, S. (2018). Disruptive technology and disruptive innovation: Ignore at your peril! *Technology Analysis and Strategic Management, 30*(11), 1247–1255. DOI: 10.1080/09537325.2018.1523384

Makwambeni, Bl., Trust, M., & Bulani, J. G. (2023). Between Utopia and Dystopia: investigating journalistic perceptions of AI deployment in community media newsrooms in South Africa. In Dralega, C. A. (Ed.), *Digitisation, AI and algorithms in African journalism and media contexts* (pp. 17–32). Emerald Publishing Limited., DOI: 10.1108/978-1-80455-135-620231002

Mampilly, A. J., & Manda, V. K. (2023). Driving Change: Securing Success through Digital Transformation. In *Modern Management Challenges: A Case Compendium* (pp. 196–205). Excel India Publishers. http://dx.doi.org/DOI: 10.13140/RG.2.2.15789.05605

Mancini, T., & Sibilla, F. (2017). Offline personality and avatar customisation. discrepancy profiles and avatar identification in a sample of mmorpg players. *Computers in Human Behavior*, *69*, 275–283. DOI: 10.1016/j.chb.2016.12.031

Manda, M., Yadav, A., & Khaliq, L. N. (2024). Role of Payment Systems in Financial Well-Being: An Examination from India. In *Emerging Perspectives on Financial Well-Being* (p. 24). IGI Global., https://www.igi-global.com/gateway/chapter/347786

Manda, V. K., Abukari, A. M., Gupta, V., & Bharathi, M. J. (2024). Revolutionizing Finance With Decentralized Finance (DeFi). In Darwish, D. (Ed.), (pp. 127–150). Advances in Web Technologies and Engineering. IGI Global., DOI: 10.4018/979-8-3693-1532-3.ch006

Manda, V. K., Christy, V., & Jitta, M. R. (2024). Ethical AI and Decision-Making in Management Leadership. In Bhattacharya, P., Hassan, A., Liu, H., & Bhushan, B. (Eds.), (pp. 197–226). Advances in Human and Social Aspects of Technology. IGI Global., DOI: 10.4018/979-8-3693-4147-6.ch009

Mandenaki, K., Sotirakou, C., Karabella, A., & Mourlas, C. (2024). *Digital Transformation in Media.* IQ Media Hub, NKUA. Research Presentation at the 18th PFJU Conference 18-20 October 2024, Samothraki.

Manin, B. (1995). *Principes du Gouvernement Représentatif.* Calmann-Lévy.

Marchetti, B. (2021). La garanzia dello human in the loop alla prova della decisione amministrativa algoritmica. *BioLaw Journal*, *2*, 367–385.

Markus, H., & Nurius, P. (1986). Possible selves. *The American Psychologist*, *41*(9), 954–969. DOI: 10.1037/0003-066X.41.9.954

Marrone, P. (2022). Democrazia epistemica, epistocrazia, algocrazia: Alcuni problemi. *Paradigmi*, *2*, 307–324.

Martínez-Caro, E., Cegarra-Navarro, J. G., & Alfonso-Ruiz, F. J. (2020). Digital transformation and customer experience in SMEs: A conceptual framework. *. *Management Decision*, *58*(8), 1602–1624. DOI: 10.1108/MD-10-2018-1160

Martínez-Estrella, C. G. (2020). Uso de personajes y metáforas en la gestión de la crisis sanitaria del COVID-19. Revisión de la comunicación de sanidad pública en España y México. *Revista Española de Comunicación en Salud*(1), 319-327. https://doi.org/doi.org/10.20318/recs.2020.5458

Martin, K. (2019). Ethical Implications and Accountability of Algorithms. *Journal of Business Ethics*, *160*(4), 835–850. DOI: 10.1007/s10551-018-3921-3

Mathijssen, J., Janssen, M., Bon-Martens, M., & Goor, I. (2012). Adolescents and alcohol: An explorative audience segmentation analysis. *BMC Public Health*, *12*, 742. DOI: 10.1186/1471-2458-12-742 PMID: 22950946

Mazzoleni, G. (1998). *La Comunicazione Politica*. il Mulino.

McDermott, V. (2021). Digital Fashion and the New Social Identity. *Journal of Virtual Fashion*, *7*(4), 97–110.

McFarland, M. (2014, September 23). The algorithm that can make fantasy football even more fun. *The Washington Post*. https://www.washingtonpost.com/news/innovations/wp/2014/09/23/the-algorithm-that-can-make-fantasy-football-even-more-fun. Retrieved 10.08.2024.

McKay, E. (2025). *Human-Computer Interaction Closes the Digital Divide: A multicultural, intergenerational ICT case study*. https://dl.acm.org/doi/pdf/10.5555/1151681.1151685

McLuhan, M. (1964). *Understanding Media. The Extension of Man*. McGraw-Hill.

McQuail, D. (1992). Media Performance. Mass Communication and the Public Interest. *Sage (Atlanta, Ga.)*.

Melnyk, L., Dehtyarova, I., Kubatko, O., Karintseva, O., & Derykolenko, A. (2019). Disruptive technologies for the transition of digital economies towards sustainability. *Economic Annals-XXI*, *179*(9-10), 22–30. DOI: 10.21003/ea.V179-02

Mena, P. (2019). Principles and boundaries of fact-checking: Journalists' perceptions. *Journalism Practice*, *13*(6), 657–672. DOI: 10.1080/17512786.2018.1547655

Mesjar, L., Cross, K., Jiang, Y., & Steed, J. (2023). The Intersection of Fashion, Immersive Technology, and Sustainability: A Literature Review. *Sustainability*, *15*(4), 3761. DOI: 10.3390/su15043761

Messinger, P. R., Ge, X., Stroulia, E., Lyons, K., Smirnov, K., & Bone, M. (2008). On the Relationship between My Avatar and Myself. *Journal of Virtual Worlds Research*, *1*(2), 1–17.

Miller, R. (2015). AP's 'robot journalists' are writing their own stories now. *The Verge*. https://www.theverge.com/2015/1/29/7939067/ap-journalism-automation-robots-financial-reporting, Retrieved, 10.08.2024.

Mobilio, G. (2020). L'intelligenza artificiale e i rischi di una "disruption" della regolamentazione giuridica. *BioLaw Journal*, *2*, 401–424.

Mong, A. (2024). "How Armenia's Hetq media fights disinformation and divisions", *DW Akademie*, April 14. Retrieved from https://akademie.dw.com/en/how-armenias-hetq-media-fights-disinformation-and-divisions/a-68771711 (Accessed 25/8/2025).

Monteros, M. (2024). Why fashion brands are inching into the gaming arena. *Modernretail.co*. Available at: https://www.modernretail.co/technology/why-fashion-brands-are-inching-into-the-gaming-arena/

Monti, M. (2019). Automated Journalism and Freedom of Information: Ethical and Juridical Problems Related to AI in the Press Field. *Opinio Juris in Comparatione*, 1/2018. Retrieved from https://ssrn.com/abstract=3318460 (Accessed 25/8/2025).

Moreno-Gil, V., Ramon, X., & Rodríguez-Martínez, R. (2021). Fact-Checking Interventions as Counteroffensives to Disinformation Growth: Standards, Values, and Practices in Latin America and Spain. *Media and Communication*, *9*(1), 251–263. DOI: 10.17645/mac.v9i1.3443

Morton, T., & Duck, J. (2001). Communication and health beliefs. Mass and interpersonal influences on perceptions of risk to self and others. *Communication Research*, *28*(5), 602–626. DOI: 10.1177/009365001028005002

Muñiz, C., & Corduneanu, V. I. (2020). Percepción de riesgo y consumo mediático durante el inicio de la pandemia de COVID-19 en México. *Más poder local*(41), 44-47. Retrieved from http://maspoderlocal.es/files/articulos/mexico-percepcion-riesgo-pandemia-mpl41.pdf

Muñiz, C., & Corduneanu, L. E. (2020). Audience segmentation and climate change: The Chilean case. *Environmental Communication*, *14*(7), 941–957. DOI: 10.1080/17524032.2020.1725084

Munoriyarwa, A., Chiumbu, S., & Motsaathebe, G. (2023). Artificial intelligence practices in everyday news production: The case of South Africa's mainstream newsrooms. *Journalism Practice*, *17*(7), 1374–1392. DOI: 10.1080/17512786.2021.1984976

Murphy, R. R. (1999). Case studies of applying Gibson's ecological approach to mobile robots. *IEEE Transactions on Systems, Man, and Cybernetics*, *29*, 105–111.

Murschetz, P. (2024). Business Models in the Media Industry: A Synoptic Overview. *Handbook of Media and Communication Economics*. https://doi.org/DOI: 10.1007/978-3-658-39909-2_19425

Nagy, P., & Neff, G. (2015). Imagined Affordance: Reconstructing a Keyword for Communication Theory. *Social Media + Society*, *1*(2). Advance online publication. DOI: 10.1177/2056305115603385

Nagy, P., & Neff, G. (2023). Rethinking affordances for human-machine communication research. In Guzman, A. L., McEwen, R., & Jones, S. (Eds.), *Rethinking affordances for human-machine communication research* (Vol. 0, pp. 273–279). SAGE Publications Ltd., DOI: 10.4135/9781529782783

Nakamura, L. (2002). *Cybertypes: Race, Ethnicity, and Identity on the Internet*. Routledge.

Nambisan, S., Wright, M., & Feldman, M. (2019). The digital transformation of innovation and entrepreneurship: Progress, challenges and key themes. *. *Research Policy*, *48*(8), 103773. DOI: 10.1016/j.respol.2019.03.018

Napoli, P. M., & Roepnack, A. (2018). Big data and media management. In Albarran, A. B., Mierzejewska, B. I., & Jung, J. (Eds.), *Handbook of media management and economics* (2nd ed., pp. 410–423). Routledge.

Neelakandan, B. V., Gopalakrishnan, S., Ramachandran, H. R. A., & Gopichandran, V. (2018). Health Beliefs Regarding Uptake of Hypertension Screening in an Urban Setting: A Population Segmentation Strategy. *International Journal of Medicine and Public Health*, *8*(1), 11-17. https://https://doi.org/DOI: <ALIGNMENT.qj></ALIGNMENT>10.5530/ijmedph.2018.1.3

Newman, N. (2024, January 9). Journalism, media, and technology trends and predictions 2024. *Reuters Institute*https://reutersinstitute.politics.ox.ac.uk/journalism-media-and-technology-trends-and-predictions-2024#header--0, Retrieved 10/08/2024.

Newman, N., Fletcher, R., Schulz, A., Simge, A., & Nielsen, R. K. (2020). Journalism, media, and technology trends and predictions 2020. *Reuters Institute Digital News Report*. https://www.digitalnewsreport.org/publications/2020/journalism-media-and-technology-trends-and-predictions-2020

Newman, N., Fletcher, R., Robertson, C. T., Ross Arguedas, A. A., & Nielsen, R. K. (2024). *Reuters Institute Digital News Report 2024*. Reuters Institute for the Study of Journalism.

Nichols, T., Shabir, N., & Nielsen, R. K. (2016). *Digital- Born News Media in Europe. Digital News Project 2016*. Reuters Institute for the Study of Journalism. University of Oxford., DOI: 10.60625/risj-d7k0-sk74

Nimmo, D. D., & Combs, J. E. (1990). *Mediated Political Realities*. Longman.

Noar, S. M. (2006). A 10-Year Retrospective of Research in Health Mass Media Campaigns: Where Do We Go From Here? *Journal of Health Communication: International Perspectives, 11*(1), 21–42. DOI: 10.1080/10810730500461059 PMID: 16546917

Noar, S. M., Grant Harrington, N., & Shemanski Aldrich, R. (2009). The Role of Message Tailoring in the Development of Persuasive Health Communication Messages. *Annals of the International Communication Association, 33*(1), 73–133. DOI: 10.1080/23808985.2009.11679085

Norman, D. A. (1988). *The psychology of everyday things*. Basic Books.

Norman, D. A. (1999). Affordance, conventions, and design. *Interaction, 6*(3), 38–43.

Norman, D. A. (2013). *The design of everyday things: Revised and Expanded Edition*. Basic Books.

Norris, P. (2000). *A Virtuous Circle. Political Communications in Postindustrial Societies*. Cambridge University Press.

Ntoutsi, E., Fafalios, P., Gadiraju, U., Iosifidis, V., Nejdl, W., Vidal, M., Ruggieri, S., Turini, F., Papadopoulos, S., Krasanakis, E., Kompatsiaris, I., Kinder-Kurlanda, K., Wagner, C., Karimi, F., Fernandez, M., Alani, H., Berendt, B., Kruegel, T., Heinze, C., & Staab, S. (2020). Bias in data-driven artificial intelligence systems—An introductory survey. *Wiley Interdisciplinary Reviews. Data Mining and Knowledge Discovery, 10*(3), e1356. DOI: 10.1002/widm.1356

Nunnally, J. C., & Bernstein, I. H. (1994). *Psychometric Theory* (3rd ed.). McGraw-Hill.

Nurelmadina, N., Hasan, M. K., Memon, I., Saeed, R. A., Ariffin, K. A. Z., Ali, E. S., Mokhtar, R. A., Islam, S., Hossain, E., & Hassan, M. A. (2021). A Systematic Review on Cognitive Radio in Low Power Wide Area Network for Industrial IoT Applications. [Switzerland]. *Sustainability, 13*, 338. Advance online publication. DOI: 10.3390/su13010338

Nyhan, B., & Reifler, J. (2012). *Misinformation and Fact-Checking: Research Findings from Social Science*, 28. Washington, D.C.: New America Foundation. https://www.issuelab.org/resources/15316/15316.pdf

Ogola, G. (2025). Between Fear and Hope: Generative AI, ChatGPT and Journalism. In Mutsvairo, B., & Orgeret, K. S. (Eds.), *The Palgrave Handbook of Global Digital Journalism*. Palgrave Macmillan., DOI: 10.1007/978-3-031-59379-6_13

Omenugha, N. O. (2019). Contextualizing Disruptive Technologies in Media Practice and Management. *Global Journal of Archaeology & Anthropology*, *8*(4). Advance online publication. DOI: 10.19080/GJAA.2019.08.555743

Orellana-Centeno, J. (2020). El coronavirus 19 (COVID-19) en México. *Revista Chilena de Salud Pública*, *24*(1), 72–73. DOI: 10.5354/0719-5281.2020.57588

Ornell, F., Schuh, J. B., Sordi, A. O., & Kessler, F. H. (2020). "Pandemic fear" and COVID-19: Mental health burden and strategies. *The British Journal of Psychiatry*, *42*(3), 232–235. DOI: 10.1590/1516-4446-2020-0008 PMID: 32267343

Osterwalder, A., & Pigneur, Y. (2010). *Business Model Generation. A Handbook for Visionaries, Game Changers, and Challengers*. John Wiley & Sons.

Otranto, P. (2021). Riflessioni in tema di decisioni amministrative, intelligenza artificiale e legalità. *Federalismi.it, 7*, 187-204.

Otterbacher, J., Bates, J., & Clough, P. (2017). "Competent men and warm women: Gender stereotypes and backlash in image search results," in *Proceedings of the 2017 chi conference on human factors in computing systems*. pp. 6620–6631.

Panagopoulos, A. M., Triantafyllou, S., & Kapos, P. (2024, June 29-30). *AI usage in the Greek Newsrooms*. CCLabs Conference, (In Greek) https://www.youtube.com/watch?v=w3qbeP205Lo&t=4235s 1.05.07 – 1.20.50.

Papa, A. (2022). Intelligenza Artificiale e decisioni pubbliche tra tecnica, politica e tutela dei diritti. *Federalismi.it, 22*, 101-114.

Papadima-Sophocleous, S. (2003). A Web of Intercultural Communication. *Informatologia*, *3*(36), 166–241.

Park, J. H., Cheong, H.-K., Son, D.-Y., Kim, S.-U., & Ha, C.-M. (2010). Perceptions and behaviors related to hand hygiene for the prevention of H1N1 influenza transmission among Korean university students during the peak pandemic period. *BMC Infectious Diseases*, *10*, 222. DOI: 10.1186/1471-2334-10-222 PMID: 20663229

Park, S.-M., & Kim, Y.-G. (2022). A metaverse: Taxonomy, components, applications, and open challenges. *IEEE Access : Practical Innovations, Open Solutions*, *10*, 4209–4251. DOI: 10.1109/access.2021.3140175

Parvanta, C. (2011). Introduction to Public Health Communication and Informatics. In Parvanta, C. F., Nelson, D. E., Parvanta, S. A., & Harner, R. N. (Eds.), *Essentials of Public Health Communication* (pp. 3–18). Jones & Bartlett Learning.

Parvanta, C. F., & Freimuth, V. (2000). Health Communication at the Centers for Disease Control and Prevention. *American Journal of Health Behavior, 24*(1), 18–25. DOI: 10.5993/AJHB.24.1.4

Passey, D. (2015). Inclusive Technologies and Learning: Research, Practice and Policy. *Social Inclusion (Lisboa), 3*(6), 1–6. DOI: 10.17645/si.v3i6.543

Patel, A. A., & Soni, S. J. (2015). A Novel Proposal for Defending against Vampire Attack in WSN. *2015 Fifth International Conference on Communication Systems and Network Technologies*, 624–627. https://doi.org/DOI: 10.1109/CSNT.2015.94

Patient, G. (2023). The Influence of Pop Culture on Fashion Trends. *Merchmates.co.uk.* Available at: https://www.merchmates.co.uk/blog/culture/influence-of-pop-culture-on-fashion/

Păvăloaia, V.-D., & Necula, S.-C. (2023). Artificial Intelligence as a Disruptive Technology—A Systematic Literature Review. *Electronics (Basel), 12*(5), 1102. DOI: 10.3390/electronics12051102

Pavithra, J., & Samy, S. S. (2022). A Comparative Study on Detection of Malware and Benign on the Internet Using Machine Learning Classifiers. *Mathematical Problems in Engineering, 2022*(1), 4893390. DOI: 10.1155/2022/4893390

Pavleska, T., Školkay, A., Zankova, B., Ribeiro, N., & Bechmann, A. (2020). Performance Analysis of Fact-Checking Organisations and Initiatives in Europe: A Critical Overview of Online Platforms Fighting Fake News. In Terzis, G., Kloza, D., Kużelewska, E., & Trottier, D. (Eds.), *Disinformation and Digital Media as a Challenge for Democracy* (pp. 217–246). Chapter, Intersentia., DOI: 10.1017/9781839700422.014

Pavlik J.V. (2023). Collaborating with ChatGPT: considering the implications of generative artificial intelligence for journalism and media education. *Journal Mass Communication Education* 78(1):84–93. https://doi.org/ to ref 2023 in articleDOI: 10.1177/10776958221149577Return

Pedersen, M. J., & Favero, N. (2020). Social Distancing during the COVID-19 Pandemic: Who Are the Present and Future Noncompliers? *Public Administration Review, 5*, 805–814. DOI: 10.1111/puar.13240 PMID: 32836442

Peng, Y. (2018). Same candidates, different faces: Uncovering media bias in visual portrayals of presidential candidates with computer vision. *Journal of Communication, 68*(5), 920–941.

Peng, Y. (2021). What makes politicians' Instagram posts popular? Analysing social media strategies of candidates and office holders with computer vision. *The International Journal of Press/Politics*, *26*(1), 143–166.

Peng, Y., & Lu, Y. (2023). Computational visual analysis in political communication. In Lilleker, D., & Veneti, A. (Eds.), *Research Handbook on Visual Politics* (pp. 42–54). Edwin Elgar.

Petzold, M. B., Bendau, A., Plag, J., Pyrkosch, L., Maricic, L. M., Betzler, F., & Strölle, A. (2020). Risk, resilience, psychological distress, and anxiety at the beginning of the COVID-19 pandemic in Germany. *Brain and Behavior*, *10*, e01745. DOI: 10.1002/brb3.1745 PMID: 32633464

Picard, R. G. (2003). Cash Cows or Entrecôte: Publishing Companies and Disruptive Technologies. *TIC. Trends in Communication*, *11*(2), 127–136. DOI: 10.1207/S15427439TC1102_04

Plasser, F., & Plasser, G. (2002). *Global Political Campaigning: A Worldwide Analysis of Campaign Professionals and Their Practices*. Praeger.

Poesy (2024). PFJU establishes a Committee for a Code of Ethics on AI. In Greek. https://www.poesy.gr/poesy-idryei-epitropi-gia-ti-systasi-k/ Retrieved 15.11,2024.

Porter, M. E. (1985). *Competitive Advantage. Creating and Sustaining Superior Performance*. Free Press.

Porter, M. E., & Kramer, M. R. (2006). Strategy and society: The link between competitive advantage and corporate social responsibility. *. *Harvard Business Review*, *84*(12), 78–92. PMID: 17183795

Postman, N. (1985). *Amusing Ourselves to Death: Public Discourse in the Age of Show Business*. Penguin.

Pressman, L. (2017, June, 14). The State of Artificial Intelligence in 2017. *Automated Insights*. https://automatedinsights.com/blog/the-state-of-artificial-intelligence/ , Retrieved 06.06.2024.

Prior, M. (2007). *Post-Broadcast Democracy: How Media Choice Increases Inequality in Political Involvement and Polarised Elections*. Cambridge University Press.

Protothema.gr. (2024a, July 4). The new AI-powered English version of Proto Thema with the support of Google. *Proto Thema*. https://en.protothema.gr/2024/07/04/the-new-ai-powered-english-version-of-proto-thema-with-the-support-of-google/ Retrieved 10.08.2024.

Protothema.gr. (2024b, July 22). How Artificial Intelligence is changing the comments on protothema.gr. *Proto Thema*https://en.protothema.gr/2024/07/22/how-artificial-intelligence-is-changing-the-comments-on-protothema-gr/ Retrieved 10.08.2024.

Protothema.gr. (2024c, November, 15). ProtoThema Introduces AI Semantic Search. https://en.protothema.gr/2024/11/15/protothema-introduces-ai-semantic-search/ Retrieved 20/11/2024.

Pucer, P., Trobec, I., & Žvanut, B. (2014). An information communication technology based approach for the acquisition of critical thinking skills. *Nurse Education Today*, *34*(6), 964–970. DOI: 10.1016/j.nedt.2014.01.011 PMID: 24581890

Rangone, N. (2022). Intelligenza artificiale e pubbliche amministrazioni: Affrontare i numerosi rischi per trarne tutti i vantaggi. *BioLaw Journal*, *2*, 473–488.

Rath, M. P., Peterson, J., Greensley, P., & Gill, P. (1994). *Introduction to fashion merchandising*. Delmar Pub.

Reddick, C. G., Enriquez, R., Harris, R. J., & Sharma, B. (2020). Determinants of broadband access and affordability: An analysis of a community survey on the digital divide. *Cities (London, England)*, *106*, 102904. DOI: 10.1016/j.cities.2020.102904 PMID: 32921864

Restubog, S. L., Ocampo, A. C., & Wang, L. (2020). Taking control amidst the chaos: Emotion regulation during the COVID-19 pandemic. *Journal of Vocational Behavior*, *119*, 103440. DOI: 10.1016/j.jvb.2020.103440 PMID: 32390659

Reynolds, S., & Nolan, J. (2025). Ethical Considerations in AI Journalism: Bias Detection and Mitigation. *ITSI Transactions on Electrical and Electronics Engineering*, *12*(1), 18–23. Retrieved from https://journals.mriindia.com/index.php/itsiteee/article/view/144

Rhpoz, L. (2019). Framing communication in Mexico: Media, politics, and public perception. *Communicatio Socialis*, *13*(1), 45–65. DOI: 10.15581/003.13.1.45-65

Rimal, R. N., & Lapinski, M. K. (2009). Why health communication is important in public health. *Bulletin of the World Health Organization*, *87*, 247. DOI: 10.2471/BLT.08.056713 PMID: 19551226

Risser, L., Picard, A., Hervier, L., & Loubes, J. (2022). A survey of Identification and mitigation of Machine Learning algorithmic biases in Image Analysis. *10.48550/arXiv.2210.04491.*

Ritter, E. H., & Conrad, C. R. (2016). Preventing and responding to dissent: The observational challenges of explaining strategic repression. *The American Political Science Review*, *110*(1), 85–99.

Robonews (n.d.) https://robonews.gr/%cf%80%ce%bf%ce%b9%ce%bf%ce%b9-%ce%b5%ce%b9%ce%bc%ce%b1%cf%83%cf%84%ce%b5/ (In Greek) Assessed 10.11.2024.

Rodríguez-Hidalgo, C. (2020). Me and my robot smiled at one another: The process of socially enacted communicative affordance in Human-Machine Communication. *Human-Machine Communication*, *1*, 55–69. DOI: 10.30658/hmc.1.4

Rojas, M., David, M., Todorov, A., & Vitria, J. (2011). Automatic prediction of facial trait judgments: Appearance vs. structural models. *PLoS One*, *6*(8). PMID: 21858069

Rondán Cataluña, F. J., Navarro García, A., Díez de Castro, E. C., Rodríguez Rad, C., & Guisado González, M. (2010). Estudio del performance de los grupos estratégicos en el sistema de franquicia español. *Investigaciones Europeas de Dirección y Economía de la Empresa*, *16*(2), 43–62. DOI: 10.1016/S1135-2523(12)60111-7

RSF (2024). Index, Greece. https://rsf.org/en/country/greece, Retrieved 15/10/2024.

Rubino, F., Agostino, D., & Spallazzo, D. (2025). Strategizing blockchain adoption in public cultural services: A comprehensive scoping review. *International Journal of Public Sector Management*, *38*(1), 77–97. DOI: 10.1108/IJPSM-12-2023-0383

Ruggiero, C., Karadimitriou, A., Lo, W. H., Núñez-Mussa, E., Bomba, M., & Sallusti, S. (2022) The professionalisation of journalism: global trends and the challenges of training and job insecurity. In: Trappel J, Tomaz T (eds) *Success and failure in news media performance: comparative analysis in the media for democracy monitor 2021*. (pp 309–335) Nordicom, University of Gothenburg, Gothenburg, https://doi.org/DOI: 10.48335/9789188855589-15

Sabedini, M. (2017, October 28). Social Media, their Platform, Respectively the Users, Ethics, Advantages and Disadvantages. *2017 UBT International Conference*. University for Business and Technology International Conference, Durres, Albania. DOI: 10.33107/ubt-ic.2017.212

Şahin, E., Çakmak, M., Doğar, M. R., Uğur, E., & Üçoluk, G. (2007). To Afford or Not to Afford: A New Formalization of Affordances Toward Affordance-Based Robot Control. *Adaptive Behavior*, *15*(4), 447–472. DOI: 10.1177/1059712307084689

Salzmann, A., Guribye, F., & Gynnild, A. (2021). Adopting a mojo mindset: Training newspaper reporters in mobile journalism. *Journalism, 24* (2), 346-362. https://doi.org/ (Original work published 2023)DOI: 10.1177/1464884921996284

Sarr, S. F. S. (2024). Artificial Intelligence and Fact Checking in Africa: Between the Logic of Dependency and the Limits of Automation. [ESJ]. *European Scientific Journal*, *20*(8), 91. DOI: 10.19044/esj.2024.v20n8p91

Sartori, G. (1997). *Homo Videns. Televisione e Post-Pensiero*. Laterza.

Schiavo, R. (2014). *Health Communication. From theory to practice* (2nd ed.). Jossey-Bass.

Schill, D. (2012). The Visual Image and the Political Image: A Review of Visual Communication Research in the Field of Political Communication. *The Review of Communication*, *12*(2), 118–142. DOI: 10.1080/15358593.2011.653504

Schjøtt, A., & Bengtsson, M. (2024). De- and recoding algorithmic systems: The case of fact-checkers and fact-checked users. *Convergence*, *0*(0). Advance online publication. DOI: 10.1177/13548565241289233

Schmeichel, M. (2012). Good Teaching? An examination of culturally relevant pedagogy as an equity practice. *Journal of Curriculum Studies*, *44*(2), 211–231. DOI: 10.1080/00220272.2011.591434

Schmidt, R. C. (2007). Scaffolds for social meaning. *Ecological Psychology*, *19*(2), 137–151.

Schreder, S. (2023). Why the metaverse needs fashion to feel human. *Culture3.com*. Available at: https://www.culture3.com/posts/why-the-metaverse-needs-digital-fashion-to-feel-human

Schrock, A. R. (2015). Communicative Affordances of Mobile Media: Portability, Availability, Locatability, and Multimediality. *International Journal of Communication*, *9*(18), 1229–1246.

Schwartz, G. (2008). Virtual Worlds Research: Global X Local Agendas. In *Virtual Worlds Research: Consumer Behavior in Virtual Worlds,* 1(2), ISSN: 1941-8477.

Scott, S. V., Van Reenen, J., & Zachariadis, M. (2017). The long-term economic impact of digital transformation. *. *Economic Journal (London)*, *127*(603), 1203–1236. DOI: 10.1111/ecoj.12459

Segran, E. (2022). Karlie Kloss wants to dress your avatar. *Fast Company*. Available at: https://www.fastcompany.com/90768159/karlie-kloss-wants-to-dress-your-avatar

Semple, K. (2020, julio 14). El gobierno de México llama a la calma y adopta medidas mesuradas frente al virus. *The New York Times*. Retrieved febrero 23, 2021, from https://www.nytimes.com/es/2020/03/16/espanol/deportes/mexico-coronavirus-amlo-futbol.html

Serino, L., Meleleo, C., Maurici, M., Bagnato, B., Sorbara, D., Zaratti, L., & Franco, E. (2011). Knowledge and worry as basis for different behaviors among university students: The case of Pandemic Flu H1N1v. *Journal of Preventive Medicine and Hygiene*, *52*, 144–147. DOI: 10.15167/2421-4248/jpmh2011.52.3.269 PMID: 22010546

Serrano, R. M. (2023). *Disinformation Landscape in Germany, EU DisinfoLab*, March (Project funded by the Friedrich Naumann Foundation for Freedom). Retrieved from www.disinfo.eu/wp-content/uploads/2023/03/20230224_GER_DisinfoFS.pdf

Seymore, C. (2024). Dress to Impress offers new era of gaming, fashion fusion. *Thepostathens.com*. Available at: https://www.thepostathens.com/article/2024/09/dress-to-impress-roblox#google_vignette

Shao, Y., & Crook, C. (2015). The Potential of a Mobile Group Blog to Support Cultural Learning Among Overseas Students. *Journal of Studies in International Education*, *19*(5), 399–422. DOI: 10.1177/1028315315574101

Shaukat, K., Luo, S., & Varadharajan, V. (2023). A novel deep learning-based approach for malware detection. *Engineering Applications of Artificial Intelligence*, *122*, 106030. DOI: 10.1016/J.ENGAPPAI.2023.106030

Shaw, A. (2017). Encoding and decoding affordances: Stuart Hall and interactive media technologies. *Media Culture & Society*, *39*(4), 592–602. DOI: 10.1177/0163443717692741

Sibley, C. S., Greaves, L. M., Satherley, N., Wilson, M. S., Overall, N. C., Lee, C. H., & Barlow, F. K. (2020). Effects of the COVID-19 Pandemic and Nationwide Lockdown on Trust, Attitudes Toward Government, and Well-Being. *The American Psychologist*, *75*(5), 618–630. DOI: 10.1037/amp0000662 PMID: 32496074

Siegrist, M., & Zingg, A. (2014). The role of public trust during pandemics: Implications for crisis communication. *European Psychologist*, *19*(1), 23–32. DOI: 10.1027/1016-9040/a000169

Simoncini, A. (2019). L'algoritmo incostituzionale: Intelligenza artificiale e il futuro delle libertà. *BioLaw Journal*, *1*, 63–89.

Simoncini, A. (2023). Il linguaggio dell'intelligenza artificiale e la tutela costituzionale dei diritti. *Rivista AIC*, *2*, 1–19.

Şimşek, C. (2021, May). Algorithmic transparency in the EU. Retrieved from https://www.sciencespo.fr/public/sites/sciencespo.fr.public/files/SIMSEK%20Can.pdf

Singer, J. B. (2021). Border patrol: The rise and role of fact-checkers and their challenge to journalists' normative boundaries. *Journalism, 22*(8), 1929–1946. DOI: 10.1177/1464884920933137

Školkay, A., & Filin, J. (2019). A Comparison of Fake News Detecting and Fact-Checking AI-Based Solutions. *Studia Medioznawcze, 20*(4), 365–383. DOI: 10.33077/uw.24511617.ms.2019.4.187

Slater, M. D., & Flora, J. A. (1991). Health Lifestyles: Audience Segmentation Analysis for Public Health Interventions. *Health Education Quarterly, 18*(2), 221–233. DOI: 10.1177/109019819101800207 PMID: 2055779

Smith, M. D., & Telang, R. (2016). *Streaming, sharing, stealing: Big data and the future ofentertainment*. MIT Press.

Smith, R. (2021). Commodifying Self-Expression: The Economics of Digital Fashion in Gaming Environments. *Journal of Digital Economies, 14*(4), 400–418.

Sobolev, A., Chen, M. K., Joo, J., & Steinert-Threlkeld, Z. C. (2020). News and geolocated social media accurately measure protest size variation. *The American Political Science Review, 114*(4), 1343–1351.

Somit, A., & Peterson, S. A. (1998). Biopolitics After Three Decades: A Balance Sheet. *Journal of Political Science, 25*, 59–571.

Soni, S. J., & Nayak, S. D. (2013). Enhancing security features & performance of AODV protocol under attack for MANET. *2013 International Conference on Intelligent Systems and Signal Processing (ISSP)*, 325–328. https://doi.org/DOI: 10.1109/ISSP.2013.6526928

Spanaki, K., Papadopoulos, T., Jayawickrama, U., Olan, F., & Liu, S. (2023). Editorial: fake news, misinformation, and supply chain disruptions: the role of emerging technologies. *Annals of Operations Research, 327*, 601–604. DOI: 10.1007/s10479-023-05498-w

Spiller, E. (2021). Il diritto di comprendere, il dovere di spiegare. Explainability e intelligenza artificiale costituzionalmente orientata. *BioLaw Journal, 2*, 419–432.

Stahl, B. C., Antoniou, J., & Bhalla, N.. (2023). A systematic review of artificial intelligence impact assessments. *Artificial Intelligence Review, 56*, 12799–12831. DOI: 10.1007/s10462-023-10420-8 PMID: 37362899

Stanton, A. (2023). How Does The Color Of Your Clothing Affect Your Mood? *The good trade*. Available at: https://www.thegoodtrade.com/features/clothing-color-psychology/

Stanyer, J. (2013). *Intimate Politics: Publicity, Privacy and the Personal Lives of Politicians in Media Saturated Democracies*. John Wiley & Sons.

Statera, G. (1986). *La Politica Spettacolo. Politici e Mass Media nell'era dell'immagine*. Mondadori.

Statista (2024). Distribution of video game users in the United States in 2023, by gender. Available at: https://www.statista.com/forecasts/494867/distribution-of-gamers-by-gender-usa

Steinert-Threlkeld, Z. C. (2017). Spontaneous collective action: Peripheral mobilisation during the Arab Spring. *The American Political Science Review, 111*(2), 379–403.

Stellefson, M. L., Hanik, B. W., Chaney, B. H., & Chaney, D. J. (2008). Challenges for Tailored Messaging in Health Education. *Health Education. American Journal of Health Education, 39*(5), 303–313. DOI: 10.1080/19325037.2008.10599054

Stephenson, N. (1992). *Snow crash*. Bantam Books.

Sterpa, A., De Vivo, I., & Capasso, C. (2023). L'ordine giuridico dell'algoritmo: La funzione regolatrice del diritto e la funzione ordinatrice dell'algoritmo. *NAD, 2*, 1120–1155.

Stray, J. (2021). Making artificial intelligence work for investigative journalism. *Algorithms, Automation, and News*, 97–118.

Street, J. (2004). Celebrity Politicians: Popular Culture and Political Representation. *British Journal of Politics and International Relations, 6*(4), 435–452.

Strömbäck, J. (2008). Four Phases of Mediatization: An Analysis of the Mediatization of Politics. *The International Journal of Press/Politics, 13*(3), 228–246.

Stypinska, J. (2023). AI ageism: A critical roadmap for studying age discrimination and exclusion in digitalized societies. *AI & Society, 38*(2), 665–677. DOI: 10.1007/s00146-022-01553-5 PMID: 36212226

Subroto, V., Kusumajaya, R., & Tobing, W. (2024). The Business Model of the Mass Media Industry in the Era of Artificial Intelligence (AI) Development in Indonesia. *Journal of Management and Informatics, 3*, 230–249. DOI: 10.51903/jmi.v3i2.31

Sulistyono, I. B., & Purwasito, A. Warto, & Pitana, T. (2023). *Epistemology of Cultural Study*. DOI: 10.5281/ZENODO.7747729

Svensson, P. (2016). *Big Digital Humanities: Imagining a Meeting Place for the Humanities and the Digital*. University of Michigan Press., DOI: 10.2307/j.ctv65sx0t

Tadhani, J. R., & Vekariya, V. (2024b). A Survey of Deep Learning Models, Datasets, and Applications for Cyber Attack Detection. *AIP Conference Proceedings, 3107*(1). Advance online publication. DOI: 10.1063/5.0208470/3287908

Tajfel, H., & Turner, J. C. (1979). An integrative theory of intergroup conflict. In Austin, W. G., & Worchel, S. (Eds.), *The social psychology of intergroup relations* (pp. 33–37). Brooks/Cole.

Takano, M., & Taka, F. (2022). Fancy avatar identification and behaviors in the virtual world: Preceding avatar customization and succeeding communication. *Computers in Human Behavior Reports, 6*, 100176.

Tapiador, F. J., Avelar, S., Tavares-Corrêa, C., & Zah, R. (2011). Deriving fine-scale socioeconomic information of urban areas using very high-resolution satellite imagery. *International Journal of Remote Sensing, 32*(21), 6437–6456.

Thackeray, R., & McCormack Brown, K. (2005). Social Marketing's Unique Contributions to Health Promotion Practice. *Health Promotion Practice, 6*(4), 365–368. DOI: 10.1177/1524839905278848 PMID: 16247912

Thatta, S., Manda, V. K., & Athota, V. S. (2024). Research and Applications of Neurotechnologies for Leadership. In Saluja, S., Kukreja, J., & Sharma, S. (Eds.), (pp. 36–62). Advances in Business Strategy and Competitive Advantage. IGI Global., DOI: 10.4018/979-8-3693-1785-3.ch004

Thomas, R. K. (2006). *Health Communication.* New York: Springer Science+Business Media Inc.

Thompson, J. B. (1995). *The Media and Modernity: A Social Theory of the Media.* Stanford University Press.

Thompson, J. B. (2010). *Merchants of culture. The Publishing Business in the Twenty First Century.* Polity Press.

Toffler, A. (1980). *The Third Wave.* Bantam.

Tooher, R., Collins, J. E., Street, J. M., Braunack-Mayer, A., & Marshall, H. (2013). Community knowledge, behaviours and attitudes about the 2009 H1N1 Influenza pandemic: A systematic review. *Influenza and Other Respiratory Viruses, 7*(6), 1316–1327. DOI: 10.1111/irv.12103 PMID: 23560537

Tornatzky, L. G., & Fleischer, M. (1990). *The Processes of Technological Innovation.* Lexington Books.

tovima (2024, July 1). Τα άρθρα του Γιάννη Πρετεντέρη στο ΒΗΜΑ αποκτούν τη φωνή του μέσω τεχνολογίας AI. *To Vima*https://www.tovima.gr/2024/07/01/media/ta-arthra-tou-gianni-pretenteri-sto-vima-apoktoun-ti-foni-tou-meso-texnologias-ai/ retrieved 10.08.2024.

Towner, T. L., & Dulio, D. A. (2012). New Media and Political Marketing in the United States: 2012 and Beyond. *Journal of Political Marketing*, *1*(1-2), 95–119.

Triantafyllou, S. (2020). *Media and journalism in transition. The changing working conditions of journalists in the digital age*. Salonika: Disigma publications. In Greek.

Triantafyllou, S., Panagopoulos, A. M., & Kapos, P. (2025). AI Pioneers and Stragglers in Greece: Challenges, Gaps, and Opportunities for Journalists and Media. *Societies (Basel, Switzerland)*, *15*(8), 209. DOI: 10.3390/soc15080209

Trigueiro, I., Molina, D., Poyatos, J., Del Ser, J., & Herrera, F. (2024). General purpose artificial intelligence systems (GPAIS): Properties, definition, taxonomy, societal implications and responsible governance. *Information Fusion*, *103*, 1–16.

Tsipi, L., Vouyioukas, D., Loumos, G., Kargas, A., & Varoutas, D. (2023). Digital Repository as a Service (D-RaaS): Enhancing Access and Preservation of Cultural Heritage Artifacts. *Heritage*, *6*(10), 6881–6900. DOI: 10.3390/heritage6100359

Turkle, S. (1995). *Life on the Screen: Identity in the Age of the Internet*. Simon & Schuster.

Turkle, S. (1997). Computational Technologies and Images of the Self. *Social Research*, *64*(3), 1093–1111. https://www.jstor.org/stable/40971200

Turulja, L., & Bajgoric, N. (2019). Information technology, knowledge management and their role in the innovation process: A review and research framework. *. *Journal of Knowledge Management*, *23*(3), 548–563.

UNB. (2022). *Malware Dataset*. Https://Www.Unb.ca/Cic/Datasets/Malmem-2022.Html

Urooj, B., Shah, M. A., Maple, C., Abbasi, M. K., & Riasat, S. (2022). Malware Detection: A Framework for Reverse Engineered Android Applications Through Machine Learning Algorithms. *IEEE Access : Practical Innovations, Open Solutions*, *10*, 89031–89050. DOI: 10.1109/ACCESS.2022.3149053

Valsecchi, A. (2020, September 14). Algoritmo, discrezionalità amministrativa e discrezionalità del giudice. Retrieved from https://www.iusinitinere.it/algoritmo-discrezionalita-amministrativa-e-discrezionalita-del-giudice-nota-a-cons-stato-sez-vi-sent-del-4-febbraio-2020-n-881-30630

Van Aelst, P., Sheafer, T., & Stanyer, T. (2012). The Personalization of Mediated Political Communication: A Review of Concepts, Operationalisations and Key Findings. *Journalism, 13*(2), 203–220.

van Dalen, A. (2024). Revisiting the Algorithms Behind the Headlines. How Journalists Respond to Professional Competition of Generative AI. *Journalism Practice*, •••, 1–18. DOI: 10.1080/17512786.2024.2389209

Van Dijck, J., Poell, T., & De Waal, M. (2018). *The Platform Society: Public Values in a Connective World.* Oxford University Press.

Van Zoonen, L. (1998). "Finally, I have my mother back": Politicians and Their Families in Popular Culture. *The Harvard International Journal of Press/Politics, 3*(1), 48–64.

Vasalou, A., & Joinson, A. N. (2009). Me, Myself and I: Avatars and the Idealized Self in Virtual Worlds. *Psychology of Popular Media Culture, 18*(3), 183–200.

Vaughan, E., & Tinker, T. (2009). Effective Health Risk Communication About Pandemic Influenza for Vulnerable Populations. *American Journal of Public Health, 99*(2), 324–331. DOI: 10.2105/AJPH.2009.162537 PMID: 19797744

Verhoef, P. C., Lemon, K. N., Parasuraman, A., Roggeveen, A., Tsiros, M., & Schlesinger, L. A. (2009). Customer experience creation: Determinants, dynamics and management strategies. *. *Journal of Retailing, 85*(1), 31–41.

Vernon, R. J., Sutherland, C. A., Young, A. W., & Hartley, T. (2014). Modelling first impressions from highly variable facial images. *Proceedings of the National Academy of Sciences of the United States of America, 111*(32), E3353–E3361. PMID: 25071197

Vesa, M.. (2021). Virtual Selfhood and Digital Identity Construction in Online Gaming. *New Media & Society, 23*(6), 1045–1063.

Vladutiu, C. J., Nansel, T. R., Weaver, N. L., Jacobsen, H. A., & Kreuter, M. W. (2006). Differential strength of association of child injury prevention attitudes and beliefs on practices: A case for audience segmentation. *Injury Prevention : Journal of the International Society for Child and Adolescent Injury Prevention, 12*, 35–40. DOI: 10.1136/ip.2004.007153 PMID: 16461418

Vyas, D., Chisalita, C. M., & Van Der Veer, G. C. (2006). Affordance in interaction. In Proceedings of the *13th European Conference on Cognitive Ergonomics: Trust and Control in Complex Socio-technical Systems*. New York: Association for Computing Machinery (pp. 92-99).

Walter, N., Cohen, J., Holbert, R. L., & Morag, Y. (2019). Fact-Checking: A Meta-Analysis of What Works and for Whom. *Political Communication, 37*(3), 350–375. DOI: 10.1080/10584609.2019.1668894

Wang, H. (2022). Transparency as manipulation? Uncovering the disciplinary power of algorithmic transparency. *Philosophy & Technology, 3*, 1–25.

Wang, H. (2023). Why should we care about the manipulative power of algorithmic transparency? *Philosophy & Technology, 1*, 1–6.

Wang, X., Zeng, M. L., Gao, J., & Zhao, K. (2024). *Intelligent Computing for Cultural Heritage: Global Achievements and China's Innovations* (1st ed.). Routledge., DOI: 10.4324/9781032707211

Wang, X., Zhang, J., Zhang, A., & Ren, J. (2019). TKRD: Trusted kernel rootkit detection for cybersecurity of VMs based on machine learning and memory forensic analysis. *Mathematical Biosciences and Engineering, 16*(4), 2650–2667. DOI: 10.3934/MBE.2019132 PMID: 31137231

Ward, C. (2023). Augmented Identities: Exploring the Intersection of Fashion and Selfhood in Digital Spaces. *Journal of Digital Studies, 10*(1), 52–75.

Washington Post (2024, Nov, 7). The Washington Post Launches "Ask The Post AI," a New Search Experience. https://www.washingtonpost.com/pr/2024/11/07/washington-post-launches-ask-post-ai-new-search-experience/ Retrieved 27.11.2024.

Watson, D., Clark, L. A., & Tellegen, A. (2001). Development and Validation of Brief Measures of Positive and Negative Affect: The PANAS Scales. *Journal of Personality and Social Psychology, 54*(6), 1063–1070. DOI: 10.1037/0022-3514.54.6.1063 PMID: 3397865

Watson, H. J., & Nations, C. (2019). Addressing the growing need for algorithmic transparency. *Communications of the Association for Information Systems, 1*(45), 488–510.

Weber, H. (2014, Jun, 30). Associated Press backs Automated Insights to automate boring earnings reports. http://venturebeat.com/2014/06/30/associated-press-backs-automated-insights-to-automate-boring-earnings-reports/ Retrieved 10.09.2024.

Weber, D. M., & Kauffman, R. J. (2011). What drives global ICT adoption? Analysis and research directions. *Electronic Commerce Research and Applications, 10*(6), 683–701. DOI: 10.1016/j.elerap.2011.01.001

Weidmann, N. B., & Schutte, S. (2017). Using night light emissions for the prediction of local wealth. *Journal of Peace Research, 54*(2), 125–140.

Wesner, S. (2022). Cultural values as policy directives: Techniques, technologies and symbolic work. *International Journal of Cultural Policy*, *28*(7), 858–874. DOI: 10.1080/10286632.2022.2137153

Westerman, G., Bonnet, D., & McAfee, A. (2014). *Leading Digital: Turning Technology into Business Transformation*. Harvard Business Review Press.

Williams, J. E., & Flora, J. A. (1995). Health Behavior Segmentation and Campaign Planning to Reduce Cardiovascular Disease Risk Among Hispanics. *Health Education Quarterly*, *22*(1), 36–48. DOI: 10.1177/109019819502200105 PMID: 7721600

Williams, R. (1974). *Television: Technology and Cultural Form*. HarperCollins.

Winschiers-Theophilus, H., Blake, E., Maasz, D., Muashekele, C., Gallert, P., Stanley, C., & Koruhama, A. K. (2021). Embracing Indigenous Knowledge Systems in ICT-Enabled Education. In Lubin, I. A. (Ed.), *ICT and International Learning Ecologies* (1st ed., pp. 132–157). Routledge., DOI: 10.4324/9780429345463-9

Wolfe, R., & Caliskan, A. (2022). American == White in Multimodal Language-and-Image AI. *Proceedings of the 2022 AAAI/ACM Conference on AI, Ethics, and Society*, 800–812. DOI: 10.1145/3514094.3534136

Wong, L. P., & Sam, I.-C. (2010). Public Sources of Information and Information Needs for Pandemic Influenza A(H1N1). *Journal of Community Health*, *35*, 676–682. DOI: 10.1007/s10900-010-9271-4 PMID: 20411411

World Economic Forum. (2024). *The Global Risks Report 2024*. Geneva: World Economic Forum. https://www3.weforum.org/docs/WEF_The_Global_Risks_Report_2024.pdf

Wu, Y., & Wang, H. (2022). Innovation Model of Journalism Education Based on Information Technology under the Background of New Media. *Mobile Information Systems*, *2022*, 1–13. DOI: 10.1155/2022/8367001

Xi, N., Ma, D., Liou, M., Steinert-Threlkeld, Z. C., Anastasopoulos, J., & Joo, J. (2020). Understanding the political ideology of legislators from social media images. In *Proceedings of the International AAAI Conference on Web and Social Media*, (14(1), pp. 726–37), https://doi.org/DOI: 10.1609/icwsm.v14i1.7338

Xudoyqulov, X., & Abdurazzoqov, F. (2022). *Speech culture development technologies among youth*. DOI: 10.5281/ZENODO.7271962

Yadav, M., Kumar, M., Sahoo, A., & Rathnasiri, M. S. H. (2024). AI and the Evolution of Artistic Expression: Impacts on Society and Culture. In Kilinç, U. (Ed.), (pp. 15–36). Advances in Media, Entertainment, and the Arts. IGI Global., DOI: 10.4018/979-8-3693-3916-9.ch002

Yamanobea, N., Wana, W., Ramirez-Alpizar, I. G., Petitb, D., Tsuji, T., Akizukid, S., Hashimotoe, M., Nagataa, K., & Harada, K. (2017). A brief review of affordance in robotic manipulation research. *Advanced Robotics*, *31*(19–20), 1086–1101. DOI: 10.1080/01691864.2017.1394912

Yam, K. C., Tan, T., Jackson, J. C., Shariff, A., & Gray, K. (2023). Cultural Differences in People's Reactions and Applications of Robots, Algorithms, and Artificial Intelligence. *Management and Organization Review*, *19*(5), 859–875. DOI: 10.1017/mor.2023.21

Yang, Z. (2024, Oct.,9). How The New York Times incorporates editorial judgment in algorithms to curate its home page. https://www.niemanlab.org/2024/10/how-the-new-york-times-incorporates-editorial-judgment-in-algorithms-to-curate-its-home-page/ Retrieved 13.10.2024.

Yankelovich, D., & Meer, D. (2006). Rediscovering Market Segmentation. *Harvard Business Review*, *84*(2), 122–131. https://pubmed.ncbi.nlm.nih.gov/16485810/ PMID: 16485810

Yee, N., & Bailenson, J. (2007). The Proteus Effect: The effect of transformed self-representation on behavior. *Human Communication Research*, *33*(3), 271–290. DOI: 10.1111/j.1468-2958.2007.00299.x

Yoon, J., Joo, J., Park, E., & Han, J. (2020). *Cross-domain classification of facial appearance of leaders. Social Informatics, SocInfo 2020* (Vol. 12467). Lecture Notes in Computer Science. Springer., DOI: 10.1007/978-3-030-60975-7_32

Young, D. G., Jamieson, K. H., Poulsen, S., & Goldring, A. (2018). Fact-Checking Effectiveness as a Function of Format and Tone: Evaluating FactCheck.org and FlackCheck.org. *Journalism & Mass Communication Quarterly*, *95*(1), 49–75. DOI: 10.1177/1077699017710453

Zachariev, Z. (2006). Éducation, Dialogue Interculturel et Société de L'information. *International Review of Education*, *52*(5), 425–442. DOI: 10.1007/s11159-006-9002-3

Zambonelli, F. (2023). *Algocrazia. Il governo degli algoritmi e dell'intelligenza artificiale*. Scienza Express.

Zayani, M. (2020). Digital Journalism, Social Media Platforms, and Audience Engagement: The Case of AJ+. *Digital Journalism (Abingdon, England)*, *9*(1), 24–41. DOI: 10.1080/21670811.2020.1816140

Zhai, L., Yang, L., & Chan, H. (2022). The role of digital transformation in customer experience and business performance. *. *Journal of Business Research*, *141*, 569–582. DOI: 10.1016/j.jbusres.2022.06.019

Zhang, X., & Zhou, M. (2023). Information and digital technology-assisted interventions to improve intercultural competence: A meta-analytical review. *Computers & Education*, *194*, 104697. DOI: 10.1016/j.compedu.2022.104697

Zhan, X. (2025). Artificial Intelligence in the Cultural Industry: Applications and Challenges. *Innovation in Science and Technology*, *4*(1), 62–68. DOI: 10.56397/IST.2025.01.06

Zichermann, G., & Linder, J. (2013). *The Gamification Revolution: How Leaders Leverage Game Mechanics to Crush the Competition* (1st ed.). McGraw-Hill.

Zwart, O., Veldhuijzen, I. K., Elam, G., Aro, A. R., Abraham, T., Bishop, G. D., & Brug, J. (2009). Perceived Threat, Risk Perception, and Efficacy Beliefs Related to SARS and Other (Emerging) Infectious Diseases: Results of an International Survey. *International Journal of Behavioral Medicine*, *16*, 30–40. DOI: 10.1007/s12529-008-9008-2 PMID: 19125335

About the Contributors

Kostas Karpouzis is an assistant professor at the Department of Communication, Media and Culture at Panteion University. His research deals with algorithms that make computer systems more adaptive to how people interact with them and how digital games and gamification, as cultural artifacts, can be used to teach conventional and social skills. At the same time, he explores the ethical and social values embedded in and influenced by AI systems. Since 1998 he has been involved in over twenty R&D projects, with prominent examples being the Siren project (the project was voted Best Learning Game in Europe in 2013 by the Games and Learning Alliance), the H2020 iRead project, which produced Navigo, the winner of the GALA Serious Games competition in 2018, and the ongoing Road-STEAMer project, which is designing the European roadmap and proposing policies for teaching STEAM topics in schools and universities. He has given three talks at TEDx events and his TED-Ed video "Can machines read your emotions?" has been viewed over 670,000 times and translated into 26 languages.

Yannis Skarpelos studied Sociology and holds a PhD in Visual Cultural Studies. He is teaching at the Department of Communication, Media and Culture since 1996. His interests are focusing on the sociological and semiotic visual analysis. The last decade his research is focused on computational analysis of big visual data, videogames and social network analysis. His research activities include the creation of Panteion University eclass, the project TRIPMENTOR Interactive Tourist (funded by the "Research-Create-Innovate" program), INTERMEDS: Intercultural Mediators (funded by Erasmus+), and RIPEC: Research for Innovative Practices in Emergency Management of Erasmus Community (funded by Erasmus+). He is the founder and head of the Cultural Informatics Lab of the Department, and has been associate chair and chair of the Department, member of the Panteion Senate, and of the Research Committee.

***

Maria Romana Allegri graduated cum laude both in humanities and in law, since September 2021 has been an associate professor in Public Law at the Department of Communication and Social Research, Sapienza University of Rome, where she has worked as a researcher since March 2001. She is currently teaching "Public Law and Information & Communication Law" and "Fundamentals of Public Law" in bachelor's courses. In the past years she also taught "International Organisation", "EU Law" and "Ethics and Law of Journalism" in bachelor's and master's courses . She is a member of the University Association of European Studies (AUSE) and of the Italian Association of public and institutional communication (Compubblica). Her most recent publications concern the right to be forgotten, the right to access the Internet, the regulation of social media, with particular reference to political propaganda and disinformation.

Dina Darwish is the Vice dean, faculty of computer science and information technology, Ahram Canadian university, Egypt. She obtained my Ph.D from Cairo university at 2009, engineering faculty. She is a professor since 2020, also my special domain is artificial intelligence and has several publications in the areas of Artificial intelligence, wireless ad hoc networks, Internet of things, Big data analytics, Blockchain applications, Cloud computing applications, Web 3 technologies, Chatbots development and others.

Cesario Armando Flores-Villanueva holds a PhD in International Relations, Business and Diplomacy at the Faculty of Political Science (FACPYRI) of the Universidad Autónoma de Nuevo León (U.A.N.L.). Master in International Relations at FACPYRI of the UANL. Throughout 24 years as a teacher at FACPYRI I have taught classes related to international marketing, business and quantitative research. At the master's level I teach international negotiations, leadership and diversity at the Master's level. The line of research is focused on the internationalization of Mexican companies with a focus on franchising, social responsibility and consumer behavior I have published 10 articles in national and international journals, 6 book chapters, I have a book coordination. I have participated in 6 national and international conferences. HAVE BEEN directed 3 bachelor's theses, 2 master's theses and 4 doctoral theses. I participated as Coordinator of the book Panorama de comercio internacional en México. Currently I belong to the NATIONAL SYSTEM OF RESEARCHERS LEVEL I The line of research is focused on the internationalization of Mexican companies with a focus on franchising, social responsibility and consumer behavior as I believe they are vital for job creation,

Stavros D. Kaperonis is assistant professor in the Department of Media, Communication and Culture at the Panteion University of Social and Political Sciences, Athens, Greece. He holds a BSc (Hons) in Natural Sciences from the School of Science and Technology (Hellenic Open University), an MSc in Services Management from the Department of Business Administration at Athens University of Economics and Business and a PhD in Design effect (aesthetics) from the Department of Public Administration at the Panteion University of Athens-Greece. Dr. Kaperonis has participated in many research projects and he has been teaching at undergraduate and postgraduate programs of Media, Communication and Design in Greece, since 2015. He has been invited to give guests lectures in Italy and China. Main research interests of Dr. Kaperonis include web design, UX, aesthetics, Atmospherics, Communication management, Social media, new media tools in journalism, strategic information systems planning, services customization, business process modeling and management. His research work has been published in Conferences and chapter of edited book.

Panagiotis Kapos was born in Athens. He holds a PhD degree from Panteion University (Department of Communication, Media and Culture). His thesis examined the digital transition of the book publishing industry in the new ecosystem of value and networking focusing on the Greek case. He is professionally active in publishing books and in the cultural management and communication. He has collaborated as a faculty member with Panteion University teaching courses on books and written culture in the digital age, publishing entrepreneurship and the digital transition of the Media and Content Industries. He also collaborates with Darmstadt University of Applied Sciences (h-da) in Germany on courses related to the Expanded Media Master's programme and with Ionian University in the Master's programme named Management and Promotion of Cultural Information teaching the course entitled Information Behaviour, Creativity and Cultural Industries. He has collaborated with Greek organizations of written culture as a scientific-research associate. He is Vice President of the Hellenic Foundation for Books and Culture, an organization supervised by the Greek Ministry of Culture. His research interests lie in the fields of digital transition of the publishing industry, media economics and digital communication.

Vijaya Kittu Manda is a multi-dimensional personal. He is a Researcher at PBMEIT, India. He is an Advocate, a Technocrat, an Academician, a Book-writer, Editor, Peer-Reviewer and a stock market enthusiast. He has 11 University Postgraduate Degrees in various disciplines ranging from Arts and Humanities to Sustainability to Regenerative Economy, and from Management to Engineering. He is a Ph.D. in Financial Management. His thesis on Mutual Funds and their Market

Competition won the prestigious NSE-IEA Best Thesis Award in 2023. He is currently pursuing his second Ph.D. in Computer Science with focus on Blockchain and Artificial Intelligence. He has nearly 13+ years of experience in capital markets, financial planning, and investing He writes Research Papers and Case Studies, and sits on Editorial Boards of various publishers. He has over 121+ Google Scholar publications and 785+ magazine article publications so far. He is the Chief Editor for a Management Book Series and the Editor of several Scopus-indexed books. He is a Peer Reviewer and is a Certified Peer Review Supervisor. He frequently reviews for leading publishers including IEEE, Springer, Emerald, Sage, amongst others. He sits as a Session Chair and an Advisor for various academic and industrial conferences. He is a guest speaker for colleges and universities. He serves as the host for a research talk show called PBMEIT Expert Insights.

José T. Olague is originally from Monterrey, Mexico. He holds a Ph.D. in Philosophy with a focus on International Relations, Business, and Diplomacy from UANL, and a Master's in Tourism from the University of Las Palmas de Gran Canaria. With over twenty years of academic experience, he has worked in teaching, research, and academic management in tourism and marketing. He is currently a professor-researcher at UABC and a member of Mexico's National System of Researchers. His research focuses on tourist consumer behavior. He has advised public and private institutions on tourism planning and collaborates with academic networks nationally and internationally.

Andreas M. Panagopoulos holds a Ph.D from the Journalism and Media Dept. of the Aristotle University of Thessaloniki. He is an experienced News manager and journalist in Broadcast and Digital Media. His research interests are in the fields of Web Metrics, AI in Journalism, Ethics and Journalism Prctices. He has published articles in International and Greek journals and in edited volumes. He has been granted with the Ath. Botsis Foundation Journalism Award. He is elected to the Board of PFJU and serves as Executive Secretary.

Nguyen Phuc Nguyen is associate professor in Management at University of Economics, The University of Danang. He holds a Ph.D. degree with a concentration in management sciences from Kobe University, Japan. His areas of research expertise cover a wide range of management and organizational behavior issues such as Inter-firm relationship, trust, competitive capability, innovation, CSR and leadership.

Ilia Evangelia (Sybil) Sdralli is a Phd Candidate at the Department of Communication, Media, and Culture at Panteion University, content strategist, and journalist specializing in the intersection of dress, fashion, pop culture, and social media. She is the founder of The StyleTitle fashion project, dedicated to researching and documenting the trends shaping the current fashion ecosystem. Her Phd thesis focuses on the visual analysis of the use of the color black in mainstream cinema costumes as well as in contemporary costume.

Sunil Soni completed an M.E. in 2008 from Sardar Patel University, earning a prestigious gold medal, and went on to achieve a Ph.D. in 2017 from RK University, Rajkot. With over 20 years of dedicated teaching experience, Dr. Sunil Soni currently serves as a Lecturer in the Information Technology Department at Government Polytechnic, Rajkot. Passionate about advancing knowledge and innovation, their primary areas of interest include machine learning, deep learning, and cloud computing, where they continue to inspire students and contribute actively to academic and research advancements.

Bikkina Srinivas is a Director at Vingas Industries Private Limited, Visakhapatnam, India. He has over two decades of industrial experience in dealing with industrial gases and related establishments. His areas of speciality involves generating unique ideas and transforming them into practically implementable models. He is a Ph.D. with specialization in Operations Management. He is the author the several journal articles and actively engages in both Industrial and Academic events. He has special interest in Computer Science and Information Technology and has several initiatives that help in the industrial gases industry.

Jaydeep R. Tadhani completed his Bachelor of Engineering in Information Technology from VGEC Chandkheda in 2011 and obtained his Master's degree in Computer Engineering from GTU in 2014. With 12 years of teaching experience, he has made significant contributions to the field of cybersecurity. In 2024, he successfully completed his PhD from GTU, further solidifying his expertise. His extensive knowledge and research in cybersecurity have positioned him as a key figure in the domain. His work has been recognized internationally, with his research paper published in Nature Scientific Reports, underscoring his contributions to the advancement of cybersecurity research.

Theodore Tarnanidis is a marketing scholar, adjunct professor at the International Hellenic University and researcher in Applications of D.Sc. and MCDA. Theodore has six years experience as a marketing and decision making practitioner. He made his post-doc research in the area of sustainable entrepreneurship from the University of Macedonia. He obtained a Ph.D. from the University of London Met., UK. He received his M.B.A from Liverpool University, UK and is a graduate from the University of Macedonia (Business Administration) and Alexander Technological Educational Institute (Marketing). His research focuses on International Marketing, Multi-cultural Marketing, Business Marketing,

Consumer science, Marketing Management, Methodology, Advanced Quantitative Methods, Decision Science, Multiple-Criteria Decision Analysis, Management of Education, Sustainable Entrepreneurship. His work has been published in various internationally renowned scientific conferences (Academy of Marketing, European Marketing Academy, PROMETHEE Days, Hellenic Operational Research) and in the Journal of Business Ethics, World Review of Entrepreneurship, Management and Sustainable Development, Journal of Retailing and Consumer Services, Current Issues in Tourism, Marketing Science & Inspirations, European Journal of Educational Management, Management Science Letters, Administrative Sciences and other journals.

Sotirios Triantafyllou was born in Athens, Greece. He holds a PhD Degree from Panteion University (Department of Communication, Media and Culture). He is teaching staff in in the postgraduate program of Sports Journalism in University of Peloponnese, department of Sports Management from 2021. From 2019 until 2023 he was teaching staff in the University of Patras and from 2015-2019 was teaching staff in Technological Institute of Western Greece. Professional journalist since 1994, present (since October 2021), he has worked in sports and political newspapers, television and radio stations and websites, having covered major sports events (Olympic Games, World Cups and European Football Championships). He is president of the Panhellenic Federation of Journalist's Unions (POESY) from October 2022 and member of the board of Athens-Macedonian News Agency form January 2023. He was president of the Greek Sports Journalists Association (2012-2018), member of the Executive Committee of the European Union of Sports Jounalists (2017-2022) and member of the Executive Committee of the Union of Sports Journalists of Southeastern Europe (2014-2016).

Adinarayana Rao U. V. is a Professor at the Department of Operations at the GITAM School of Business, GITAM (Deemed to be University), Visakhapatnam, India. He is an award-winning teacher and has worked at Addis Ababa University before turning to be a Professor at GITAM. He has numerous publications in reputed journals, published books, resents papers at seminars and conferences, and is guiding research scholars in the areas of Business and Industrial Operations.

Thomai Voulgari is a postdoctoral researcher at the Department of Communication, Media, and Culture at Panteion University and a Member of cultural informatics, data & computational cultural studies laboratory. Her postdoctoral research focuses on the topic of the impact of fashion culture as reflected in gaming world through an interactive experience. Her research interests focuses on visual studies, data analysis, semiotics, gaming environment, digital fashion, social network analysis, anti-immigration discourse and hate speech in digital era. Her Phd dissertation specialized on manipulation through Social Media Platforms involving data collection methods.

Index

A

B

C

D

E

F

G

I